UV-B Radiation and Plant Life

UV-B Radiation and Plant Life

Molecular Biology to Ecology

Edited by

Brian R. Jordan

Professor of Plant Biotechnology
Lincoln University
New Zealand

CABI

CABI is a trading name of CAB International

CABI
Nosworthy Way
Wallingford
Oxfordshire OX10 8DE
UK

Tel: +44 (0)1491 832111
Fax: +44 (0)1491 833508
E-mail: info@cabi.org
Website: www.cabi.org

CABI
745 Atlantic Avenue
8th Floor
Boston, MA 02111
USA

Tel: +1 (617)682-9015
E-mail: cabi-nao@cabi.org

British Library Cataloguing-in-Publication Data

A catalogue record for this book is available from the British Library.

Library of Congress Cataloging-in-Publication Data

Names: Jordan, Brian R., editor.
Title: UV-B radiation and plant life : molecular biology to ecology / edited by Brian R. Jordan, Professor of Plant Biotechnology, Lincoln University, New Zealand.
Description: Wallingford, Oxfordshire, UK ; Boston, MA : CABI, [2017] | Includes bibliographical references and index.
Identifiers: LCCN 2017016226 (print) | LCCN 2017032358 (ebook) | ISBN 9781780648606 (ePDF) | ISBN 9781780648613 (ePub) | ISBN 9781780648590 (hbk : alk. paper)
Subjects: LCSH: Plants--Effect of ultraviolet radiation on. | Ultraviolet radiation. | Plant molecular biology. | Plant ecology.
Classification: LCC QK757 (ebook) | LCC QK757 .U96 2017 (print) | DDC 571.4/56--dc23
LC record available at https://lccn.loc.gov/2017016226

ISBN: 978 1 78064 859 0 (hardback)
978 1 78064 860 6 (e-book)
978 1 78064 861 3 (e-pub)

Commissioning editor: David Hemming
Editorial assistant: Emma McCann
Production editor: Alan Worth

Typeset by SPi, Pondicherry, India
Printed and bound in the UK by CPI Group (UK) Ltd, Croydon, CR0 4YY

Contents

Preface

It is a great pleasure to introduce this book on the role of ultraviolet-B radiation (UV-B) in plant growth and development. UV-B is highly energetic radiation that can have a profound effect on all biological systems, and terrestrial plants being sessile are consistently exposed to this radiation. UV-B is absorbed by a wide range of compounds, including DNA, proteins and lipids and can potentially lead to damage to these critical molecules, such as the formation of dimeric lesions in DNA. Consequently, UV-B radiation was historically considered to be purely damaging and with the discovery of the ozone hole over the Antarctic in the early 1980s, UV-B was seen as a major threat to the biosphere. Research was focused on harmful aspects of UV-B and unfortunately, unrealistic UV-B environments were used to study the responses. Recently however, the understanding of how UV-B affects plants has changed dramatically, from both ecological and molecular perspectives.

For this book, I have brought together the undoubted world authorities on UV-B plant research. The book covers a range of topics involving UV-B effects on plants from: the UV-B environment and ecosystem impact; UV-B changes in plant physiology and secondary metabolism; the biochemistry and molecular biology of UV-B responses; isolation and characterisation of the UV-B photoreceptor; signal transduction and finally the application of UV-B in agriculture and horticulture. Through these chapters a number of aspects emerge. The first aspect is how the understanding of UV-B responses has evolved from essentially just being considered damaging, to a more substantive regulatory role in photomorphogenesis. The second aspect is the important breakthrough that has taken place to characterize a UV-B photoreceptor and the molecular mechanism of action. This major scientific discovery came in 2011 with the identification of a putative UV-B photoreceptor, UVR8. Importantly all this new knowledge has allowed a rethink of the potential to use UV-B in a more positive way to manipulate horticultural and agricultural plants.

From a personal perspective the study of UV-B responses has been a large part of my career. I have always been involved in 'light and plant development' research. However, in 1990 I was invited to work on molecular aspects of UV-B responses in the laboratory of Professor Jan Anderson in Canberra, Australia. I was fortunate to be working with Åke Strid, Fred Chow and other great colleagues. This period etched deeply my interest in all aspects of UV-B research and now through my UV-B colleagues I have been able to invite this eminent assembly of authors to write the chapters in this book. Although there is still a lot more to learn about the responses of plants to UV-B, I am certain the information in this book will provide a foundation of knowledge for many years to come.

I would like to acknowledge and thank all the authors for their time and effort to produce an excellent portfolio of authoritative and comprehensive reviews. In a time of constant work pressure with a variety of demands, I am very appreciative of their willingness to provide their knowledge and expertise to this book. It is very telling that within one week of my invitation, all the authors had responded positively, the overwhelming message was 'I have to be part of this project'. I think that says it all!

Thanks to all my colleagues in the UV-B field, particularly Marcel Jansen, Jason Wargent and Rainer Hoffman. I would especially like to thank Professor Åke Strid for his friendship to myself and family, and academic support since our time in Canberra together. Thanks also to David Hemming and all the staff at CABI who have been very supportive in the preparation of the book.

Finally, I would like to dedicate my contribution in this book to Gillian Barbara Jordan who passed away on the 25th of December 2012. Gill was the 'love of my life' for 35 years and supported my career aspirations throughout.

Brian Jordan
Professor of Plant Biotechnology
Lincoln University
New Zealand
February 2017

Contributors

Pedro J. Aphalo, Viikki Plant Science Centre, Department of Biosciences, P.O. Box 65, 00014 University of Helsinki, Finland. E-mail: pedro.aphalo@helsinki.fi

Amy T. Austin, IFEVA, Consejo Nacional de Investigaciones Científicas y Técnicas (CONICET),Universidad de Buenos Aires, Avenida San Martín 4453, C1417DSE, Buenos Aires, Argentina and Instituto de Investigaciones Biotecnológicas, Consejo Nacional de Investigaciones Científicas y Técnicas, Universidad Nacional de San Martín, B1650HMP San Martín, Buenos Aires, Argentina. E-mail: austin@ifeva.edu.ar

Carlos L. Ballaré, IFEVA, Consejo Nacional de Investigaciones Científicas y Técnicas (CONICET),Universidad de Buenos Aires, Avenida San Martín 4453, C1417DSE, Buenos Aires, Argentina and Instituto de Investigaciones Biotecnológicas, Consejo Nacional de Investigaciones Científicas y Técnicas, Universidad Nacional de San Martín, B1650HMP San Martín, Buenos Aires, Argentina. E-mail: ballare@ifeva.edu.ar

Paul W. Barnes, Department of Biological Sciences and Environment Program, Loyola University New Orleans, 6363 St. Charles Avenue, New Orleans, Louisiana 70118. E-mail: pwbarnes@loyno.edu

Susanne Baldermann, Leibniz Institute of Vegetable and Ornamental Crops, Großbeeren, Germany. E-mail: baldermann@igzev.de

Melanie Binkert, Department of Botany and Plant Biology, University of Geneva, Geneva, Switzerland. E-mail: melanie.binkert@unige.ch

Janet F. Bornman, International Institute of Agri-Food Security, Curtin Business School, Curtin University, PO Box U1987, Perth WA 6845, Western Australia. E-mail: Janet.Bornman@Curtin.edu.au

Isabelle N. Bottger, Department of Biological Sciences and Environment Program, Loyola University New Orleans, 6363 St. Charles Avenue, New Orleans, Louisiana 70118. E-mail: issybo@gmail.com

Stephan D. Flint, Department of Forest, Rangeland and Fire Sciences, UIPO 441135, University of Idaho, Moscow, Idaho 83844-1135. E-mail: sflint@uidaho.edu

Franziska S. Hanschen, Leibniz Institute of Vegetable and Ornamental Crops, Großbeeren, Germany. E-mail: hanschen@igzev.de

Éva Hideg, Institute of Biology, University of Pécs, Ifjúság útja 6, H-7624 Pécs, Hungary. E-mail: ehideg@gamma.ttk.pte.hu

Marcel A.K. Jansen, University College Cork, School of Biological, Earth and Environmental Science & Environmental Research Institute, North Mall, Cork, Ireland. E-mail: m.jansen@ucc.ie

Gareth I. Jenkins, Institute of Molecular, Cell and Systems Biology, College of Medical, Veterinary and Life Sciences, Bower Building, University of Glasgow, Glasgow G12 8QQ, UK. E-mail: Gareth.Jenkins@glasgow.ac.uk

Brian R. Jordan, Faculty of Agriculture and Life Sciences, Department of Wine, Food and Molecular Biosciences, Centre for Viticulture & Oenology, PO Box 84, Lincoln University, Lincoln 7647, Canterbury, New Zealand. E-mail: Brian.Jordan@lincoln.ac.nz

Karel Klem, Global Change Research Institute CAS, v.v.i., Bělidla 4a, CZ 60300 Brno, Czech Republic. E-mail: klem.k@czechglobe.cz

Susanne Neugart, Leibniz Institute of Vegetable and Ornamental Crops, Grossbeeren, Germany. E-mail: neugart@igzev.de

T. Matthew Robson, Department of Biosciences, Viikki Plant Science Center, University of Helsinki, Helsinki 00014, Finland. E-mail: matthew.robson@helsinki.fi

Monika Schreiner, Leibniz Institute of Vegetable and Ornamental Crops, Grossbeeren, Germany. E-mail: schreiner@igzev.de

Åke Strid, School of Science and Technology, Örebro University, SE-70182 Örebro, Sweden, E-mail: ake.strid@oru.se

Mark A. Tobler, Department of Biological Sciences and Environment Program, Loyola University New Orleans, 6363 St. Charles Avenue, New Orleans, Louisiana 70118. E-mail: tobler@loyno.edu

Roman Ulm, Department of Botany and Plant Biology, University of Geneva, Geneva, Switzerland. E-mail: roman.ulm@unige.ch

Otmar Urban, Global Change Research Institute CAS, v.v.i., Bělidla 4a, CZ 60300 Brno, Czech Republic. E-mail: urban.o@czechglobe.cz

Jason Wargent, Institute of Agriculture and Environment, Massey University, Palmerston North, New Zealand. E-mail: J.Wargent@massey.ac.nz

Melanie Wiesner-Reinhold, Leibniz Institute of Vegetable and Ornamental Crops, Grossbeeren, Germany. E-mail: wiesner@igzev.de

1 Towards an Understanding of the Implications of Changing Stratospheric Ozone, Climate and UV Radiation

Janet F. Bornman*
Curtin University, Perth, Australia

Introduction

Changing profiles of ultraviolet radiation

The stratospheric ozone layer, located *c.* 10 to 50 km above the Earth's surface (Fig. 1.1), makes up approximately 90% of the world's ozone. The remaining ozone is located in the troposphere closest to Earth. Although ozone is an effective filter against transmission of ultraviolet (UV) radiation to the Earth's surface, even a small amount of the short wavelengths can have environmental effects. UV radiation is conventionally defined as UV-C (< 280 nm), UV-B (280–315 nm) and UV-A (315–400 nm). About 97–99% of UV radiation in the wavelength range of 200–300 nm is absorbed by ozone with little or no filtering effect on UV-A radiation (NASA, 2016). Thus, as the UV radiation passes through the atmosphere to Earth, all UV-C radiation and most of the UV-B radiation is absorbed. Other factors influencing the amounts of UV radiation reaching the Earth's surface include altitude, latitude, sun angle, clouds, aerosols, ground reflectivity, depth and quality of water bodies, as well as climate-induced changes.

More than 40 years ago scientists contemplated the likely cause of a decreasing stratospheric ozone layer (Molina and Roland, 1974) and the consequent threat of increased amounts of UV radiation. Thirty-two years ago, the Antarctic 'ozone hole' was discovered (Farman *et al.*, 1985). Research has since shown that substances used in many applications such as air conditioners, fire extinguishers, refrigerators, foams, aerosol sprays and agricultural fumigants as well as certain solvents, were ozone-depleting substances (ODS). Most were also contributors to the warming greenhouse effect. These ODS include chlorofluorocarbons, methyl bromide, methyl chloroform, halons, hydrochlorofluorocarbons, and carbon tetrachloride. Subsequently, several of the substances used as substitutes for the ODS have also been found to add to global warming. The Montreal Protocol and its Amendments have successfully controlled further production of the ODS, preventing catastrophic exposures to UV radiation (Newman *et al.*, 2009; Newman and McKenzie, 2011; Chipperfield *et al.*, 2015; United Nations Environment Programme Environmental Effects Assessment Panel, 2016). These evolving events and human activities demonstrate the intricate interrelationship of ozone dynamics, UV radiation and climate change,

*Corresponding author: Janet.Bornman@Curtin.edu.au

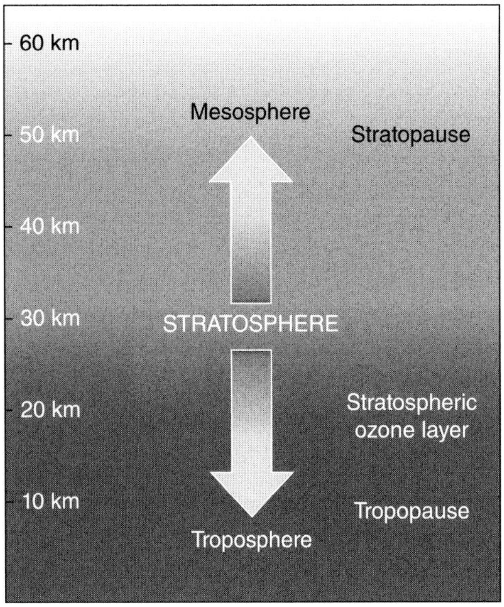

Fig. 1.1. Diagrammatic sketch of the stratosphere and its boundaries.

which in turn affect the environment and life on Earth in complex ways.

Environmental and health implications

Projections involving the dynamics of UV radiation, climate and ozone have important implications for the environment and human health. In areas with reduced UV radiation, vitamin D levels may drop below the recommended concentrations, and the positive effects of the UV radiation on certain autoimmune diseases, cancers and infections (Lucas *et al.*, 2015) may become lessened. However, behavioural patterns towards sun exposure among diverse population groups will largely determine the amount of UV radiation and levels of vitamin D acquired. At the same time, reduced levels of UV radiation would mean decreased incidences of skin cancers and cataracts. In natural ecosystems and agricultural systems, low exposure to UV radiation may favour pathogens and herbivores as a consequence of decreased levels of UV-induced phenolic compounds, which would otherwise function as deterrents against attack. Depending on the amount of UV radiation received, crop quality may be affected due to changes in the amounts and profiles of plant phenolics (many of which are effective antioxidants), nutritional composition, general plant fitness and morphology (Wargent and Jordan, 2013; Bornman *et al.*, 2015; Robson *et al.*, 2015). These patterns of change also offer opportunities in crop management (Raviv and Antignus, 2004; Paul *et al.*, 2005; Wargent and Jordan, 2013).

Complexities of Ozone Dynamics, UV Radiation and Climate Change

Shaping of the current and future environment

Annual ozone depletions are still occurring in the polar regions, especially in Antarctica because of the long atmospheric lifetimes (close to 100 years) of some of the ODS such as chlorofluorocarbons (CFCs; 'freons') and halons containing chlorine and bromine. Substantially smaller ozone depletions occur also at mid-latitudes, with periodic large depletions due to volcanic eruptions and the resultant sulphate emissions, which enhance activation of chlorine that in turn catalyses the loss of ozone. Over the tropics, the stratospheric ozone layer is always naturally thinner than in other regions, and variations in the concentration of the ozone layer here are so far small.

There now appear to be indications of initial recovery (Fig. 1.2) of the stratospheric ozone layer (Solomon *et al.*, 2016) as a consequence of the regulations put in place by the Montreal Protocol and its Amendments. However, predicting future changes in the ozone layer is difficult because of the confounding influence of rapid climate change. The Montreal Protocol has been instrumental in stimulating research and production of substitutes for many of the ODS. Among these substitutes are the typical hydrofluorocarbons (HFCs), which are used in refrigeration and air conditioning. However, HFCs have a large global warming potential and long atmospheric lifetimes (Hurwitz

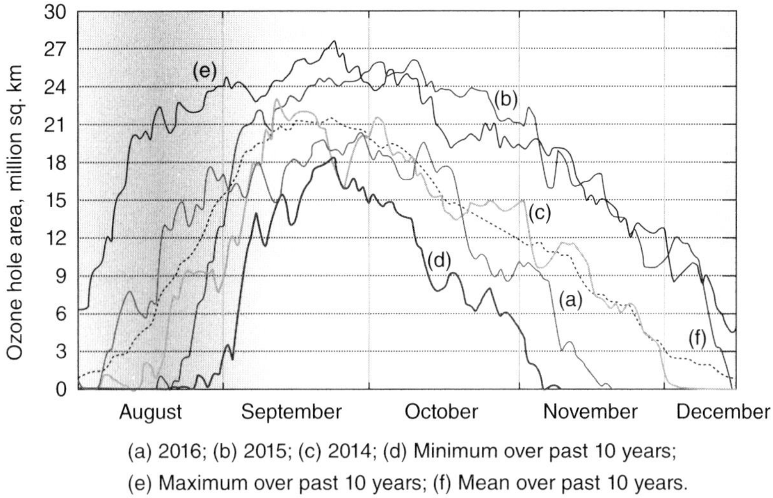

Fig. 1.2. Progression of the ozone 'hole' area in millions of sq. km. The shaded area during August depicts decreasing uncertainty in the size of the ozone 'hole' as the polar region becomes sunlit. (NASA Ozone Hole Watch.)

et al., 2016). For example, HFC-23 has a lifetime of *c.* 228 years, and a global warming potential thousands of times greater than carbon dioxide (Chipperfield, 2015). Despite their potential to contribute to global warming, HFCs did not come under the Montreal Protocol since they have a negligible effect on the ozone layer. However, because they were produced as a result of the agreements to phase out the major ODS under the Protocol, much effort finally culminated in a decision by 197 countries in Kigali, Rwanda (Kigali Amendment, 2016) to phase out the use of HFCs. This is expected to have profound and positive effects on mitigating climate warming.

Ozone itself absorbs heat and, therefore, decreases or increases in ozone concentration can have a cooling or warming effect. This effect also depends on altitude. Since ozone absorbs heat at relatively low altitudes, it cools the lower stratosphere over Antarctica (Thompson and Solomon 2002; Hartmann *et al.*, 2013; Bais *et al.*, 2015), contributing favourable conditions for the formation of polar stratospheric clouds that form a catalytic ice crystal surface for ozone-depleting chlorine free radicals.

As the environment changes, so too will the levels of exposure to UV radiation and the ecosystem's responses to the interactive effects of multiple climate factors (Bornman *et al.*, 2015; Robinson and Erickson, 2015), including temperature, water availability and soil nutrients. Thus the effects of ozone depletion on climate change – and impacts of climate change events less directly dependent on ozone dynamics – will very probably continue to further modify the amount of UV radiation reaching the Earth. Some of these UV-modifying conditions due to climate change include variations in cloud cover, UV-absorbing tropospheric gases, and changes in reflectivity from melting snow and ice as temperatures increase (Bais *et al.*, 2015). In regions outside the polar areas, cooling of the middle and upper stratosphere from increasing amounts of greenhouse gases is predicted to decrease the catalytic destruction of ozone and reduce levels of UV radiation outside the tropics (Eyring *et al.*, 2007; Shepherd, 2008; Waugh, *et al.*, 2009; Bais *et al.*, 2015). However, this may be partly offset by the highly reactive nitrogen oxides (NO_x) from nitrous oxide (N_2O) that catalyse the destruction of the upper stratospheric ozone. Emissions of N_2O come from biomass burning, industry, agriculture and also natural sources (e.g. soils) but human activity is set to account for substantially increased emissions by the middle of the 21st century unless mitigating actions are taken (Ravishankara *et al.*, 2009; Davidson and Kanter, 2014; Revell *et al.*, 2015). In contrast to regions outside the tropics, UV radiation in the tropics is likely to increase slightly because of large-scale circulation changes in the upper atmosphere brought about by the increase in greenhouse gases (Butchart, 2014; Bais *et al.*, 2015).

Ozone affects climate and vice versa

There is further emerging evidence of the way in which stratospheric ozone is influencing climate change and vice versa (Thompson and Solomon, 2002; Shepherd, 2008; Nowack *et al.*, 2015; Iglesias-Suarez *et al.*, 2016), and how these two factors modify the amount of UV radiation received by ecosystems, humans and other animals (Williamson *et al.*, 2014). Thus several consequences of current and future climate change are becoming apparent through both observation and modelling. One such example is the effect on climate by ozone depletion in the Southern hemisphere (Thompson *et al.*, 2011; Turner *et al.*, 2014; Bais *et al.*, 2015). It is predicted that the cooling of the lower stratosphere will intensify, and that stronger winds (Li *et al.*, 2016; Gent, 2016) will increase the meridional overturning – a circulation system of deep ocean and surface currents resulting in the transport and storage of large quantities of water, heat and carbon – thus playing a major role in climate change and in modifying the environment.

Ozone level variation and increasing climate change are highly dynamic processes, and consequently there is some uncertainty in the way in which they will play out as the Earth's climate evolves and as research unravels more interacting factors. Global climate is perturbed by stratospheric ozone through temperature changes from radiative forcings (Myhre *et al.*, 2013) and also by changes in tropospheric and stratospheric circulations (WMO, 2015). Radiative forcing refers to the changes in the radiative or energy balance from differences between incoming solar radiation and outgoing infrared radiation, which can modify climatic conditions. Since ozone is itself a greenhouse gas, where increases occur, there is a warming effect (positive radiative forcing), and consequently a depletion in ozone generally results in a cooling effect (negative radiative forcing). Therefore, after 2050, projected climate change will probably become the dominant driver of future stratospheric ozone dynamics, affecting also the UV radiation environment, as the amounts of ozone depleting substances gradually decrease (Eyring *et al.*, 2007; IPCC, 2013).

Ecological consequences of ozone depletion

Only recently has attention turned to considering the consequences for ecosystems of the impact of the dynamics of ozone depletion *per se* on climate change (Villalba *et al.*, 2012; Bornman *et al.*, 2015; Gutt *et al.*, 2015; Robinson and Erickson, 2015). There are already indications that the complex events arising from ozone depletion are altering ecosystems in the Southern hemisphere through changes in precipitation, wind circulation patterns and wind speed, leading in some instances to increased aridity, thereby impacting plant habitats (Clarke *et al.*, 2012) and altering growth response of, for example, forest ecosystems (Villalba *et al.*, 2012).

Nitrous oxide and the future

One of the intriguing conundrums is the idea that future environmental change may require consideration of some policy intervention with respect to the ozone-depleting nitrous oxide (N_2O) (Butler *et al.*, 2016), to prevent what has been termed 'super recovery' of stratospheric ozone. If CO_2 and methane (CH_4) levels continue to increase, they will contribute to ozone recovery due to the temperature effects in the stratosphere of these greenhouse gases (GHGs). On the other hand, curbing CO_2 and CH_4 would also have obvious beneficial environmental effects with respect to global warming.

However, if N_2O is reduced against a background of rising CO_2 and CH_4, stratospheric ozone is projected to increase beyond its historical values – i.e. the so-called super recovery (Portmann and Solomon, 2007; Iglesias-Suarez *et al.*, 2016; Maycock, 2016). As a consequence, a reduction in UV radiation exceeding pre-1980s values would intuitively be a positive outcome for some human diseases such as skin cancer and cataracts, but may be detrimental for other diseases, e.g. where UV-induced vitamin D is involved, as well as for other health conditions benefitting from appropriate exposure to UV radiation (Lucas *et al.*, 2015). Ecosystems and plant development would be affected by a lowered UV radiation regime which would probably also decrease plant tolerance to pathogen and insect attack (see below: *UV radiation: environmental stress or regulatory factor?*).

A significant reduction in UV radiation reaching Earth as a result of ozone super recovery also has implications for the chemical composition of the atmosphere, since it would result in reduced action by UV radiation in 'cleaning' or oxidising the troposphere through the generation of hydroxyl radicals (·OH) (Levy, 1971; Madronich *et al.*, 2015). These radicals control atmospheric lifetimes of many pollutants such as nitrogen oxides, methane, halocarbons, and sulphur dioxide (Madronich *et al.*, 2015), which have consequences for climate change, ozone concentration and possible further reductions in UV radiation reaching the Earth's surface. Some of these effects may be partly counterbalanced by global measures to reduce air pollutants (McKenzie *et al.*, 2011; Watanabe *et al.*, 2011), which would result in higher levels of UV radiation reaching the Earth's surface. Thus, trends in air quality, important for ecosystems and health, will be modulated by UV radiation. Post-2050, it is likely that we will see CO_2 and N_2O becoming progressively important in determining the future of the ozone layer (Stolarski *et al.*, 2015) and the UV radiation environment. It is therefore becoming very clear that increasing climate change will influence the recovery of stratospheric ozone and modulate the penetration of UV radiation to the Earth's surface. It is also becoming apparent that apart from the effects of ozone on climate, and vice versa, climate changes can modify exposure to UV radiation, independently of ozone. By way of human adaptation strategies and opportunism, these rapidly changing environmental conditions can also be exploited for practical purposes, as reviewed by Wargent and Jordan (2013), to improve the nutritional quality of agricultural crops through UV-induced enhancement of antioxidants and other health-promoting compounds (see above).

UV Radiation: Environmental Stress or Regulatory Factor?

Early on, it was recognised that UV radiation was part of the environmental cue for plants and fungi that shaped their morphology (Kumagai, 1988; Ensminger, 1993; Kim *et al.*, 1998; Paul and Gwynn-Jones, 2003), growth and biochemistry (Klein, 1978). Early work also raised the question whether UV-B radiation posed a threat to photosynthesis. The finding was that inhibition was generally only seen at high UV-B irradiances and that even these could be compensated for by acclimation mechanisms (Allen *et al.*, 1998). However, in the wake of increasing evidence of ozone depletion, most of the research quickly centred around damage, giving in many instances an unbalanced interpretation due to unrealistic experimental conditions of UV radiation and visible light (Searles *et al.*, 2001). This trend has slowly reversed and consequently our understanding has broadened regarding the diversity of response in an increasingly complex and rapidly changing environment (assessed in Ballaré *et al.*, 2011; Jansen and Bornman, 2012; Williamson *et al.*, 2014; Bornman *et al.*, 2015). It has also highlighted the need for a strong interdisciplinary approach in order to gain a comprehensive, whole-systems perspective of the plant environment. Similarly, evaluation of the role of UV radiation at plant and ecosystem levels, under multi-environmental conditions (e.g. water availability, temperature,

CO_2, and soil nutrients (assessed in Caldwell *et al.*, 2007; Ballaré *et al.*, 2011; Bornman *et al.*, 2015)) is important for obtaining realistic outcomes and determining potential interacting effects.

Increasingly, more information on the regulatory and acclimatory role of UV radiation has been facilitated by molecular studies that have demonstrated some of the mechanisms underlying plant genetic, biochemical, physiological and morphological modifications. These mechanistic studies have included investigation of the way in which UV-B radiation is perceived by the plant through the UV-B photoreceptor, UV RESISTANCE LOCUS8 (UVR8), which mediates photomorphogenic response to UV-B radiation (Jenkins, 2009, 2014).

Research on some of the indirect responses to UV radiation, in particular, UV-B radiation, of individual plants and terrestrial ecosystems has also contributed to the shift in focus from UV radiation as mainly a stress issue to one of a modifying or regulatory factor. The indirect effects are often manifested by a response not directly induced by a current stressor, but through a series of interactions (Paul and Gwynn-Jones, 2003; Miller and TerHorst, 2012). Typical indirect effects are exemplified by changes in plant chemistry leading to plant tolerance against pathogens and herbivores due to UV-induced plant polyphenolics (Ballaré *et al.*, 2011; Ferreyra *et al.*, 2012) at toxic concentrations or at levels that deter pathogen or herbivore attack. These polyphenolics, e.g. flavonoids, function as chemical defence compounds and also contribute to antioxidant activity. Other indirect modifications by UV radiation occur below the soil surface, although penetration by UV is minimal. Rather, the response appears to be mainly mediated through flavonoids in plant root exudates as a result of exposure to UV radiation of the above-ground plant parts (Zaller *et al.*, 2002; Avery *et al.*, 2003; Caldwell *et al.*, 2007; Cesco *et al.*, 2010; Bornman *et al.*, 2015).

Although the research emphasis on damaging effects of UV radiation on plants and ecosystems has lessened, potential deleterious effects can still occur under certain environmental situations. These effects are largely dependent on genotype, co-occurring stress factors, regional location, season and duration of the stress(es). Importantly, in light of the projected changes in the UV radiation environment (as a consequence of the diverse interactive effects of changes in ozone and climate, compounded by human activities) detrimental modifications may increase if plant defence systems become less effective under harsh conditions (Williamson *et al.*, 2014).

Conclusions

Thus, although stratospheric ozone levels are projected to recover or super-recover, future exposure to UV radiation will be strongly influenced by the interactive processes involving ozone dynamics and climate change, either singly or together. With the projected increase and complexity of climate change, ozone dynamics and land-use changes, research on the effects of UV radiation will continue to be relevant.

References

Allen, D.J., Nogués, S. and Baker, N.R. (1998) Ozone depletion and increased UV-B radiation: Is there a real threat to photosynthesis? *Journal of Experimental Botany* 49, 1775–1788.

Avery, L.M., Lewis Smith, R.I. and West, H.M. (2003) Response of rhizosphere microbial communities associated with Antarctic hairgrass (*Deschampsia antarctica*) to UV radiation. *Polar Biology* 26, 525–529. DOI:10.1007/s00300-003-0515-y

Bais, A.F., McKenzie, R.L., Bernhard, G., Aucamp, P.J., Ilyas, M. *et al.* (2015) Ozone depletion and climate change: impacts on UV radiation. *Photochemical & Photobiological Sciences* 14, 19–52.

Ballaré, C.L., Caldwell, M.M., Flint, S.D., Robinson, S.A. and Bornman, J.F. (2011) Effects of solar ultraviolet radiation on terrestrial ecosystems: patterns, mechanisms, and interactions with climate change. *Photochemical & Photobiological Sciences* 10, 226–241. DOI:10.1039/C0PP90035D

Bornman, J.F., Barnes, P.W., Robinson, S.A., Ballare, C.L., Flint, S.D. *et al.* (2015) Solar ultraviolet radiation and ozone depletion-driven climate change: effects on terrestrial ecosystems. *Photochemical & Photobiological Sciences* 14, 88–107. DOI:10.1039/C4PP90034K

Butchart, N. (2014) The Brewer-Dobson circulation. *Reviews of Geophysics* 52, 157–184.

Butler, A., Daniel, J.S., Portmann, R.W., Ravishankara, A.R., Young, P.J. *et al.* (2016) Diverse policy implications for future ozone and surface UV in a changing climate. *Environmental Research Letters* 11, 064017. DOI:10.1088/1748-9326/11/6/064017

Caldwell, M.M., Bornman, J.F., Ballaré, C.L., Flint, S.D. and Kulandaivelu, G. (2007) Terrestrial ecosystems, increased solar ultraviolet radiation, and interactions with other climate change factors. *Photochemical & Photobiological Sciences* 6, 252–266. DOI:10.1039/B700019G

Cesco, S., Neumann, G., Tomasi, N., Pinton, R. and Weisskopf, L. (2010) Release of plant-borne flavonoids into the rhizosphere and their role in plant nutrition. *Plant and Soil* 329, 1–25.

Clarke, L.J., Robinson, S.A., Hua, Q., Ayre, D.J. and Fink, D. (2012) Radiocarbon bomb spike reveals biological effects of Antarctic climate change. *Global Change Biology* 18, 301–310.

Chipperfield, M.P. (2015) Global atmosphere – the Antarctic ozone hole. In: Harrison, R.M. and Hester, R.E. (eds) *Still Only One Earth: Progress in the 40 Years Since the First UN Conference on the Environment.* Royal Society of Chemistry, UK, pp. 1–33. DOI:10.1039/9781782622178-00001

Chipperfield, M.P., Dhomse, S.S., Feng, W., McKenzie, R.L., Velders, G.J.M. *et al.* (2015) Quantifying the ozone and ultraviolet benefits already achieved by the Montreal Protocol. *Nature Communications* 6, 7233. DOI:10.1038/ncomms8233

Davidson, E.A. and Kanter, D. (2014) Inventories and scenarios of nitrous oxide emissions. *Environmental Research Letters* 9, 105012. DOI:10.1088/1748-9326/9/10/105012

Ensminger, P.A. (1993) Control of development in plants and fungi by far-UV radiation. *Physiologia Plantarum* 88, 501–508.

Eyring, V., Waugh, D.W., Bodeker, G.E., Cordero, E., Akiyoshi, H. *et al.* (2007) Multimodel projections of stratospheric ozone in the 21st century. *Journal of Geophysical Research: Atmospheres* 112, D16303. DOI:10.1029/2006JD008332

Farman, J.C., Gardiner, B.G. and Shanklin, J.D. (1985) Large losses of total ozone in Antarctica reveal seasonal ClOx/NOx interaction. *Nature* 315, 207–210.

Ferreyra, M.F., Rius, S.P. and Casati, P. (2012) Flavonoids: biosynthesis, biological functions, and biotechnological applications. *Frontiers in Plant Science* 3, 222. DOI:10.3389/fpls.2012.00222

Gent, P.R. (2016) Effects of Southern Hemisphere wind changes on the meridional overturning circulation in ocean models. *Annual Review of Marine Science* 8, 79–94.

Gutt, J., Bertler, N., Bracegirdle, T.J., Buschmann, A., Comiso, J. *et al.* (2015) The Southern Ocean ecosystem under multiple climate change stresses: an integrated circumpolar assessment. *Global Change Biology* 21, 1434–1453. DOI:10.1111/gcb.12794

Hartmann, D.L., Tank, A.K., Rusticucci, M., Alexander, L.V., Brönnimann, S. *et al.* (2013) Observations: atmosphere and surface. In: Stocker, T.F., Qin, D., Plattner, G.-K., Tignor, M.B.M., Allen, S.K. *et al.* (eds) *Climate Change 2013: The Physical Science Basis. Contribution of Working Group I to the Fifth Assessment Report of the Intergovernmental Panel on Climate Change.* Cambridge University Press, Cambridge, UK and New York. pp. 159–254. DOI:10.1017/CBO9781107415324.008

Hurwitz, M.M., Fleming, E.L., Newman, P.A., Li, F. and Liang, Q. (2016) Early action on HFCs mitigates future atmospheric change. *Environmental Research Letters* 11, 114019.

Iglesias-Suarez, F., Young, P.J. and Wild, O. (2016) Stratospheric ozone change and related climate impacts over 1850–2100 as modelled by the ACCMIP ensemble. *Atmospheric Chemistry and Physics* 16, 343–363. DOI:10.5194/acp-16-343-2016

IPCC (2013) *Climate Change 2013: The Physical Science Basis. Contribution of Working Group I to the Fifth Assessment Report of the Intergovernmental Panel on Climate Change* (Stocker, T.F., Qin, D., Plattner, G.-K., Tignor, M.B.M., Allen, S.K. *et al.* (eds)). Cambridge University Press, Cambridge, UK and New York. DOI:10.1017/CBO9781107415324

Jansen, M.A. and Bornman, J.F. (2012) UV-B radiation: from generic stressor to specific regulator. *Physiologia Plantarum* 145, 501–504. DOI:10.1111/j.1399-3054.2012.01656.x

Jenkins, G.I. (2009) Signal transduction in responses to UV-B radiation. *Annual Review of Plant Biology* 60, 407–431.

Jenkins, G.I. (2014) The UV-B photoreceptor UVR8: from structure to physiology. *The Plant Cell* 26, 21–37.

Kigali Amendment (2016) Available at: https://treaties.un.org/doc/Publication/CN/2016/CN.872.2016-Eng.pdf (accessed 29 March 2017).

Kim, B.C., Tennessen, D.J. and Last, R.L. (1998) UV-B-induced photomorphogenesis in *Arabidopsis thaliana*. *The Plant Journal* 15, 667–674.

Klein, R.M. (1978) Plants and near-ultraviolet radiation. *The Botanical Review* 44, 1–127.

Kumagai, T. (1988) Photocontrol of fungal development. *Photochemistry and Photobiology* 47, 889–896.

Levy, H. (1971) Normal atmosphere: large radical and formaldehyde concentrations predicted. *Science* 173, 141–143.

Li, F., Vikhliaev, Y.V., Newman, P.A., Pawson, S., Perlwitz *et al.* (2016) Impacts of interactive stratospheric chemistry on Antarctic and Southern Ocean climate change in the Goddard Earth Observing System, version 5 (GEOS-5). *Journal of Climate* 29, 3199–3218.

Lucas, R.M., Norval, M., Neale, R.E., Young, A.R., de Gruijl, F.R. *et al.* (2015) The consequences for human health of stratospheric ozone depletion in association with other environmental factors. *Photochemical & Photobiological Sciences* 14, 53–87.

Madronich, S., Shao, M., Wilson, S.R., Solomon, K.R., Longstreth, J.D. *et al.* (2015) Changes in air quality and tropospheric composition due to depletion of stratospheric ozone and interactions with changing climate: implications for human and environmental health. *Photochemical & Photobiological Sciences* 14, 149–169.

Maycock, A.C. (2016) The contribution of ozone to future stratospheric temperature trends. *Geophysical Research Letters* 43, 4609–4616.

McKenzie, R.L., Aucamp, P.J., Bais, A.F., Björn, L.O., Ilyas, M. *et al.* (2011) Ozone depletion and climate change: impacts on UV radiation. *Photochemical & Photobiological Sciences* 10, 182–198. DOI:10.1039/C0PP90034F

Miller, T.E. and TerHorst, C.P. (2012) Indirect effects in communities and ecosystems. *Ecology (Oxford Bibliographies)*. Oxford University Press, Oxford, UK. DOI:10.1093/OBO/9780199830060-0020

Molina, M.J. and Rowland, F.S. (1974) Stratospheric sink for chlorofluoromethanes: chlorine atom-catalysed destruction of ozone. *Nature* 249, 810–812.

Myhre, G., Shindell, D., Bréon, F.M., Collins, W., Fuglestvedt, J. *et al.* (2013) Anthropogenic and natural radiative forcing. In: Stocker, T.F., Qin, D., Plattner, G.-K., Tignor, M.B.M., Allen, S.K. *et al.* (eds) *Climate Change 2013: The Physical Science Basis. Contribution of Working Group I to the Fifth Assessment Report of the Intergovernmental Panel on Climate Change*. Cambridge University Press, Cambridge, UK and New York, pp. 659–740. DOI:10.1017/ CBO9781107415324.018

NASA (2016) *Ozone Hole Watch*. Available at: https://ozonewatch.gsfc.nasa.gov/facts/ozone.html (accessed 29 March 2017).

Newman, P.A. and McKenzie, R. (2011) UV impacts avoided by the Montreal Protocol. *Photochemical & Photobiological Sciences* 10, 1152–1160.

Newman, P.A., Oman, L.D., Douglass, A.R., Fleming, E.L., Frith, S.M. *et al.* (2009) What would have happened to the ozone layer if chlorofluorocarbons (CFCs) had not been regulated? *Atmospheric Chemistry and Physics* 9, 2113–2128.

Nowack, P.J., Abraham, N.L., Maycock, A.C., Braesicke, P., Gregory, J.M. *et al.* (2015) A large ozone-circulation feedback and its implications for global warming assessments. *Nature Climate Change* 5, 41–45.

Paul, N.D. and Gwynn-Jones, D. (2003) Ecological roles of solar UV radiation: towards an integrated approach. *Trends in Ecology & Evolution* 18, 48–55.

Paul, N.D., Jacobson, R.J., Taylor, A., Wargent, J.J. and Moore, J.P. (2005) The use of wavelength-selective plastic cladding materials in horticulture: understanding of crop and fungal responses through the assessment of biological spectral weighting functions. *Photochemistry and Photobiology* 81, 1052–1060.

Portmann, R.W. and Solomon, S. (2007) Indirect radiative forcing of the ozone layer during the 21st century. *Geophysical Research Letters* 34. DOI:10.1029/2006GL028252

Ravishankara, A.R., Daniel, J.S. and Portmann, R.W. (2009) Nitrous oxide (N_2O): the dominant ozone-depleting substance emitted in the 21st century. *Science* 326, 123–125.

Raviv, M. and Antignus, Y. (2004) UV radiation effects on pathogens and insect pests of greenhouse-grown crops. *Photochemistry and Photobiology* 79, 219–226.

Revell, L.E., Tummon, F., Salawitch, R.J., Stenke, A. and Peter, T. (2015) The changing ozone depletion potential of N_2O in a future climate. *Geophysical Research Letters* 42, 10047–10055. DOI:10.1002/2015GL065702

Robinson, S.A. and Erickson, D.J. (2015) Not just about sunburn – the ozone hole's profound effect on climate has significant implications for Southern Hemisphere ecosystems. *Global Change Biology* 21, 515–527.

Robson, T., Klem, K., Urban, O. and Jansen, M.A. (2015) Re-interpreting plant morphological responses to UV-B radiation. *Plant, Cell & Environment* 38, 856–866.
Searles, P.S., Flint, S.D. and Caldwell, M.M. (2001) A meta-analysis of plant field studies simulating stratospheric ozone depletion. *Oecologia* 127, 1–10.
Shepherd, T.G. (2008) Dynamics, stratospheric ozone, and climate change. *Atmosphere-Ocean* 46, 117–138. DOI:10.3137/ao.460106
Solomon, S., Ivy, D.J., Kinnison, D., Mills, M.J., Neely, R.R. and Schmidt, A. (2016) Emergence of healing in the Antarctic ozone layer. *Science* 353, 269–274.
Stolarski, R.S., Douglass, A.R., Oman, L.D. and Waugh, D.W. (2015) Impact of future nitrous oxide and carbon dioxide emissions on the stratospheric ozone layer. *Environmental Research Letters* 10, 034011. DOI:10.1088/1748-9326/10/3/034011
Thompson, D.W. and Solomon, S. (2002) Interpretation of recent Southern Hemisphere climate change. *Science* 296, 895–899. DOI:10.1126/science.1069270
Thompson, D.W., Solomon, S., Kushner, P.J., England, M.H., Grise, K.M. *et al.* (2011) Signatures of the Antarctic ozone hole in Southern Hemisphere surface climate change. *Nature Geoscience* 4, 741–749.
Turner, J., Barrand, N.E., Bracegirdle, T.J., Convey, P., Hodgson, D.A. *et al.* (2014) Antarctic climate change and the environment: an update. *Polar Record* 50, 237–259.
United Nations Environment Programme, Environmental Effects Assessment Panel (2016) Environmental effects of ozone depletion and its interactions with climate change. Progress report, 2015. *Photochemical & Photobiological Sciences* 15, 141–174. DOI:10.1039/C6PP90004F
Villalba, R., Lara, A., Masiokas, M.H., Urrutia, R., Luckman, B.H. *et al.* (2012) Unusual Southern Hemisphere tree growth patterns induced by changes in the Southern Annular Mode. *Nature Geoscience* 5, 793–798.
Wargent, J.J. and Jordan, B.R. (2013) From ozone depletion to agriculture: understanding the role of UV radiation in sustainable crop production. *New Phytologist* 197, 1058–1076.
Waugh, D.W., Oman, L., Kawa, S.R., Stolarski, R.S., Pawson, S. *et al.* (2009) Impacts of climate change on stratospheric ozone recovery. *Geophysical Research Letters* 36, L03805. DOI:10.1029/2008GL036223
Williamson, C., Zepp, R., Lucas, R., Madronich, S., Austin, A.R. *et al.* (2014) Solar ultraviolet radiation in a changing climate. *Nature Climate Change* 4, 434–441.
Watanabe, S., Sudo, K., Nagashima, T., Takemura, T., Kawase, H. and Nozawa, T. (2011) Future projections of surface UV-B in a changing climate. *Journal of Geophysical Research: Atmospheres* 116, D16118. DOI:10.1029/2011JD015749
WMO (2015) *Scientific Assessment of Ozone Depletion: 2014.* Global Ozone Research and Monitoring Project Report No. 55. World Meteorological Organization, Geneva, Switzerland.
Zaller, J.G., Caldwell, M.M., Flint, S.D., Scopel, A.L., Salo, O.E. and Ballaré, C.L. (2002) Solar UV-B radiation affects below-ground parameters in a fen ecosystem in Tierra del Fuego, Argentina: implications of stratospheric ozone depletion. *Global Change Biology* 8, 867–871.

2 Quantification of UV Radiation

Pedro J. Aphalo*
Department of Biosciences, University of Helsinki, Finland

Introduction

The accuracy needed in the quantification of exposure for research on the effects of UV-B radiation is similar to that required for visible radiation, but it requires much more effort to achieve (Aphalo, 2016). When measuring the UV-B component of solar radiation at ground level, the main difficulty is that this component is only a very small fraction of the global irradiance. Based on a standardized 1.5-air-mass global radiation spectrum for middle latitudes (ASTM G173), 0.015% of photons are in the UV-B region. Even if we use photosynthetically active radiation PAR (400–700 nm) instead of global radiation (280–4000 nm) as a reference, less than 0.1% of photons are in the UV-B region (computed with the R for photobiology suite of packages, see Aphalo *et al.*, 2016). If we consider the spread across the whole day or wintertime, the contribution of UV-B is even smaller. On the other hand, UV-B radiation is very effective in eliciting responses in organisms. Taking both things together, an error in the quantification of UV-B irradiance that is extremely small compared to global or PAR photon irradiance can be biologically highly relevant. Even under a clear sky at noon in summer, when UV-B irradiance is at its maximum, a measurement error equivalent to erroneously detecting only 0.05% of the PAR photons impinging on a UV-B sensor as UV-B photons can lead to an overestimation of UV-B irradiance by 20% or more. In this chapter I highlight the difficulties inherent in the quantification of UV-B radiation and discuss practical ways of obtaining reliable estimates of UV-B irradiance and exposure in spite of these difficulties. I keep the presentation simple and avoid complicated maths and physics. A deeper and more detailed discussion of these issues as well as those related to other aspects of research on the effects of UV radiation on plants is available in the publication *Beyond the Visible: A Handbook of Best Practice in Plant UV Photobiology* (Aphalo *et al.*, 2012a; Björn *et al.*, 2012).

I analyse the quantification of UV-B exposure and irradiance as a process that includes all the different steps affecting the final estimate. This is similar to approaches used in metrology; however, I discuss the uncertainties introduced at each step in the quantification process qualitatively, as the exact size of errors introduced at each step varies. The sections below discuss the different sources

*Corresponding author: pedro.aphalo@helsinki.fi

of errors and uncertainties and describe the typical data flow involved in the quantification of radiation.

The Definition of UV-B

According to ISO and CIE standards, ultraviolet-B radiation is the region of the spectrum between 280 nm and 315 nm (Björn, 2015). Other definitions are sometimes used, such as those having 320 nm as the boundary between UV-A and UV-B regions (Zeman, 2016). For the quantification of UV-B radiation in sunlight, this difference of only 5 nm makes a huge difference in the 'UV-B' irradiance estimate. The use of 320 nm as the boundary can yield a value that is more than twice that obtained with the standard definition, and this discrepancy depends on solar elevation. Consequently, stating 'UV-B radiation' alone is ambiguous and introduces uncertainty. To avoid this problem, whenever feasible, the ISO definition should be used and its usage stated when describing methods. When the only calibration available for a broadband instrument is based on a non-standard definition of the UV-B spectral band, the definition used should be clearly indicated and the implications with respect to comparison to other studies discussed.

Spectral Irradiance versus UV-B Irradiance

There are good practical reasons to use spectroradiometers in preference to broadband sensors. A very important one is that, for a spectrometer, a single calibration is valid for measuring any light source; broadband UV sensors require a separate calibration for each light source to be measured. The exception are thermopile-based pyranometers when measuring global radiation in energy-based units.

In the case of broadband UV-B sensors, calibration factors for sunlight and fluorescent UV-B lamps can differ by a factor of two or more. Ignoring the need for separate calibrations causes huge errors that are difficult to detect unless methods are described in enough detail with respect to the type, number and distance to lamps, plus any filtering.

It is regrettable that many suppliers of broadband UV sensors do not highlight this requirement in their brochures and user guides. One positive exception is Irradian Ltd (Tranent, Scotland, UK; formerly Macam), a company that sells readout devices that can be programmed with up to five different calibration constants. With these UV radiometers, it is easy to choose the calibration used based on the type of lamp or light source being measured at the time of measurement. The several different types of calibrations available are also clearly explained in the documentation.

Any broadband UV sensor can be calibrated to match different light sources by comparison to a properly calibrated spectroradiometer with a stray-light specification good enough for reliable measurement of the light source in question. In the case of the study of long-term trends or geographical variation in UV-B radiation, intercomparisons of instruments may be needed to obtain reliable data (Leszczynski *et al.*, 1998; Huelsen *et al.*, 2008).

Factors that affect the validity of a calibration are various: in the short term, difference in sensor temperature between the calibration condition and the measurement condition can be an important source of errors. This is especially relevant for some traditional types of UV-B broadband sensors based on fluorescent pigments, for which only types with thermostatic heating can be used reliably in the field. Some sensors do have built-in temperature sensors allowing the correction of readings within a more restricted range of temperatures. Readings from semiconductor-based sensors, e.g. based on silicon photodiodes and SiC devices, are much less affected by temperature changes.

Spectrometer readings can also be affected by temperature changes, unless the detector itself or the whole device is cooled to a constant temperature. The main effect of increased spectrometer temperature is an increase in dark noise, which can be to some extent corrected by frequent dark readings and averaging of multiple measurements. Another effect of changes in temperature on array spectrometers is a shift of the wavelength calibration, an effect that depends on individual instrument design.

Several reports from the World Meteorological Organization give recommendations for the use of different types of instruments in the quantification of UV-B radiation (Seckmeyer *et al.*, 2001, 2005, 2010a, b).

Signal and Noise

Different types of errors can affect the baseline reading of a light sensor: drift of the sensor signal in darkness as a result of electrical noise and/or thermal radiation, and stray light, which can be thought as light that has strayed from its path due to optical imperfections like reflections on the inside of a spectrometer, or 'leakage' of unwanted wavelengths by the optical filters used in broadband sensors. The dark signal is relatively easy to correct, as it can be measured by blocking radiation from reaching the sensor. Stray light is more difficult to measure because it cannot be easily separated from the radiation to be measured. Furthermore, its contribution to the instrument reading can depend on the spectrum of the light source being measured. Problems with filters can be characterized by measuring the sensor response to monochromatic light using a spectrograph, and comparing readings to those of a thermopile. Some broadband UV sensors are based on light detection devices that are inherently 'blind' to visible radiation, such as silicon carbide (SiC) devices, and are immune to errors caused by strong visible light. The principles of operation of broadband sensors are discussed by Björn *et al.* (2012), as well as of dosimeters which we do not discuss here.

To be able to reject enough visible light when measuring UV-B radiation in sunlight, so as to achieve a usable estimate, say an overestimation of not more than 10%, we need a sensor with a sensitivity at least four orders of magnitude higher in the UV-B band than in the visible and infrared regions. This is enough only when the sun elevation is high. To achieve precise estimates when the sun is lower in the sky, four orders of magnitude may be too little. Single monochromator spectrometers achieve a stray light level that is only approximately three orders of magnitude, making them unsuitable for measuring UV-B radiation in sunlight. It should be remembered that all spectrometers using array detectors have to be by necessity built with a single monochromator in the light pass. Unless very special measuring protocols that include the measurement and subtraction of the stray light are used, array spectrometers should never be used to measure the UV-B component of sunlight (Ylianttila *et al.*, 2005; Kreuter and Blumthaler, 2009). Of course, when measuring artificial light sources, in the absence of strong visible radiation, these instruments are perfectly capable of giving reliable estimates of UV-B irradiance. For measuring UV-B radiation in sunlight, or of supplementation of sunlight with UV-B lamps, a double monochromator scanning spectroradiometer is needed. In the case of supplementation, a workaround is to measure the lamp output in the absence of sunlight, e.g. by blocking sunlight or at night. If this approach is used, one must be careful not to expose the target plants to UV-B radiation in the absence of 'normal' levels of visible radiation.

Workflow for UV-B Quantification through Measurement

A crucial step is the calibration of the instruments to be used. How frequently an instrument needs to be calibrated depends on how fast its responsivity changes in time and the maximum allowable error size we set. In addition to steady drift, calibration of instruments can change as a result of non-catastrophic malfunctions. When faced with such a malfunction, all data acquired between a good and a bad calibration has to be discarded, adding another reason to prefer frequent recalibration, or at least checking by comparison to another instrument or detector. One needs to consider the cost of repeating the measurements, sometimes requiring whole experiments to be redone when deciding on the frequency of recalibration. Another consideration is the harshness of the environment under which the instrument is used, as for example, prolonged exposure of broadband sensors to strong UV-B or UV-C radiation can cause deterioration of optical filters and in some cases the detectors themselves, leading to changes in overall sensitivity and even spectral response characteristics. A reasonable rule of thumb is to recalibrate instruments used for measuring UV-B radiation at least once per year. It is also important to keep in mind that calibration in most cases is done against standard lamps or calibrated instruments that are not primary standards. In other words, the calibration is usually transferred several times before calibration of a user's instrument. As a consequence of this, errors accumulate, and calibrations of UV-B instruments, even if done with utmost care and the best equipment, are subject to considerable errors. In practice accumulated calibration uncertainties can be as large as ±10%.

Once we have suitably calibrated instruments, UV-B quantification involves several steps as presented in Fig. 2.1. Depending on the instrument and software used for data acquisition the first few steps of the workflow (dashed boxes) will be applied automatically, but it is important for the user to understand what is happening behind the scenes as the validity of the acquired data depends on these steps.

Acquisition of a raw reading from a detector (column (a) in Fig. 2.1) is the first step in any radiation measurement. The acquisition step depends on the type of detector (e.g. reading of counts in the case of digital detectors or measurement of an analogue current or voltage in the case of analogue detectors). The most basic test is to check that the detector is not saturated, i.e. that the acquired signal is not clipped and that the reading is being obtained using an integration time or sensitivity setting that provides enough resolution. Frequently, CCD and CMOS arrays, have a few malfunctioning pixels, e.g. 'hot', 'cold' and 'dead' pixels that

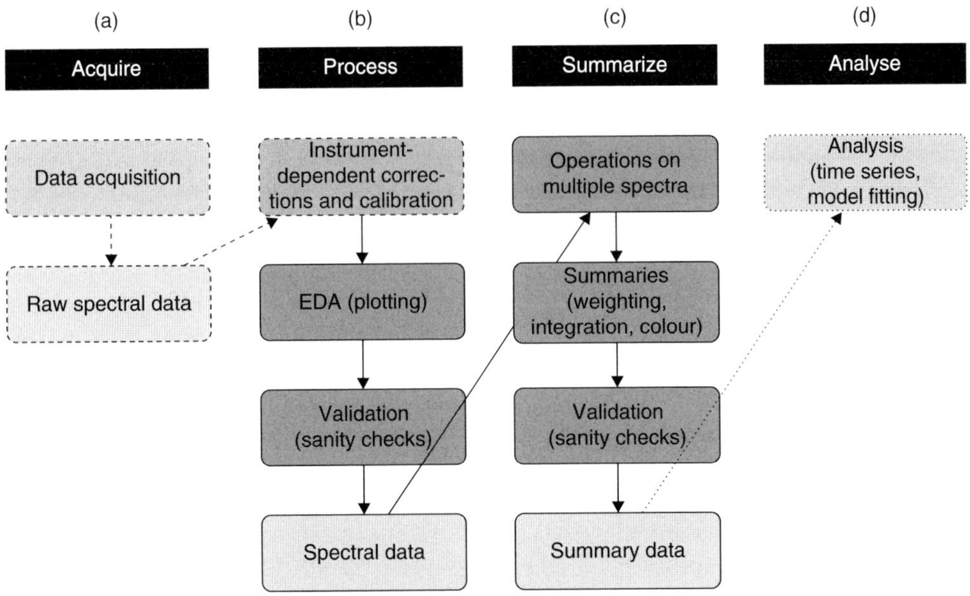

Fig. 2.1. Data flow for the quantification of radiation by spectroscopy. Four main steps (a)–(d) can be distinguished, in the flow of spectral data from acquisition to their use in the interpretation of experimental results. Based on Aphalo *et al.* (2016).

produce bad data that needs to be culled before further calculations.

The second step (column (b) in Fig. 2.1) is the conversion of the raw readings into values expressed in physical units of radiation – in other words first correcting for the dark signal, and possible non-linearity of sensor response, and then multiplying by a calibration constant. In the case of an array spectrometer, these calculations need to be done individually for each pixel in the array detector, and a wavelength value matched to each pixel in the array. In other words, the calibration of an array spectrometer is recorded as a vector of up to a few thousands calibration multipliers, plus coefficients of a function relating pixel positions in the detector array to wavelengths. In this step we also include the very important process of data 'sanity' checks. It is quite easy, with experience or by searching the literature, to estimate what irradiance to expect from a certain setup of filters and/or lamps and compare these values to what one has measured. The overall shape of the emission spectrum of a source can be also checked in this way. This step, if done carefully, will guarantee that no *gross* errors in quantification remain undetected and end up being reported.

The third step (column (c) in Fig. 2.1) consists in summarizing spectral data as totals by region of the spectrum, either not weighted or weighted with a biological spectral weighting function (BSWF). In many cases each of these observations may be the average of several instrument readings, used to reduce uncertainties caused by instrument instability. Radiation conditions will vary in time and space, even in controlled environments, so even after having taken care of uncertainties related to instrumentation, the variation inherent to the plants exposure or growing conditions needs to be quantified, leading to an additional step of data analysis.

The fourth step (column (d) in Fig. 2.1) consists mostly of the statistical analysis of the data obtained by replicated measurements in time and/or space. For example, in a controlled environment we usually need to map the spatial variation in irradiance; or in a canopy we may be interested in the change in irradiance and spectral composition with depth in the canopy. This type of measurement is frequently needed even when variation in the radiation environment is not the subject under study. As an example, Fig. 2.2 shows a map of solar UV-B attenuation under a polyester filter.

Geometry Considerations

The first consideration is the heterogeneity due to light sources, and after that, the hetereogeneity dependent on plant morphology and optical properties. When the light source is the sun, the relatively small yearly variation in the Sun–Earth distance, has only a minor effect on solar irradiance at ground level. In contrast, the solar elevation angle determines the length of the path of solar radiation through the atmosphere. The effect of solar elevation is the main cause of the well-known daily and yearly time courses of solar UV-B irradiance. When we consider artificial light sources located usually some tens of centimetres from the top of the plants, the varying distance to the light source as plants grow can significantly affect irradiance at the top of the plants. For a point source, like a single light bulb, irradiance will vary proportionally to the square of the distance because the radiation beam 'spreads' in two dimensions. With tubular lamps and arrays of many small light sources like LEDs, when located in an enclosed space with reflective walls, the decay with distance will be less abrupt but still important. In practice, this means that either (i) the distance between lamps and the top of the plants must be maintained approximately constant all the way through an experiment, or (ii) that the irradiance or spectral irradiance must be repeatedly measured throughout an experiment's duration. The biggest problems are in those cases when the radiation treatment affects the height of the plant(s), and this change in height in turn affects the radiation treatment. Another case is when the different genotypes being compared grow to different heights. Outdoors, the variations in growth do not affect the irradiance the plants are exposed to, as long as they grow

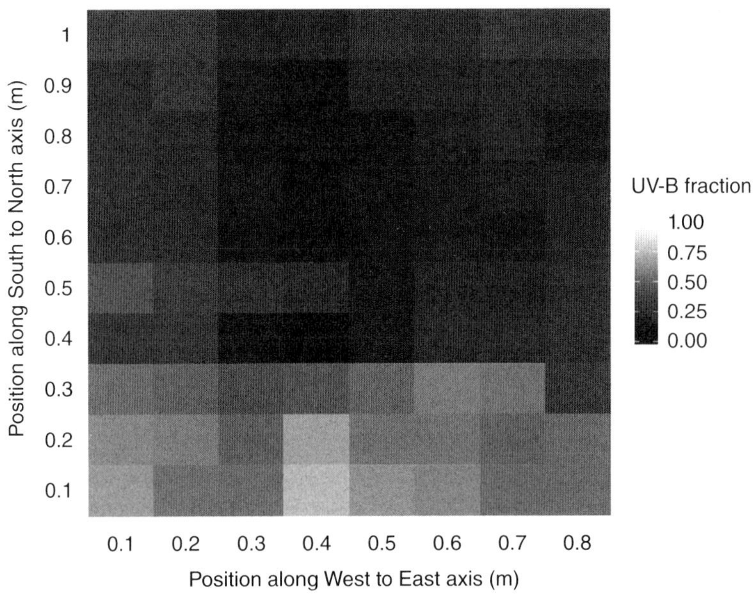

Fig. 2.2. A UV-B relative irradiance map under a 1 x 0.8 m polyester filter set outdoors. Measured at noon, under clear sky conditions, at 0.4 m below the filter. (Based on Fig. 2.31C from Aphalo *et al.* (2012b) (data of T. Matthew Robson, Saara Hartikainen and Oriane Loiseau).)

in a pure canopy. In a controlled environment, however, such variable growth could drastically affect the levels of exposure.

When quantifying the radiation received by a plant, we need to consider its three-dimensional structure. The morphology of plants, including positions of organs, affects the irradiance at the surface of these organs. There are two aspects to this question: (i) shading and (ii) the angle between the radiation beam and the plant surface. Inside the canopy these two effects are interlinked. Surfaces perpendicular to the radiation beam will be exposed to higher irradiances than those surfaces displayed at a shallower angle, so heliotropism and other nastic movements affect the UV-B irradiance received (Bawhey *et al.*, 2003). Unless plants are very small, of the same height and growing apart from each other, and display leaves at similar angles, the irradiance and spectrum impinging on different plants and on different parts of the shoot will vary also as a consequence of shading, or radiation attenuation, and/or reflection by other organs of the same and neighbouring plants. When interpreting results for genotypes of different morphology, or when using treatments that affect morphology, the differences in the effective UV-B exposure on the organs of plants with different morphologies must be taken into account. Some, but not all, of these shading and reflection dependent effects can be decreased by having ample spacing between individual plants.

We should also be aware of obstacles in the path of radiation. Obstacles such as beams supporting the roof of a greenhouse, or frames supporting lamps or filters affect the UV and visible radiation field the plants are exposed to. They cause shaded patches whose position depends on the position of the obstacle relative to the plants and on the position of the sun. Obstacles, like walls, or even more frequently the instrument operator, can function either as reflectors or shades depending on their position. Interference by operators should be avoided, and that of obstacles permanently positioned near the plants quantified when it cannot be avoided. Measurements should capture temporal and spatial variation, and in many cases this requires measurements at different solar elevations and under both clear sky and overcast conditions. A corollary to this is that filter and lamp frames should be identical, and located at exactly the same distance from plants in all the treatments being compared, i.e. irradiance should never be adjusted by having different distances between lamp frames and plants in different treatments (Flint *et al.*, 2009).

The Spectrum of UV-B Light Sources

Ideally when experimentally studying the effect of UV-B radiation one should use a source emitting radiation only in this region of the spectrum. Such sources exist but are for most research projects prohibitively expensive or only able to illuminate a space that is too small. UV lasers, even tuneable UV lasers, are available, but expensive, and of relatively low power and consequently can only illuminate a small area. LEDs emitting UV-B are also available, but have low output, are very inefficient compared to LEDs emitting at longer wavelengths, relatively short-lived and very expensive. Once again, they are useful for illuminating at most a few plants for a relatively short period of time. The third possibility is the use of spectrographs, based on a Xenon-arc lamp and a monochromator. They are expensive and have relatively high running costs, and in most cases illuminate a rather small area but have good wavelength resolution and permit simultaneous irradiation of different samples, each exposed to a different narrow range of wavelengths.

In research with plants, the most frequently used UV-B radiation sources are low pressure mercury lamps with special 'phosphors' as coating of the ampoule. They differ from household fluorescent tubes and lamps in the composition of the 'phosphor'. Notwithstanding the trade names under which they are sold, only a fraction of the total radiation emitted by these lamps is in the UV-B region of the spectrum. This adds several complications to the interpretation of experimental results unless all the required control treatments are included in an experiment. Dark controls are not useful, as they just reveal the effect of lamps as a whole, rather than what we are usually interested in: the effect of the UV-B portion of the radiation emitted. Adding a second type of control, with lamps optically filtered to block UV-B wavelengths but pass longer wavelengths, allows us to separate the effect of UV-B from all other lamp effects bulked together. To separate the effects of different portions of the emitted radiation, further types of filters could be used. Finally, a filter blocking all emitted UV and visible radiation would allow us to observe other possible effects of energized lamps, like those of electrical fields and thermal radiation.

An example of the dangers of using dark controls is the apparent enhancement of seed germination by UV-B radiation, that in the example given in Fig. 2.3 could be almost fully explained by the small amount of orange radiation emitted by UV-B lamps. This example highlights that even photon fluences that may be difficult to measure or would seem biologically irrelevant can, under some conditions, be confounded with effects of UV-B radiation unless adequate controls are included in experiments.

A problem extensively discussed in the literature is that of simulating the effect of stratospheric ozone depletion on UV-B exposure by means of lamps that cannot reproduce the change in spectrum that actual ozone depletion causes (Caldwell *et al.*, 1986; McLeod, 1997). This has led to the extensive use of different BSWFs in an attempt to adapt lamp-based UV-B treatments to mimic the predicted increase in solar UV-B radiation triggered by stratospheric ozone depletion. One can think of BSWFs as 'transfer functions' for predicting effects or expressing treatments in terms of different conditions that are difficult to achieve experimentally. Conceptually they are similar to quantities used to quantify illumination using the spectral response of the human eye as BSWF, or PAR using a rather arbitrary BSWF that nonetheless 'works well' across many different plant species. The use of BSWFs in UV-B research has become less important as the focus has shifted to other areas more closely related to photoperception and ecological roles under 'normal' conditions and away from the effects of ozone depletion.

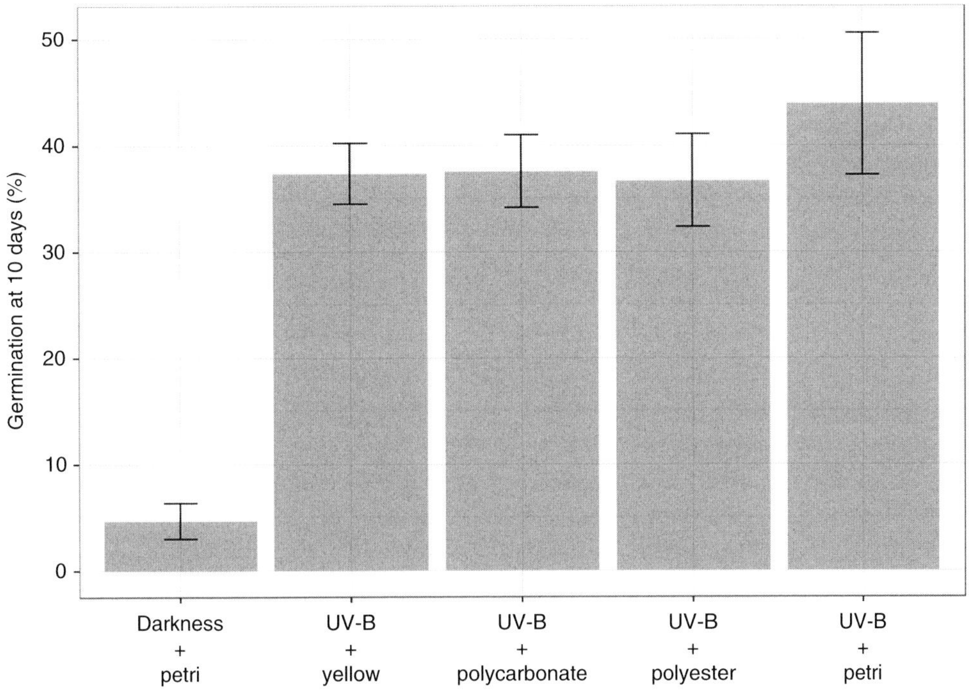

Fig. 2.3. Germination of seeds of silver birch (*Betula pendula*) exposed to radiation from UV-B lamps. Seeds were irradiated in plastic petri dishes with lids transmitting approximately 70% of the UV-B radiation from Q-Panel UVB313 lamps filtered with cellulose acetate plus yellow, polycarbonate or polyester as long-pass filters blocking all UV and blue radiation, all UV radiation and UV-B radiation, respectively. The dark controls were located on the same bench of the dark room, beside the irradiated ones, but wrapped in several layers of black polythene film. Irradiation time was 12 h, and seeds were not exposed to other light except very briefly to a very weak green safelight. (Unpublished data of Xavier Anguera and Pedro J. Aphalo.)

Simulation of the Solar Spectrum for Estimation of UV-B Exposure

The accuracy of simulations of the solar spectrum at ground level can be surprisingly good, in some cases better than actual measurements with ordinary broadband sensors (Lindfors *et al.*, 2007, 2009). This good accuracy depends, however, on the availability of good estimates of cloudiness. Such good estimates can be obtained from hourly global radiation data at ground level, which is frequently available from meteorological stations. Cloudiness is affected by topography and the presence of large bodies of water, so global radiation data should be acquired at a nearby location with a topography similar to that of the site of interest. Especially under difficult measuring conditions, low solar elevations, or extreme temperature, simulations based on local global radiation measurements can outperform routine measurements with normal broadband instruments. The other big advantage is that if local global radiation data is available together with a regional estimate of stratospheric ozone concentration, the simulations can be used to estimate exposure to UV-B (or any other region of the solar spectrum) retrospectively. Estimates based purely on satellite data are not usually as good because the cloud imaging data from orbiting satellites is sparser over time because it is available once per satellite overpass rather than, at least in theory, continuously for a stationary ground based sensor. The radiation transfer model most frequently used for ultraviolet radiation estimation is currently libRadtran (Emde *et al.*, 2016). The core program of the libRadtran package is the radiative transfer tool uvspec.

Uvspec was originally designed to calculate spectral irradiance and actinic flux in the ultraviolet and visible parts of the spectrum, where the name uvspec stems from. Over the years, uvspec has undergone numerous revisions that added extensions and led to many improvements. The current version of uvspec covers the full solar spectrum (UV, visible and infrared regions) from 120 nm to 100 μm. Sasha Madronich's TUV model is also frequently used.

For this type of estimate (i.e. UV-B) libRadtran is especially suitable. While TUV can also provide good estimates in the ultraviolet region, the range of wavelengths covered is smaller. Although TUV simulates in more detail the atmospheric chemistry, when interested only in the solar spectrum at ground level, being able to simulate a broader range of wavelengths can make the use of libRadtran preferable.

Software for Calculations

When quantifying UV-B radiation, many biologists find that the biggest stumbling blocks are related to processing and summarizing of spectral data, including the computation of meaningful summaries. In addition, when doing data sanity checks, or planning experiments, access to suitable example data to use for comparisons can be limited.

A suite of R packages has been developed (Aphalo, 2015; Aphalo *et al.*, 2016) with two aims in mind. One aim is to provide tools for teaching and learning about VIS and UV radiation physics and photobiology. The other aim is to make it easier for researchers in the field of photobiology to do calculations required for the description of irradiation conditions and for simulations useful for data validation and/or when designing experiments. The suite is a collection of classes, methods and functions, accompanied by data sets. In particular, the large sets of example data will make it easy to carry out sanity checks of newly acquired and/or already published data.

Given the expected audience of both students and biologists, rather than data analysts, or experienced programmers, we have aimed at designing a consistent and easy to understand paradigm for the analysis of spectral data. The design is based on my own experience as a user, and on feedback from our students and 'early adopters'.

Most elements of the framework are used by all packages in the suite:

- Spectral objects are containers for different types of spectral data, data which is referenced to wavelength. These data normally originate in measurements with spectrometers or simulations with models.
- Containers for spectral objects are used to store collections of spectral objects, such as time series of spectral objects or spectral images, or other sets of related spectral data.
- Waveband objects are containers of instructions for the quantification of spectral data. In addition to the everyday definition as a range of wavelengths, we include the spectral weighting functions used in the calculation of what are frequently called weighted or effective exposures and doses.
- Maths operators and functions for operations on spectral objects are used to combine and/or transform spectral data, and in some cases to apply weights defined by wavebands. They handle possible mismatches in the wavelength values between the operands or arguments automatically by interpolation.
- Apply methods are used to apply functions to each individual spectrum stored in collections of spectra. They replace explicit iteration loops such as for-loops with faster code with a more convenient syntax.
- Summary methods and functions are defined for calculation of summary quantities such as irradiance and transmittance. Different summary functions return different quantities through integration over wavelengths and take as arguments spectra and wavebands.
- Plot methods for spectral objects simplify the construction of specialized

plots of spectral data. As data is stored in spectral objects using known units and these objects contain also metadata, these methods can automatically construct plots with suitable axis labels and other annotations.
- Foreign data exchange functions can be used for importing data output by diverse measuring instruments including various spectrometers and data loggers. Other functions allow two-way exchange of data with other R packages or classes defined in base R.

By using an object-oriented design, methods of the same name can be defined for objects of different classes, some of them already implemented for other classes in R itself. This reduces the number of names a user needs to remember. For example, there are many implementations of methods plot and print, but the user needs only one name for each of these operations. The order and naming of arguments is consistent throughout the suite, and also consistent with base R methods and functions whenever possible.

Here is an example using packages photobiology and photobiologyWavebands:

q_irrad(sun.spct, UVB())

Example data of the spectral irradiance of sunlight is used as the first argument (sun.spct) to summary function 'q_irrad()' so as to obtain the summary quantity photon (=quantum) irradiance. If we supply as a second argument a waveband defining the wavelength range for UV-B, we obtain UV-B photon irradiance with no weighting function applied.

e_irrad(sun.spct, UVB())

By substituting 'e_irrad' for 'q_irrad' in the code above, we obtain the summary quantity energy irradiance instead of photon irradiance.

e_irrad(sun.spct, CIE())

The same example data of the spectral irradiance of sunlight (sun.spct) is used as the first argument to summary function 'e_irrad()'. Here the second argument, 'CIE()', is a waveband object defining the BSWF corresponding to the definition of the erythemal action spectrum from the 1998 CIE (International Commission on Illumination) Standard. The code now gives us the biologically effective UV irradiance according to the standard set out by the CIE.

*e_irrad(sun.spct * polyester.spct, CIE())*

By means of the multiplication operator '*' we effect the convolution of data of spectral irradiance for sunlight (sun.spct) with data of spectral transmittance for a polyester film (polyester.spct), obtaining an approximation of the spectral irradiance of filtered sunlight. This intermediate result is then used as the first argument to summary function 'e_irrad()'. As we again supply as the second argument a waveband object defining the CIE98 BSWF, we obtain the biologically effective UV irradiance according to CIE for sunlight filtered by a polyester film. Of course, instead of using example data included in the suite, the user can use any measured spectral irradiance and spectral transmittance data of interest.

plot(sun.spct)

Generates a fully formatted and annotated plot of the solar spectrum.

plot(polyester.spct)

Generates a fully formatted and annotated plot of the spectral transmittance of a polyester film.

*plot(sun.spct * polyester.spct)*

Generates a fully formatted and annotated plot of the solar spectrum filtered by a film of polyester.

plot(sun.spct, range = c(290:410))

Generates a fully formatted and annotated plot of the solar spectrum, for wavelengths in the range 290–410 nm.

*plot(sun.spct * CIE(), range = c(290:410))*

Generates a fully formatted and annotated plot of biologically effective spectral irradiance according to the CIE98 BSWF, for wavelengths in the range 290–410 nm.

These are only a few simple examples. The suite defines and uses eight different classes of spectral objects for different types of spectral data, and corresponding classes for collections of these spectral objects. All mathematical operators and functions in base R are also defined for objects of these classes, as well as plot, print, and numerous summary methods. In the case of plotting, not only plot methods are defined, but several extensions to package 'ggplot2' are also implemented so as to facilitate the plotting of spectral data with full control of the resulting plot by users.

Up to date information on the suite is available at http://www.r4photobiology.info. A handbook of photobiological calculations with R is under preparation, with a draft already available at http://www.leanpub.com/r4photobiology/.

Checklists

This chapter concludes with a summary of the main points discussed in the form of two checklists that can be used as a guide when quantifying UV-B radiation with broadband sensors and spectrometers.

Checklist for broadband sensors

1. Is the sensor suitable for the measurement attempted? Suitable range of irradiance sensitivity, good-enough suppression of extraneous wavelengths, insensitive or corrected for the environmental variation expected (e.g. temperature), robust with respect to expected disturbances (e.g. rain).
2. Is a valid calibration available? The sensor has been calibrated under the light source to be measured. The calibration is recent enough to be valid. The calibration has been done at a similar temperature, or if not, the effect of the difference in temperature can be corrected for.
3. Sanity check at time of measurement. The sensor is in the intended position, usually perfectly horizontal. There are no extraneous reflections or shading caused by the measurement procedure or operator. Dark reading is equal to zero or if not, recorded for later correction of readings. Highest reading is not off scale.
4. Sanity check of processed data. The (photon/energy) irradiance values are similar to what could be expected for the light source being measured.
5. Do the readings provide all the information needed? Temporal and spatial variation. Enough replication for reliable estimates.
6. Are obtained summary quantities sufficient for the correct interpretation of observed plant responses?

Checklist for spectrometers

1. Is the spectrometer suitable for the measurement attempted? Stray light level low enough for meaningful spectral readings in the UV-B region. Suitable range of spectral irradiance sensitivity in the wavelength range of interest. Good-enough spectral resolution. Insensitive or corrected for the environmental variation expected (e.g. temperature), robust with respect to expected disturbances (e.g. handling).
2. Is a valid calibration available? The calibration is recent enough to be valid. The calibration has been done at a similar temperature, or if not, the effect of the difference in temperature can be corrected for.
3. Sanity check at time of measurement. The cosine diffuser used as entrance optics is in the intended position, usually perfectly horizontal. There are no extraneous reflections or shading caused by the measurement procedure or operator. Dark reading is equal to zero or if not, recorded to later correction of readings. Highest reading does not cause clipping of the spectrum near or in a region of interest.
4. Sanity check of processed data. The shape of the spectrum matches expectations for the source being measured. The (photon/

energy) irradiance values are similar to what could be expected for the light source being measured.

5. Do the readings provide all the information needed? Temporal and spatial variation. Enough replication for reliable estimates.

6. Are obtained summary quantities sufficient for the correct interpretation of observed plant responses?

Acknowledgements

I thank Titta K. Kotilainen for comments on an early version of the manuscript. Discussions over many years with Andy McLeod, Lars Olof Björn, Lasse Ylianttila, Anders Lindfors, Nigel Paul, T. Matthew Robson and others have influenced the ideas presented in this chapter.

References

Aphalo, P.J. (2015) The r4photobiology suite: spectral irradiance. *UV4Plants Bulletin* 2015, 1, 19–27.

Aphalo, P.J. (2016) Measuring solar UV-B: Why is it so difficult? *UV4Plants Bulletin* 2016, 1, 21–27.

Aphalo, P.J., Albert, A., Björn, L.O., McLeod, A.R., Robson, T.M. and Rosenqvist, E. (eds) (2012a) *Beyond the Visible: A Handbook of Best Practice in Plant UV Photobiology*. University of Helsinki, Department of Biosciences, Division of Plant Biology, Helsinki.

Aphalo, P.J., Albert, A., McLeod, A.R., Heikkilä, A., Gómez, I. *et al.* (2012b) Manipulating UV radiation. In: Aphalo, P.J., Albert, A., Björn, L.O., McLeod, A.R., Robson, T.M. and Rosenqvist, E. (eds) *Beyond the Visible: A Handbook of Best Practice in Plant UV Photobiology*. University of Helsinki, Department of Biosciences, Division of Plant Biology, Helsinki, pp. 35–70.

Aphalo, P.J., Albert, A., Robson, T.M. and Kotilainen T.K. (2016) *R for Photobiology: Theory and Recipes for Common Calculations*. Leanpub. Available at: http://www.leanpub.com/r4photobiology (accessed 31 March 2017).

Bawhey, C.I., Grant, R.H. and Gao, W. (2003) Digital measurement of heliotropic leaf response in soybean cultivars and leaf exposure to solar UVB radiation. *Agricultural and Forest Meteorology* 120, 161–175.

Björn, L.O. (2015) Ultraviolet-A, B, and C. *UV4Plants Bulletin* 2015, 17–18.

Björn, L.O., McLeod, A.R., Aphalo, P.J., Albert, A., Lindfors, A.V. *et al.* (2012) Quantifying UV radiation. In: Aphalo, P.J., Albert, A., Björn, L.O., McLeod, A.R., Robson, T.M. and Rosenqvist, E. (eds) *Beyond the Visible: A Handbook of Best Practice in Plant UV Photobiology*. University of Helsinki, Department of Biosciences, Division of Plant Biology, pp. 71–117.

Caldwell, M.M., Camp, L.B., Warner, C.W. and Flint, S.D. (1986) Action spectra and their key role in assessing biological consequences of solar UV-B radiation change. In: Caldwell, M.M., Camp, L.B., Warner, C.W., Flint, S.D. Worrest, R.C. and Caldwell, M.M. (eds) *Stratospheric Ozone Reduction: Solar Ultraviolet Radiation and Plant Life*. Springer, Heidelberg, Germany, pp. 87–111.

Emde, C., Buras-Schnell, R., Kylling, A., Mayer, B., Gasteiger, J. *et al.* (2016) The libRadtran software package for radiative transfer calculations (version 2.0.1). *Geoscientific Model Development* 9, 1647–1672.

Flint, S.D., Ryel, R.J., Hudelson, T.J. and Caldwell, M.M. (2009) Serious complications in experiments in which UV doses are effected by using different lamp heights. *Journal of Photochemistry and Photobiology, B: Biology* 97, 48–53.

Huelsen, G., Groebner, J., Bais, A., Blumthaler, M., Disterhoft, P. *et al.* (2008) Intercomparison of erythemal broadband radiometers calibrated by seven UV calibration facilities in Europe and the USA. *Atmospheric Chemistry and Physics* 8, 4865–4875.

Kreuter, A. and Blumthaler, M. (2009) Stray light correction for solar measurements using array spectrometers. *Review of Scientific Instruments, AIP* 80, 096108.

Leszczyns, K., Jokela, K., Ylianttila, L., Visuri, R. and Blumthaler, M. (1998) Erythemally weighted radiometers in solar UV monitoring: results from the WMO/STUK intercomparison. *Photochemistry and Photobiology* 67, 212–221.

Lindfors, A., Kaurola, J., Arola, A., Koskela, T., Lakkala, K. *et al.* (2007) A method for reconstruction of past UV radiation based on radiative transfer modeling: applied to four stations in northern Europe. *Journal of Geophysical Research: Atmospheres (1984–2012)*, 112, 1–15.

Lindfors, A., Heikkilä, A., Kaurola, J., Koskela, T. and Lakkala, K. (2009) Reconstruction of solar spectral surface UV irradiances using radiative transfer simulations. *Photochemistry and Photobiology* 85, 1233–1239.

McLeod, A.R. (1997) Outdoor supplementation systems for studies of the effects of increased UV-B radiation. *Plant Ecology* 128, 78–92.

Seckmeyer, G., Bais, A., Bernhard, G., Blumthaler, M., Booth, C.R. *et al.* (2001) *Instruments to Measure Solar Ultraviolet Radiation – Part 1: Spectral Instruments.* Global Atmosphere Watch Report No. 125. Research Department, Atmospheric Research and Environment Branch, World Meteorological Organization, Geneva, Switzerland.

Seckmeyer, G., Bais, A., Bernhard, G., Blumthaler, M., Booth, C.R. *et al.* (2005) *Instruments to Measure Solar Ultraviolet Radiation – Part 2: Broadband Instruments Measuring Erythemally Weighted Solar Irradiance.* Global Atmosphere Watch Report No. 164. Research Department, Atmospheric Research and Environment Branch, World Meteorological Organization, Geneva, Switzerland.

Seckmeyer, G., Bais, A., Bernhard, G., Blumthaler, M., Johnsen, B., Lantz, K. and McKenzie, R. (2010a) *Instruments to Measure Solar Ultraviolet Radiation – Part 3: Multi-channel Filter Instruments.* Global Atmosphere Watch Report No. 190. Research Department, Atmospheric Research and Environment Branch, World Meteorological Organization, Geneva, Switzerland.

Seckmeyer, G., Bais, A., Bernhard, G., Blumthaler, M., Drüke, S., *et al.* (2010b) *Instruments to Measure Solar Ultraviolet Radiation – Part 4: Array Spectroradiometers.* Global Atmosphere Watch Report No. 191. Research Department, Atmospheric Research and Environment Branch, World Meteorological Organization, Geneva, Switzerland.

Ylianttila, L., Visuri, R., Huurto, L. and Jokela, K. (2005) Evaluation of a single-monochromator diode array spectroradiometer for sunbed UV-radiation measurements. *Photochemistry and Photobiology* 81, 333–341.

Zeman, G. (2016) *Ultraviolet Radiation.* Health Physics Society. Available at: http://hps.org/hpspublications/articles/uv.html (accessed 4 May 2017).

3 UV Radiation and Terrestrial Ecosystems: Emerging Perspectives

Carlos L. Ballaré[1,2]* and Amy T. Austin[1,2]*

[1]IFEVA, Consejo Nacional de Investigaciones Científicas y Técnicas (CONICET), Universidad de Buenos Aires, Avenida San Martín 4453, C1417DSE, Buenos Aires, Argentina; [2]IIB-INTECH, CONICET, Universidad Nacional de San Martín, B1650HMP San Martín, Buenos Aires, Argentina

Introduction

Understanding the effects of ultraviolet (UV) radiation in terrestrial ecosystems has been a very active area of research in photobiology. This topic has stimulated extensive cooperation between scientists working in a broad cross-section of disciplines, from photochemistry to physiology and molecular biology, and to ecology and environmental sciences. Historically, UV research has gone through several phases, with shifting scientific foci and visibility. A major initial driver of UV research was associated with the prediction of strong negative effects of environmental pollutants on the integrity of the ozone layer (Molina and Rowland, 1974) and the discovery of the Antarctic ozone hole (Farman *et al.*, 1985), which led to heightened concerns about increasing UV levels that might have catastrophic consequences for life on Earth. Consequently, most of the initial efforts were focused on understanding the biological effects of the UV-B component of solar radiation, particularly damaging effects that could affect human health and compromise the productivity of managed ecosystems (UNEP, 1994; Caldwell *et al.*, 1995). Because ozone depletion was particularly severe in polar regions (Solomon, 1990; Madronich *et al.*, 1995), the majority of the empirical studies were carried out in high-latitude ecosystems (Smith *et al.*, 1992; Johanson *et al.*, 1995; Ballaré *et al.*, 2001). During the course of those experiments, evidence began to emerge for important roles of UV-B radiation in the regulation of plant function and ecosystem-level processes (Rozema *et al.*, 1997b; Caldwell *et al.*, 2003). These observations, along with the prospect of a recovery of the ozone layer following the implementation of the Montreal Protocol (McKenzie *et al.*, 2011; Solomon *et al.*, 2016), led to a shift in research emphasis from damage to regulation of biological processes. In plant biology, the characterization of a specific UV-B photoreceptor opened the way to mechanistic studies aimed at understanding the molecular basis of plant responses to UV-B radiation. At the ecosystem level, it became obvious that other wavelengths of the solar spectrum, not affected by ozone depletion but clearly sensitive to global change, could also play important roles

*Corresponding authors: ballare@ifeva.edu.ar; austin@ifeva.edu.ar

as regulators of ecosystem processes and biogeochemical cycles (Ballaré *et al.*, 2011; Williamson *et al.*, 2014). In this chapter, we discuss these changing perspectives and highlight some open questions in the field of UV effects on terrestrial ecosystems.

Plant Growth and Function

Predicting the consequences of ozone depletion

Highly energetic UV-B quanta can have deleterious effects on living organisms. DNA, proteins, lipids and other key macromolecules can interact with UV-B quanta, leading to the formation of non-functional photoproducts. The action spectra for these interactions usually show a sharp rise in quantum effectiveness in the UV-B region (Caldwell, 1971; Setlow, 1974; Caldwell and Flint, 1997).

The ozone layer absorbs nearly all of the UV-C and most of the UV-B in the terrestrial atmosphere. Therefore, predictions of a significant decrease in ozone levels caused by anthropogenic sources and the discovery of significant ozone depletion over Antarctica in the mid-1980s (Solomon, 1999) led to strong concerns about potential impacts of increased doses of UV-B radiation on organisms, including plants. As a result, several research programmes were launched in different parts of the globe to investigate the effects of UV-B radiation on photosynthesis and plant growth (Caldwell *et al.*, 1995).

Plant acclimation and resistance to solar UV

Given the prospect of increased UV-B exposure, it is understandable that early efforts were centred on the detection and evaluation of damaging effects of UV-B radiation, in particular the consequences for agricultural productivity. Most of the initial studies demonstrated deleterious effects of UV-B on DNA integrity, photosystem II stability, gas exchange, and plant growth (reviewed in Tevini and Teramura, 1989; Strid *et al.*, 1994; Teramura and Sullivan, 1994). Many of those pioneering studies may have suffered from a lack of realism in the simulation of potential scenarios of ozone depletion, usually employing very high doses of UV-B radiation, and unbalanced levels of UV-B relative to visible radiation (Caldwell and Flint, 1997). Those unrealistic conditions may have contributed to exaggerate the negative effects of UV-B on plants (Aphalo *et al.*, 2012).

When more realistic irradiation conditions were applied, along with improved UV dosimetry, it was found that, to a large extent, plants are able to acclimate to variations in UV-B levels and that certain processes, such as photosynthesis, are largely unaffected by natural or moderately enhanced doses of UV-B radiation (Fiscus and Booker, 1995). The fact that terrestrial plants are well equipped to deal with UV-B radiation should not be surprising, given that, as sessile photosynthetic organisms, plants are normally exposed to large variations in solar irradiance, which includes large variations in the UV-B component.

Plant acclimation to UV-B radiation has several components, with the first layer of protection being filtration. Plants produce different types of epidermal sunscreens, which block UV-B quanta before they reach sensitive cellular targets (Caldwell *et al.*, 1983; Braun and Tevini, 1993). Of these sunscreens, phenolic compounds are probably the most important, and the ones that have been studied in greatest detail (Li *et al.*, 1993; Landry *et al.*, 1995; Reuber *et al.*, 1996). Colourless flavonoids and phenylpropanoids accumulate in the vacuoles of epidermal cells and form a very effective filter, which blocks UV-B radiation without significantly affecting the penetration of photosynthetically active radiation (PAR) (Bilger *et al.*, 1997; Barnes *et al.*, 2000; Burchard *et al.*, 2000; Mazza *et al.*, 2000). Accumulation of phenolic sunscreens is often accompanied by developmental responses that contribute to increase their efficiency in protecting against UV-B. One of these responses is increased leaf thickness, which concomitantly reduces the penetration of UV-B to the inner layers of photosynthetic cells (Cen and Bornman, 1993). Another response is the inhibition of hypocotyl elongation in response to UV-B perceived by the apical hook, which retards the emergence of seedlings of certain species from the soil, thereby allowing time for accumulation of phenolic sunscreens before the cotyledons are exposed to solar radiation (Ballaré *et al.*, 1995b; Ballaré *et al.*, 1996). Sunscreen levels have also been shown to fluctuate on a daily basis, providing maximum protection around midday (Veit *et al.*, 1996; Barnes *et al.*, 2016a; Barnes *et al.*, 2016b).

A second layer of protection usually involves the accumulation of enzymatic and non-enzymatic antioxidants that scavenge toxic-free radicals generated in response to interactions of UV-B quanta with several molecules (Willekens *et al.*, 1994; Malanga and Puntarulo, 1995; Rao *et al.*, 1996; Takeuchi *et al.*, 1996). Although the majority of the experiments quantifying antioxidant responses have been carried out under very unbalanced light conditions (i.e. UV-B levels applied against very low levels of PAR), there is evidence from field studies demonstrating strong enzymatic and non-enzymatic antioxidant responses to solar UV-B radiation, which contribute to minimize oxidative stress under field conditions (Mazza *et al.*, 1999; Giordano *et al.*, 2004).

A critical additional layer of defence involves repair of UV-B-induced cellular damage. DNA repair has been studied in considerable detail in terrestrial plants. The main mechanism of DNA repair is based on the action of photolyases (photorepair), which effectively remove cyclobutane pyrimidine dimers (CPDs) and pyrimidine (6-4) pyrimidinone dimers (6-4 photoproducts) (reviewed in Britt, 2004). Natural levels of UV-B cause measurable increases in the abundance of CPDs at midday (Stapleton *et al.*, 1997; Mazza *et al.*, 1999); however, CPD repair appears to be fast (Rousseaux *et al.*, 1999). Peak levels of CPDs at noon increase with UV-B irradiance in the field (Rousseaux *et al.*, 1999; Mazza *et al.*, 2000), suggesting that DNA damage could be one of the mechanisms by which solar UV-B radiation inhibits leaf growth in terrestrial plants (Giordano *et al.*, 2004).

The Montreal Protocol, signed in 1989, was a highly successful international agreement, which led to significant reductions in the emissions of ozone-depleting substances and a drastic reduction in the rate of ozone destruction. Current projections suggest that the ozone layer could recover to historical levels during the next four decades (Bais *et al.*, 2015). Given that ozone depletion was rapidly detected and effectively attenuated by the regulations of the Montreal Protocol, the changes in stratospheric ozone were relatively small, outside of polar regions (Herman, 2010). These small changes, coupled with the fact that plants tend to be relatively resistant to variations in UV-B, has led to the conclusion that the negative effects of ozone depletion between 1980 and 2010 on plant biomass were probably modest, even in those areas where ozone depletion was maximum (i.e. high latitudes in the Southern Hemisphere) (Ballaré *et al.*, 2011).

UV-B as a signal for plants

Although the main focus of the initial research on UV-B responses was clearly the potential for damaging effects, it soon became obvious that UV-B radiation was also an important regulator of plant function. One of the first processes found to be regulated by UV-B was the accumulation of phenolic sunscreens that play a role in UV-B photoprotection. In addition, growth-related responses to UV-B revealed action spectra that were not easily explained solely on the basis of cellular damage (discussed in Stapleton, 1992; Ballaré *et al.*, 1995a). This body of accumulating evidence led support to the idea that plants have mechanisms to specifically perceive the UV-B component of solar radiation and activate adaptive physiological responses. Growth inhibition and increased accumulation of leaf phenolics could be seen as complementary parts of a protection mechanism activated in the plant in response to increased levels of UV-B radiation.

An interesting spin-off from the early work on plant responses to solar UV-B was the realization that some of the changes in plant chemistry elicited by natural levels of solar UV-B radiation involved compounds known to be important for plant interactions with other organisms. The list included phenylpropanoid compounds (Izaguirre *et al.*, 2007), isoflavonoids (Zavala *et al.*, 2015), conjugated polyamines (Demkura *et al.*, 2010); cuticular waxes (Kuhlmann and Müller, 2010), proteinase inhibitors (Stratmann *et al.*, 2000; Izaguirre *et al.*, 2003) and jasmonates (Dinh *et al.*, 2013), among others. In parallel with these findings, field studies demonstrated that, in many cases, manipulations of solar UV-B radiation resulted in large changes in the levels of invertebrate herbivory and plant interactions with microorganisms (reviewed in Caldwell *et al.*, 2003; Ballaré *et al.*, 2011).

Negative effects of solar UV-B radiation on herbivory levels are particularly well documented, and appear to be proportionally much larger than the effects of UV-B inhibiting plant growth (Ballaré *et al.*, 2011). Some insects can perceive solar UV-B radiation (Mazza *et al.*, 2002), but many of the negative effects of UV-B on levels of insect herbivory are thought to be indirect (i.e. mediated by changes in host plant traits) (Ballaré *et al.*, 2012). Shade-intolerant plants often down-regulate their defences against pathogens and pests in those leaves that are exposed to shade or shade signals (such as a low red to far-red ratios, R:FR) (Izaguirre *et al.*, 2006; Moreno *et al.*, 2009; de Wit *et al.*, 2013), presumably to save resources that could be invested in growth responses to avoid shade. A plausible interpretation of the positive effects of solar UV-B on plant defences is that plants 'interpret' UV-B as a signal of full sunlight (i.e. no shade) (Mazza and Ballaré, 2015). According to this interpretation, plants growing in patchy canopies use solar UV-B as a 'gap' signal, to adaptively regulate their growth and defence phenotypes. The interplay between shade signals (such as low R:FR perceived by phytochromes) and gap signals (such as high UV-B) will optimize the allocation of resources between growth and defence (reviewed in Ballaré, 2014).

Discovery of a UV-B photoreceptor in Arabidopsis

Plant responses to solar UV-B are highly specific. The definition of 'specific', in this context, takes into account the fact that solar UV-B represents less than 1% of the total shortwave photons received at the ground surface. Therefore, if changes in UV-B within this range result in a plant response, it must be assumed that the plant has a very specific sensory system to detect these changes (i.e. a sensory system that is not activated by other wavelengths of the solar spectrum) (discussed in Demkura *et al.*, 2010). Nucleic acids and aromatic amino acids absorb strongly in the UV-B region and could therefore act as potential components of UV-B sensors (Walker, 1984; Kim *et al.*, 1992). The cloning (Kliebenstein *et al.*, 2002) and functional characterization of UV RESISTANCE LOCUS 8 (UVR8) (Rizzini *et al.*, 2011; Christie *et al.*, 2012; Wu *et al.*, 2012) represented an important step in UV photobiology, defining the first specific UV-B photoreceptor. UVR8 is a seven-bladed β-propeller protein that forms a homodimer held together by interactions between charged tryptophans. After UV-B absorption, the UVR8 homodimer converts to monomers that enter the nucleus ultimately leading to transcriptional regulation of target genes (Brown and Jenkins, 2008; Favory *et al.*, 2009; Rizzini *et al.*, 2011). The amount of UVR8 monomer is thought to be a measure of the UV-B levels perceived by the plant (Findlay and Jenkins, 2016). The central components of the UVR8 photocycle, including the CONSTITUTIVELY PHOTOMORPHOGENIC 1 (COP1) E3 ubiquitin ligase and the REPRESSOR OF UV-B PHOTOMORPHOGENESIS (RUP) 1 and 2 proteins, have been elucidated (Heijde and Ulm, 2012; Jenkins, 2014). Interaction of UVR8 with COP1 stabilizes the ELONGATED HYPOCOTYL 5 (HY5) transcription factor and enhances the association of HY5 with target promoters, thereby activating transcription of many UV-B-responsive genes (Heijde and Ulm, 2012; Jenkins, 2014). Many of these target genes are associated with UV-B protection and UV-B damage repair; therefore, a major role of UVR8 in UV protection is inferred (Kliebenstein *et al.*, 2002; Brown *et al.*, 2005; Favory *et al.*, 2009). In fact, *uvr8* is hypersensitive to prolonged exposure to UV-B radiation (Kliebenstein *et al.*, 2002), or to treatments in which UV-B is administrated against a background of low PAR, or using very high UV-B doses.

The role of UVR8 under natural conditions remains to be established. Using realistic UV-B treatments, Demkura and Ballaré (2012) inferred that UVR8 could play a role regulating plant defence against fungal pathogens, and there is evidence from limited field studies that *uvr8* mutants are somewhat more sensitive to natural levels of solar radiation than wild-type plants (Morales *et al.*, 2013). However, additional information from field studies, measuring fitness components, is needed to establish the adaptive importance of UVR8-mediated UV-B perception. It is worth noting that some growth (Casadevall *et al.*, 2013) and defence responses (Mazza and Ballaré, 2015) to UV-B radiation appear to be activated in a UVR8-independent manner. The logical enthusiasm associated with the discovery of a specialized UV-B receptor should not obscure the fact that UV-B photons have multiple targets in plant cells, including DNA, where they cause UV-B specific photoproducts. CPDs, for example, which are specifically induced by UV-B, are known to activate transcription and cellular responses in mammalian cells (Boros *et al.*, 2015). Measurable levels of CPDs are induced by natural UV-B irradiances (Ballaré *et al.*, 1996; Stapleton *et al.*, 1997; Mazza *et al.*, 1999), and for some responses, such as leaf growth inhibition, there is a good correlation with DNA damage (Giordano *et al.*, 2004). More work should be carried out to establish the molecular mechanisms underlying adaptive plant responses under natural field conditions. Care should be taken not to repeat the initial errors of UV-B dosimetry and spectral balance that were common in the pioneering experiments aimed to evaluate the biological effects of increased UV-B radiation caused by stratospheric ozone depletion (Aphalo *et al.*, 2012).

Ecosystem Processes

UV-B radiation and litter decomposition with stratospheric ozone depletion

Ecological interest in understanding the effects of UV radiation on litter decomposition and biogeochemical cycles began with the recognition of stratospheric ozone depletion and the potentially damaging effects of UV-B on a range of ecosystem processes beyond photosynthesis and plant growth (Caldwell and Flint, 1994; Rozema *et al.*, 1997b; Ballaré *et al.*, 2001). At the time, the prevailing hypotheses were that exposure to elevated UV-B levels could affect plant litter decomposition in two ways, both negatively: i) through alteration of litter quality via chemical changes in green leaves (particularly phenolic compounds, lignin and other photoprotective compounds); and ii) through damage to soil biota on litter and soil surfaces such as bacteria, thus impeding litter decomposition, soil organic matter formation and nutrient mineralization.

There was general support for the first hypothesis in initial studies from both supplementation and attenuation experiments in a range of ecosystems. Enhanced UV-B radiation during growth typically reduced overall litter quality (e.g. Gehrke *et al.*, 1995; Rozema *et al.*, 1997a; Pancotto *et al.*, 2003; Pancotto *et al.*, 2005), although this was not universally observed, particularly under conditions of realistic changes in UV-B radiation simulating stratospheric ozone depletion (Newsham *et al.*, 1999; Cybulski III *et al.*, 2000; Hoorens *et al.*, 2004). Increases in phenolic compounds or lignin in litter were identified as the major factors contributing to reduced litter quality in these experiments, which in some cases coincided with reductions in litter decomposition.

The second hypothesis focused on the potential damage to microbial populations on leaf litter exposed to elevated levels of UV-B radiation (Sinha and Häder, 2002). In practice, reduction in litter decomposition was attributed to changes in microbial community activity, as it was very difficult to demonstrate direct damage to microbial populations under field conditions. Some disputed evidence for a strong influence of UV-B radiation on microbial community efficiency was shown in a laboratory experiment (Johnson *et al.*, 2002; Stark and Hart, 2003), but overall, effects on both litter decomposition and microbial community composition were modest or undetectable (Searles *et al.*, 2001; Robson *et al.*, 2004). In cases where significant compositional changes in litter biota were detected, fungal species appeared to be more sensitive to UV-B exposure than bacteria (Gehrke *et al.*, 1995; Pancotto *et al.*, 2003; Robson *et al.*, 2004). The difficulty in interpreting many of these studies, however, stems from the continuing challenge for soil ecologists to link microbial community composition to function (Schimel and Schaeffer, 2012), due to the fact that changes in microbial community do not necessarily translate to changes in ecosystem processes such as carbon respiration or nutrient turnover. The few studies that evaluated methane or carbon flux from microbial activity demonstrated reduced (Gehrke *et al.*, 1995; Niemi *et al.*, 2002; Pancotto *et al.*, 2003), or neutral (Rinnan *et al.*, 2003) responses, without a strong case for significant UV-B effects on gas efflux from the terrestrial environment. There is some evidence that these changes may be more important in the long term, as increased microbial enzymatic activity and reduced C respiration were observed in *Vaccinum* heathland only after two decades of exposure to elevated UV-B radiation (Jones *et al.*, 2015).

Interestingly, at this time it was also noted that lignin concentrations were reduced in litter exposed to UV-B radiation (Gehrke *et al.*, 1995); it was suggested that photochemical mineralization may be responsible for these changes in litter quality (Moorhead and Callaghan, 1994). However, the general conclusion from these studies associated with stratospheric ozone depletion was that direct photochemical breakdown of terrestrial litter was unimportant or made only a minor contribution to the decomposition process and thus carbon release from terrestrial ecosystems (Moody *et al.*, 2001). This was somewhat misleading due to the fact that the vast majority of these studies had been conducted at high latitudes where background levels of UV-B were constitutively low. In addition, the lack of direct photodegradative effects was in direct contrast to several important studies from aquatic and marine ecosystems, which indicated a key role for UV radiation affecting the biotic availability of dissolved organic matter (Kieber *et al.*, 1989; Keiber *et al.*, 1990; Amon and Benner, 1996). Given the lukewarm response of many ecosystem processes to the alterations of UV-B irradiance, it would take another decade for our attention to return to the importance of solar UV radiation on carbon turnover in terrestrial ecosystems.

Photochemical degradation of terrestrial plant litter

The curious lack of correlation between climatic parameters and litter decomposition in arid and semiarid ecosystems puzzled ecologists for some time (Austin, 2011), given that litter quality and climate (particularly actual evapotranspiration) were considered to be the most fundamental controls on carbon turnover in terrestrial ecosystems (Meentemeyer, 1978). Nevertheless, enigmatic studies demonstrated a higher than predicted decomposition in sites of very low rainfall (e.g.Whitford *et al.*, 1981; Moorhead and Reynolds, 1989), as well as little correlation with mean annual precipitation (Vanderbilt *et al.*, 2008) or water availability (Austin *et al.*, 2009).

While direct photochemical transformation resulted in volatile C emissions from degrading plant litter (Tarr *et al.*, 1995; Schade *et al.*, 1999), the connection between photodegradation and carbon turnover in terrestrial ecosystems had not been recognized. The identification of a dominant control of direct photochemical mineralization of aboveground plant litter in a semiarid terrestrial ecosystem (Austin and Vivanco, 2006) was a step forward in our understanding of this control on carbon turnover. This study demonstrated that photodegradation (conversion of organic compounds to carbon dioxide (CO_2) and other volatile C emissions with exposure to solar radiation) due to UV-B and solar radiation exposure was independent of biotic activity in the Patagonian steppe. Attenuation of UV-B radiation reduced litter mass loss by 33%, while blocking total solar radiation reduced litter decomposition by two-thirds, independently of the presence of soil biota (Austin and Vivanco, 2006). This study brought to centre stage the potential importance of UV-B and solar radiation as a control on carbon turnover in terrestrial ecosystems. As a result, there was a resurgence of interest in understanding how solar radiation affected rates of mass loss and carbon turnover (reviewed in Austin, 2011; King *et al.*, 2012; Barnes *et al.*, 2015). In addition, the incorporation of photodegradation in our conceptual framework of carbon turnover in arid lands helped to explain the rapid turnover of plant litter in these biotically impoverished ecosystems. A flurry of new studies focused on the direct production of carbon gases from exposure to solar radiation in the absence of biotic activity (Brandt *et al.*, 2009; Rutledge *et al.*, 2010; Lee *et al.*, 2012), which yielded important quantitative information on the rates of carbon efflux to the atmosphere. These studies suggested that at the ecosystem scale, carbon loss from photodegradation in semiarid ecoystems with marked seasonality could be on par with microbial respiration, although in general, fluxes from photodegradation were smaller than CO_2 microbial fluxes in most terrestrial ecosystems. Other studies demonstrated significant effects of UV-B and solar radiation on litter decomposition in semiarid zones, which were often equally or more important than changes in water availability or precipitation regime (Brandt *et al.*, 2007; Day *et al.*, 2007; Henry *et al.*, 2008; Austin *et al.*, 2009; Brandt *et al.*, 2010; Rutledge *et al.*, 2010; Lin and King, 2014; Baker and Allison, 2015). At the same time, several studies demonstrated little or no direct UV-B (or UV) effects on aboveground litter decomposition (e.g. Smith *et al.*, 2010; Uselman *et al.*, 2011; Barnes *et al.*, 2012; Lambie *et al.*, 2014), which was in agreement with the results of a meta-analysis of UV-B effects (Song *et al.*, 2013). The lack of demonstration of UV-B effects may stem from our previous focus on UV-B radiation due to stratospheric ozone depletion, which conditioned many of the experiments in terrestrial ecosystems to concentrate exclusively on the role of the UV-B component in high latitude ecosystems, rather than on the range of wavelengths that make up the solar spectrum and are potentially important for photodegradation (Austin and Ballaré, 2010). Given the increasing evidence of the importance of other (non UV-B) wavelengths of radiation in biogeochemical cycling, there clearly needs to be more focus in current and future experiments to emulate the real range of variation in solar radiation in terrestrial ecosystems (Williamson *et al.*, 2014).

Photopriming and terrestrial ecosystems

One of the most exciting new developments in our understanding of the role of solar radiation on biogeochemical cycles in terrestrial ecosystems has come from the further recognition of the indirect role of solar radiation (including UV-B) on the biotic component of litter decomposition. In contrast to modest negative effects that were previously observed for litter decomposition at high latitudes, the indirect effects of photodegradation are largely positive. Stimulation of litter decomposition and carbon turnover has been observed in various semiarid terrestrial ecosystems (Baker and Allison, 2015; Day *et al.*, 2015; Gaxiola and Armesto, 2015; Lin *et al.*, 2015b). In addition, soil respiration with pulsed rain events increased dramatically due to antecedent exposure of aboveground litter to solar radiation (Ma *et al.*, 2012). Several studies have shown that litter exposed to solar radiation subsequently decayed more rapidly (Henry *et al.*, 2008; Gallo *et al.*, 2009; Foereid *et al.*, 2010), particularly if the litter was in strong contact with surface soil (Barnes *et al.*, 2012; Hewins *et al.*, 2013). These new studies represent a shift in our understanding of the role of solar radiation affecting carbon turnover for two reasons: first is that there has been a bias over time to assume harmful or negative effects of UV radiation on biotically-mediated processes; and second, that UV-B is the only part of the solar spectrum that varies and is potentially important in affecting ecosystem processes such as carbon turnover.

A very recent, broad-scale study of the effects of solar radiation through photodegradation of terrestrial leaf litter suggests a mechanism for the stimulation of biotically-mediated decomposition (Fig. 3.1; Austin *et al.*, 2016).

Across a wide range of woody and herbaceous terrestrial plant species, previous exposure to UV and particularly blue-green (BG) solar radiation significantly increased rates of organic matter decomposition, with a 30% increase for exposure to full solar radiation (UV+BG light). The mechanistic explanation for this stimulation of biotic decomposition lies in the accessibility of the plant cell wall polysaccharides in decomposing plant litter. The first step in this observed priming of litter decomposition appears to come from the fact that exposure to solar radiation results in a photodegradation of lignin (Day *et al.*, 2007; Austin and Ballaré, 2010). This photodegradation has been documented using NMR spectroscopy (Lin *et al.*, 2015a), and is due to the absorbance of solar radiation by lignin in a broad range of wavelengths of the solar spectrum (Austin and Ballaré, 2010). This degradation of lignin in

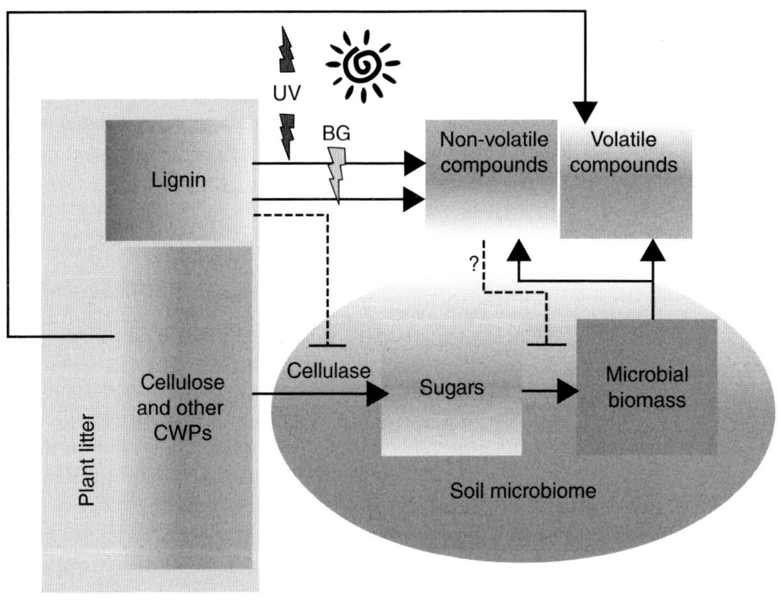

Fig. 3.1. Conceptual model of positive effects of solar radiation on lignin degradation and microbial decomposition in terrestrial ecosystems. Boxes indicate carbon pools, solid arrows indicate fluxes of carbon between pools. In the proposed mechanism, shown, the UV and blue-green (BG) photons in sunlight cause lignin photodegradation and the formation of photoproducts, some of which are released to the atmosphere (volatile compounds) and others (non-volatile compounds) are retained in litter. Cellulose and hemicelluloses, the most abundant cell wall polysaccharides (CWPs) and major components of plant litter, can also be directly photodegraded by UV photons, which contribute to the losses of volatile carbon to the atmosphere (Schade *et al.*, 1999). The biotic decomposition of CWPs is catalysed by microbial cellulases. Because cellulase activity and cellulase access to its substrate (Gressel, 2008) are both inhibited by lignin, photodegradation of the lignin component of plant litter facilitates saccharification (i.e. cellulase degradation of plant CWPs). In addition, some lignin photodegradation products (such as quinones) might have inhibitory effects on decomposer microorganisms (Müller *et al.*, 2003; Bais *et al.*, 2006).

the secondary cell walls liberates cell wall polysaccharides from the lignocellulolytic linkages, thereby enabling increased access for microbial degradation (Austin *et al.*, 2016). This mechanistic insight suggests that photopriming of plant litter may be a widespread and general phenomenon in terrestrial ecosystems that are subject to exposure to solar radiation during some part of the year. A recent meta-analysis of the effects of UV radiation on litter decomposition is consistent with the idea that UV exposure increases biodegradability of senescent plant material (Wang *et al.*, 2015), a point which has perhaps been underappreciated until now. An emerging perspective from these studies is that photodegradation and its interaction with soil biota could function as a primary regulator of litter decomposition, carbon storage and nutrient turnover in a broad range of terrestrial ecosystems.

Climate change, biogeochemical cycles and UV-B radiation in terrestrial ecosystems

We have travelled a long way in our understanding of the complexities of the controls of UV-B radiation on plant growth and ecosystem processes. An emerging perspective from the enormous effort and energy in understanding the effects of UV-B radiation in terrestrial ecosystems is clear: we can now recognize that in addition to the potentially negative effects of UV-B on terrestrial biota, UV-B can also control ecosystem processes through its natural variation and changes due to human activity or climate change. The focus on negative impacts of UV radiation needs to be recast in the light of the multiple mechanistic insights that have been highlighted in recent years due to our broadened understanding of the role of UV-B as a regulator of plant function and ecosystem processes. At the same time, we must not overestimate the depth of our mechanistic understanding of the regulation of processes with variation in UV-B radiation. The relative importance of different photosensory pathways (i.e. UVR8-, DNA-dependent, etc.) for the perception of 'natural' levels of UV-B radiation remains to be established. Future work must connect the important advances in our molecular understanding of UVR8's function with the suite of physiological responses that plants display under natural conditions. We can take a valuable lesson from the pioneering work aimed at understanding stratospheric ozone depletion in terrestrial ecosystems and take care not to repeat the initial errors of UV-B dosimetry and spectral balance that were common in those initial experiments from some decades ago.

This review highlights the need to extend our understanding of the effects of solar UV radiation beyond the wavelengths associated only with the UV-B fraction. The recent demonstration of the importance of UV-A and short-wave visible wavelengths of solar radiation in affecting carbon turnover suggests that human activity and climate change may have large and persistent effects on ecosystem functioning in ways that go far beyond stratospheric ozone depletion. For example, the relative importance of photodegradation is likely to increase under conditions of increased aridity, in particular in Mediterranean ecosystems. Carbon turnover was dramatically reduced with lower solar radiation exposure due to shading from increased vegetation cover from afforestation, and due to the combination of UV attenuation and land-use change (Throop and Archer, 2007; Bosco *et al.*, 2016). These results suggest that land-use change as it affects vegetative cover and thus solar radiation exposure may be a principal driver of changes in carbon turnover in human-modified ecosystems. This broader perspective on the role of solar radiation affecting plant function, trophic interactions (e.g. plant–insect interactions) and biogeochemical cycling should be taken into account in modelling predictions of the responses of terrestrial ecosystems to climate change and human impact (Williamson *et al.*, 2014).

Acknowledgements

We thank the members of the UNEP Environmental Effects Assessment panel for stimulating discussions regarding the role and impacts of UV-B radiation in terrestrial ecosystems. Work in our laboratories was financially supported by grants from the Agencia Nacional de Promoción de Ciencia y Tecnología (ANPCyT) of Argentina and UBACyT (To C.L.B. and A.T.A.) and the L'Oréal-UNESCO Foundation for Women in Science (to A.T.A.).

References

Amon, R.W. and Benner, R. (1996) Photochemical and microbial consumption of dissolved organic carbon and dissolved oxygen in the Amazon River system. *Geochimica et Cosmochimica Acta* 60, 1783–1792.

Aphalo, P.J., Albert, A., Björn, L.O., McLeod, A., Robson, T.M. and Rosenqvist, E. (eds) (2012) *Beyond the Visible: A Handbook of Best Practice in Plant UV Photobiology*. Helsinki University.

Austin, A.T. (2011) Has water limited our imagination for aridland biogeochemistry? *Trends in Ecology and Evolution* 26, 229–235.

Austin, A.T. and Ballaré, C.L. (2010) Dual role of lignin in litter decomposition in terrestrial ecosystems. *Proceedings of the National Academy of Sciences of the United States of America* 107, 4618–4622.

Austin, A.T. and Vivanco, L. (2006) Plant litter decomposition in a semiarid ecosystem controlled by photodegradation. *Nature* 442, 555–558.

Austin, A.T., Araujo, P.I. and Leva, P.E. (2009) Interaction of position, litter type and pulsed water events on decomposition of grasses from the semiarid Patagonian steppe. *Ecology* 90, 2642–2647.

Austin, A.T., Méndez, M.S. and Ballaré, C.L. (2016) Photodegradation alleviates the lignin bottleneck for carbon turnover in terrestrial ecosystems. *Proceedings of the National Academy of Sciences of the United States of America* 113, 4392–4397.

Bais, A.F., McKenzie, R.L., Bernhard, G., Aucamp, P.J., Ilyas, M., Madronich, S. and Tourpali, K. (2015) Ozone depletion and climate change: impacts on UV radiation. *Photochemical and Photobiological Sciences* 14, 19–52.

Bais, H.P., Weir, T.L., Perry, L.G., Gilroy, S. and Vivanco, J.M. (2006) The role of root exudates in rhizosphere interactions with plants and other organisms. *Annual Review of Plant Biology* 57, 233–266.

Baker, N.R. and Allison, S.D. (2015) Ultraviolet photodegradation facilitates microbial litter decomposition in a Mediterranean climate. *Ecology* 96, 1994–2003.

Ballaré, C.L. (2014) Light regulation of plant defense. *Annual Review of Plant Biology* 65, 335–363.

Ballaré, C.L., Barnes, P.W. and Flint, S.D. (1995a) Inhibition of hypocotyl elongation by ultraviolet-B radiation in de-etiolating tomato seedlings. I. The photoreceptor. *Physiologia Plantarum* 93, 584–592.

Ballaré, C.L., Barnes, P.W., Flint, S.D. and Price, S. (1995b) Inhibition of hypocotyl elongation by ultraviolet-B radiation in de-etiolating tomato seedlings. II. Time-course, comparison with flavonoid responses and adaptive significance. *Physiologia Plantarum* 93, 593–601.

Ballaré, C.L., Scopel, A.L., Stapleton, A.E. and Yanovsky, M.J. (1996) Solar ultraviolet-B radiation affects seedling emergence, DNA integrity, plant morphology, growth rate, and attractiveness to herbivore insects in *Datura ferox*. *Plant Physiology* 112, 161–170.

Ballaré, C.L., Rousseaux, M.C., Searles, P.S., Zaller, J.G., Giordano, C.V. *et al.* (2001) Impacts of solar ultraviolet-B radiation on terrestrial ecosystems of Tierra del Fuego (southern Argentina) – an overview of recent progress. *Journal of Photochemistry and Photobiology B:Biology* 62, 67–77.

Ballaré, C.L., Caldwell, M.M., Flint, S.D., Robinson, S.A. and Bornman, J.F. (2011) Effects of solar ultraviolet radiation on terrestrial ecosystems: patterns, mechanisms, and interactions with climate change. *Photochemical and Photobiological Sciences* 10, 226–241.

Ballaré, C.L., Mazza, C.A., Austin, A.T. and Pierik, R. (2012) Canopy light and plant health. *Plant Physiology* 160, 145–155.

Barnes, P.W., Searles, P.S., Ballaré, C.L., Ryel, R.J. and Caldwell M.M. (2000) Non-invasive measurements of leaf epidermal transmittance of UV radiation using chlorophyll fluorescence: field and laboratory studies. *Physiologia Plantarum* 109, 274–283.

Barnes, P.W., Throop, H.L., Hewins, D.B., Abbene, M.L. and Archer, S.R. (2012) Soil coverage reduces photodegradation and promotes the development of soil-microbial films on dryland leaf litter. *Ecosystems* 15, 311–321.

Barnes, P.W., Throop, H.L., Archer, S.R., Breshears, D.D., McCulley, R.L. and Tobler, M.A. (2015) Sunlight and soil–litter mixing: drivers of litter decomposition in drylands. *Progress in Botany* 76, 273–302.

Barnes, P.W., Flint, S.D., Tobler, M.A. and Ryel R.J. (2016a) Diurnal adjustment in ultraviolet sunscreen protection is widespread among higher plants. *Oecologia* 181, 55–63.

Barnes, P.W., Tobler, M.A., Keefover-Ring, K., Flint, S.D., Barkley, A.E., Ryel, R.J. and Lindroth, R.L. (2016b) Rapid modulation of ultraviolet shielding in plants is influenced by solar ultraviolet radiation and linked to alterations in flavonoids. *Plant, Cell & Environment* 39, 222–230.

Bilger, W., Veit, M., Schreiber, L. and Schreiber, U. (1997) Measurement of leaf epidermal transmittance of UV radiation by chlorophyll fluorescence. *Physiologia Plantarum* 101, 754–763.

Boros, G., Miko, E., Muramatsu, H., Weissman, D., Emri, E. *et al.* (2015) Identification of cyclobutane pyrimidine dimer-responsive genes using UVB-irradiated human heratinocytes transfected with *in vitro*-synthesized photolyase mRNA. *PLoS ONE* 10, e0131141.

Bosco, T., Bertiller, M.B. and Carrera, A.L. (2016) Combined effects of litter features, UV radiation, and soil water on litter decomposition in denuded areas of the arid Patagonian Monte. *Plant and Soil* 406, 71–82.

Brandt, L.A., King J.Y. and Milchunas, D.G. (2007) Effects of ultraviolet radiation on litter decomposition depend on precipitation and litter chemistry in a shortgrass steppe ecosystem. *Global Change Biology* 13, 2193–2205.

Brandt, L.A., Bohnet, C. and King, J.Y. (2009) Photochemically induced carbon dioxide production as a mechanism for carbon loss from plant litter in arid ecosystems. *Journal of Geophysical Research: Biogeosciences* 114, G02004. DOI:02010.01029/02008JG000772

Brandt, L.A., King, J.Y., Hobbie, S.E., Milchunas, D.G. and Sinsabaugh, R.L. (2010) The role of photodegradation in surface litter decomposition across a grassland ecosystem precipitation gradient. *Ecosystems* 13, 765–781.

Braun, J. and Tevini, M. (1993) Regulation of UV-protective pigment synthesis in the epidermal layer of rye seedlings (*Secale cereale* L-Cv Kustro). *Photochemistry and Photobiology* 57, 318–323.

Britt, A.B. (2004) Repair of DNA damage induced by solar UV. *Photosynthesis Research* 81, 105–112.

Brown, B.A. and Jenkins, G.I. (2008) UV-B signaling pathways with different fluence-rate response profiles are distinguished in mature Arabidopsis leaf tissue by requirement for UVR8, HY5, and HYH. *Plant Physiology* 146, 576–588.

Brown, B.A., Cloix, C., Jiang, G.H., Kaiserli, E., Herzyk, P., Kliebenstein, D.J. and Jenkins G.I. (2005) A UV-B-specific signaling component orchestrates plant UV protection. *Proceedings of the National Academy of Sciences of the United States of America* 102, 18225–18230.

Burchard, P., Bilger, W. and Weissenböck, G. (2000) Contribution of hydroxycinnamates and flavonoids to epidermal shielding of UV-A and UV-B radiation in developing rye primary leaves as assessed by ultraviolet-induced chlorophyll fluorescence measurements. *Plant, Cell & Environment* 23, 1373–1380.

Caldwell, M.M. (1971) Solar UV irradiation and the growth and development of higher plants. In: Giese, A.C. (ed.) *Photophysiology*. Vol. VI. *Current Topics in Photobiology and Photochemistry*. Academic Press, New York, pp. 131–177.

Caldwell, M.M. and Flint, S.D. (1994) Stratospheric ozone reduction, solar UV-B radiation and terrestrial ecosystems. *Climatic Change* 28, 375–394.

Caldwell, M.M. and Flint, S.D. (1997) Uses of biological spectral weighting functions and the need of scaling for the ozone reduction problem. *Plant Ecology* 128, 66–76.

Caldwell, M.M., Robberecht, R. and Flint, S.D. (1983) Internal filters: prospects for UV-acclimation in higher plants. *Physiologia Plantarum* 58, 445–450.

Caldwell, M.M., Teramura A.H., Tevini, M., Bornman, J.F., Bjorn, L.O., Kulandaivelu, G. (1995) Effects of increased solar ultraviolet radiation on terrestrial plants. *Ambio* 24, 166–173.

Caldwell, M.M., Ballaré, C.L., Bornman, J.F., Flint, S.D., Björn, L.O. *et al.* (2003) Terrestrial ecosystems, increased solar ultraviolet radiation and interactions with other climatic change factors. *Photochemical and Photobiological Sciences* 2, 29–38.

Casadevall, R., Rodriguez, R.E., Debernardi, J.M., Palatnik, J.F. and Casati, P. (2013) Repression of growth regulating factors by the MicroRNA396 inhibits cell proliferation by UV-B radiation in Arabidopsis leaves. *The Plant Cell* 25, 3570–3583.

Cen, Y.P. and Bornman, J.F. (1993) The effect of exposure to enhanced UV-B radiation on the penetration of monochromatic and polychromatic UV-B radiation in leaves of *Brassica napus*. *Physiologia Plantarum* 87, 249–255.

Cybulski III, W.J., Peterjohn, W.T. and Sullivan, J.H. (2000) The influence of elevated ultraviolet-B radiation (UV-B) on tissue quality and decomposition of loblolly pine (*Pinus taeda* L.) needles. *Environmental and Experimental Botany* 44, 231–241.

Christie, J.M., Arvai, A.S., Baxter, K.J., Heilmann, M., Pratt, A.J. *et al.* (2012) Plant UVR8 photoreceptor senses UV-B by tryptophan-mediated disruption of cross-dimer salt bridges. *Science* 335, 1492–1496.

Day, T.A., Zhang, E.T. and Ruhland, C.T. (2007) Exposure to solar UV-B radiation accelarates mass and lignin loss of *Larrea tridentata* litter in the Sonoran Desert. *Plant Ecology* 193, 185–194.

Day, T.A., Guénon, R., and Ruhland, C.T. (2015) Photodegradation of plant litter in the Sonoran Desert varies by litter type and age. *Soil Biology and Biochemistry* 89, 109–122.

de Wit, M., Spoel, S.H., Sanchez Perez, G.F., Gommers, C.M.M., Pieterse, C.M.J., Voesenek, L.A.C.J. and Pierik, R. (2013) Perception of low Red:Far-red ratio compromises both salicylic acid- and jasmonic acid-dependent pathogen defences in Arabidopsis. *The Plant Journal* 75, 90–103.

Demkura, P.V. and Ballaré, C.L. (2012) UVR8 mediates UV-B-induced *Arabidopsis* defense responses against *Botrytis cinerea* by controlling sinapate accumulation. *Molecular Plant* 5, 642–652.

Demkura, P.V., Abdala, G., Baldwin, I.T. and Ballaré, C.L. (2010) Jasmonate-dependent and -independent pathways mediate specific effects of solar ultraviolet-B radiation on leaf phenolics and antiherbivore defense. *Plant Physiology* 152, 1084–1095.

Dinh, S.T., Baldwin, I.T. and Galis, I. (2013) The HERBIVORE ELICITOR-REGULATED1 (HER1) gene enhances abscisic acid levels and defenses against herbivores in *Nicotiana attenuata* plants. *Plant Physiology* 162, 2106–2124.

Farman, J.C., Gardiner, B.G. and Shanklin, J.D. (1985) Large losses of total ozone in Antarctica reveal seasonal ClOx/NOx interaction. *Nature* 315, 207–210.

Favory, J.J., Stec, A., Gruber, H., Rizzini, L., Oravecz, A. *et al.* (2009) Interaction of COP1 and UVR8 regulates UV-B-induced photomorphogenesis and stress acclimation in *Arabidopsis*. *EMBO Journal* 28, 591–601.

Findlay, K.M.W. and Jenkins, G.I. (2016) Regulation of UVR8 photoreceptor dimer/monomer photoequilibrium in Arabidopsis plants grown under photoperiodic conditions. *Plant, Cell and Environment* 39, 1706–1714.

Fiscus, E.L. and Booker, F.L. (1995) Is increased UV-B a threat to crop photosynthesis and productivity? *Photosynthesis Research* 43, 81–92.

Foereid, B., Bellarby J., Meier-Augenstein, W. and Kemp, H. (2010) Does light exposure make plant litter more degradable? *Plant and Soil* 333, 275–285.

Gallo, M.E., Porras-Alfaro, A., Odenbach, K.J. and Sinsabaugh, R.L. (2009) Photoacceleration of plant litter decomposition in an arid environment. *Soil Biology and Biochemistry* 41, 1433–1441.

Gaxiola, A. and Armesto, J.J. (2015) Understanding litter decomposition in semiarid ecosystems: linking leaf traits, UV exposure and rainfall variability. *Frontiers in Plant Science* 6, 140. DOI:110.3389/fpls.2015.00140

Gehrke, C., Johanson, U., Callaghan, T.V., Chadwick, D. and Robinson, C.H. (1995) The impact of enhanced UV-B radiation on litter quality and decomposition processes in *Vaccinum* leaves from the subarctic. *Oikos* 72, 213–222.

Giordano, C.V., Galatro, A., Puntarulo, S. and Ballaré, C.L. (2004) The inhibitory effects of UV-B radiation (280–315 nm) on *Gunnera magellanica* growth correlate with increased DNA damage but not with oxidative damage to lipids. *Plant, Cell & Environment* 27, 1415–1423.

Gressel, J. (2008) Transgenics are imperative for biofuel crops. *Plant Science* 174, 246–263.

Heijde, M. and Ulm, R. (2012) UV-B photoreceptor-mediated signalling in plants. *Trends in Plant Science* 17, 230–237.

Henry, H.A.L., Brizgys, K. and Field, C.B. (2008) Litter decomposition in a California annual grassland: interactions between photodegradation and litter layer thickness. *Ecosystems* 11, 545–554.

Herman, J.R. (2010) Global increase in UV irradiance during the past 30 years (1979–2008) estimated from satellite data. *Journal of Geophysical Research: Atmospheres* 115, D04203.

Hewins, D.B., Archer, S.R., Okin, G.S., McCulley, R.L. and Throop, H.L. (2013) Soil–litter mixing accelerates decomposition in a Chihuahuan desert grassland. *Ecosystems* 16, 183–195.

Hoorens, B., Aerts, R. and Stroetenga, M. (2004) Elevated UV-B radiation has no effect on litter quality and decomposition of two dune grassland species: evidence from a long-term field experiment. *Global Change Biology* 10, 200–208.

Izaguirre, M.M., Scopel, A.L., Baldwin, I.T. and Ballaré, C.L. (2003) Convergent responses to stress. Solar ultraviolet-B radiation and *Manduca sexta* herbivory elicit overlapping transcriptional responses in field-grown plants of *Nicotiana longiflora*. *Plant Physiology* 132, 1755–1767.

Izaguirre, M.M., Mazza, C.A., Biondini, M., Baldwin, I.T. and Ballaré, C.L. (2006) Remote sensing of future competitors: impacts on plant defenses. *Proceedings of the National Academy of Sciences of the United States of America* 103, 7170–7174.

Izaguirre, M.M., Mazza, C.A., Svatosm, A., Baldwin, I.T. and Ballaré, C.L. (2007) Solar ultraviolet-B radiation and insect herbivory trigger partially overlapping phenolic responses in *Nicotiana attenuata* and *Nicotiana longiflora*. *Annals of Botany* 99, 103–109.

Jenkins, G.I. (2014) The UV-B photoreceptor UVR8: from structure to physiology. *Plant Cell* 26, 21–37.

Johanson, U., Gehrke, C., Bjorn, L.O., Callaghan, T.V. and Sonesson, M. (1995) The effects of enhanced UV-B radiation on a subarctic heath ecosystem. *Ambio* 24, 106–111.

Johnson, D., Campbell, C.D., Lee, J.A. and Callaghan, T.V. (2002) Arctic microorganisms respond more to elevated UV-B radiation than CO_2. *Nature* 416, 82–83.

Jones, A.G., Bussell, J., Winters, A., Scullion, J. and Gwynn-Jones, D. (2015) The functional quality of decomposing litter outputs from an Arctic plant community is affected by long-term exposure to enhanced UV-B. *Ecological Indicators* 60, 8–17.

Keiber, R.J., Zhou, X. and Mopper, K. (1990) Formation of carbonyl compounds from UV-induced photodegradation of humic substances in natural waters: fate of riverine carbon in the sea. *Limnology and Oceanography* 35, 1503–1515.

Kieber, D.J., McDaniel, J. and Mopper, K. (1989) Photochemical source of biological substrates in sea water: implications for carbon cycling. *Nature* 341, 637–639.

Kim, S.T., Li, Y.F. and Sancar, A. (1992) The third chromophore of DNA photolyase: Trp-277 of *Escherichia coli* DNA photolyase repairs thymine dimers by direct electron transfer. *Proceedings of the National Academy of Sciences* 89, 900–904.

King, J.Y., Brandt, L.A. and Adair, E.C. (2012) Shedding light on plant litter decomposition: advances, implications and new directions in understanding the role of photodegradation. *Biogeochemistry* 111, 57–81.

Kliebenstein, D.J., Lim, J.E., Landry, L.G. and Last, R.L. (2002) Arabidopsis UVR8 regulates ultraviolet-B signal transduction and tolerance and contains sequence similarity to human regulator of chromatin condensation. *Plant Physiology* 130, 234–243.

Kuhlmann, F. and Müller, C. (2010) UV-B impact on aphid performance mediated by plant quality and plant changes induced by aphids. *Plant Biology* 12, 676–684.

Lambie, S.M., Kirschbaum, M.U.F. and Dando, J. (2014) No photodegradation of litter and humus exposed to UV-B radiation under laboratory conditions: no effect of leaf senescence or drying temperature. *Soil Biology and Biochemistry* 69, 46–53.

Landry, L.G., Chapple, C.C. and Last, R.L.(1995) *Arabidopsis* mutants lacking phenolic sunscreens exhibit enhanced ultraviolet-B injury and oxidative damage. *Plant Physiology* 109, 1159–1166.

Lee, H., Rahn, T. and Throop, H. (2012) An accounting of C-based trace gas release during abiotic plant litter degradation. *Global Change Biology* 18, 1185–1195.

Li, J., Ou-Lee, T.M., Raba, R., Amundson, R.G. and Last, R.L. (1993) Arabidopsis flavonoid mutants are hypersensitive to UV-B irradiation. *Plant Cell* 5, 171–179.

Lin, Y., and King, J.Y. (2014) Effects of UV exposure and litter position on decomposition in a California grassland. *Ecosystems* 17, 158–168.

Lin, Y., King, J.Y., Karlen, S.D. and Ralph, J. (2015a) Using 2D NMR spectroscopy to assess effects of UV radiation on cell wall chemistry during litter decomposition. *Biogeochemistry* 125, 427–436.

Lin, Y., Scarlett, R.D. and King, J.Y. (2015b) Effects of UV photodegradation on subsequent microbial decomposition of *Bromus diandrus* litter. *Plant and Soil* 395, 263–271.

Ma, S., Baldocchi, D.D., Hatala, J.A., Detto, M. and Curiel Yuste, J. (2012) Are rain-induced ecosystem respiration pulses enhanced by legacies of antecedent photodegradation in semi-arid environments? *Agricultural and Forest Meteorology* 154–155, 203–213.

Madronich, S., McKenzie, R.L., Caldwell, M.M. and Bjorn, L.O. (1995) Changes in ultraviolet radiation reaching the earth's surface. *Ambio* 24, 143–152.

Malanga, G. and Puntarulo, S. (1995) Oxidative stress and antioxidant content in *Chlorella vulgaris* after exposure to ultraviolet-B radiation. *Physiologia Plantarum* 94, 672–679.

Mazza, C.A. and Ballaré, C.L. (2015) Photoreceptors UVR8 and phytochrome B cooperate to optimize plant growth and defense in patchy canopies. *New Phytologist* 207, 4–9.

Mazza, C.A., Battista, D., Zima, A.M., Szwarcberg-Bracchitta, M., Giordano C.V. *et al.* (1999) The effects of solar ultraviolet-B radiation on the growth and yield of barley are accompanied by increased DNA damage and antioxidant responses. *Plant, Cell & Environment* 22, 61–70.

Mazza, C.A., Boccalandro, H.E., Giordano, C.V., Battista, D., Scopel, A.L. and Ballaré, C.L. (2000) Functional significance and induction by solar radiation of ultraviolet-absorbing sunscreens in field-grown soybean crops. *Plant Physiology* 122, 117–125.

Mazza, C.A., Izaguirre, M.M., Zavala, J., Scopel, A.L. and Ballaré, C.L. (2002) Insect perception of ambient ultraviolet-B radiation. *Ecology Letters* 5, 722–726.

McKenzie, R.L., Aucamp, P.J., Bais, A.F., Bjorn, L.O., Ilyas, M. and Madronich, S. (2011) Ozone depletion and climate change: impacts on UV radiation. *Photochemical and Photobiological Sciences* 10, 182–198.

Meentemeyer, V. (1978) Macroclimate and lignin control of litter decomposition rates. *Ecology* 59, 465–472.

Molina, M.J. and Rowland, F.S. (1974) Stratospheric sink for chlorofluoromethanes: chlorine atom catalysed destruction of ozone. *Nature* 249, 810–812.

Moody, S.A., Paul, N.D., Björn, L.O., Callaghan, T.V., Lee, J.A. *et al.* (2001) The direct effects of UV-B radiation on *Betula pubescens* litter decomposing at four European field sites. *Plant Ecology* 154, 29–36.

Moorhead, D.L. and Callaghan, T.V. (1994) Effects of increasing UV-B radiation on decomposition and soil organic matter dynamics: a synthesis and modeling study. *Biology and Fertility of Soils* 18, 19–26.

Moorhead, D.L. and Reynolds, J.F. (1989) Mechanisms of surface litter mass loss in the northern Chihuahuan desert: a reinterpretation. *Journal of Arid Environments* 16, 157–163.

Morales, L.O., Brosché, M., Vainonen, J., Jenkins, G.I., Wargent, J.J. *et al.* (2013) Multiple roles for UV RESISTANCE LOCUS8 in regulating gene expression and metabolite accumulation in arabidopsis under solar ultraviolet radiation. *Plant Physiology* 161, 744–759.

Moreno, J.E., Tao, Y., Chory, J. and Ballaré, C.L. (2009) Ecological modulation of plant defense via phytochrome control of jasmonate sensitivity. *Proceedings of the National Academy of Sciences of the United States of America* 106, 4935–4940.

Müller, U., Rätzsch, M., Schwanninger, M., Steiner, M. and Zöbl, H. (2003) Yellowing and IR-changes of spruce wood as result of UV-irradiation. *Journal of Photochemistry and Photobiology B: Biology* 69, 97–105.

Newsham, K.K., Greenslade, P.D., Kennedy, V.H. and McLeod, A.R. (1999) Elevated UV-B radiation incident on *Quercus robur* leaf canopies enhances decomposition of resulting leaf litter in soil. *Global Change Biology* 5, 403–409.

Niemi, R., Martikainen, P.J., Silvola, J., Wulff, A., Turtola, S. and Holopainen, T. (2002) Elevated UV-B radiation alters fluxes of methane and carbon dioxide in peatland microcosms. *Global Change Biology* 8, 361–371.

Pancotto, V.A., Sala, O.E., Cabello, M., López, N., Ballaré, C.A., Caldwell, M.M. and Scopel, A.L. (2003) Solar UV-B decreases decomposition in herbaceous plant litter in Tierra del Fuego, Argentina: potential role of an altered decomposer community. *Global Change Biology* 9, 1465–1474.

Pancotto, V.A., Sala, O.E., Robson, T.M., Caldwell, M.M. and Scopel, A.L. (2005) Direct and indirect effects of solar ultraviolet-B radiation on long-term decomposition. *Global Change Biology* 11, 1982–1989.

Rao, M.V., Paliyath, G. and Ormrod, D.P. (1996) Ultraviolet-B- and ozone-induced biochemical changes in antioxidant enzymes of *Arabidopsis thaliana*. *Plant Physiology* 110, 125–136.

Reuber, S., Bornman, J.F. and Weissenböck, G. (1996) A flavonoid mutant of barley (*Hordeum vulgare* L.) exhibits increased sensitivity to UV-B radiation in the primary leaf. *Plant, Cell & Environment* 19, 593–601.

Rinnan, R., Impiö, M., Silvola, J., Holopainen, T. and Martikainen, P.J. (2003) Carbon dioxide and methane fluxes in boreal peatland microcosms with different vegetation cover: effects of ozone or ultraviolet-B exposure. *Oecologia* 137, 475–483.

Rizzini, L., Favory, J.J., Cloix, C., Faggionato, D., O'Hara, A. *et al.* (2011) Perception of UV-B by the Arabidopsis UVR8 protein. *Science* 332, 103–106.

Robson, T.M., Pancotto, V.A., Ballaré, C.L., Sala, O.E., Scopel, A.L. and Caldwell, M.M. (2004) Reduction of solar UV-B mediates changes in the *Sphagnum capitulum* microenvironment and the peatland microfungal community. *Oecologia* 140, 480–490.

Rousseaux, M.C., Ballaré, C.L., Giordano, C.V., Scopel, A.L., Zima, A.M. *et al.* (1999) Ozone depletion and UVB radiation: impact on plant DNA damage in southern South America. *Proceedings of the National Academy of Sciences of the United States of America* 96, 15310–15315.

Rozema, J., Tosserams, M., Nelissen, H.J.M., van Heerwaarden, L., Broekman, R.A. and Flierman, N. (1997a) Stratospheric ozone reduction and ecosystem processes: enhanced UV-B radiation affects chemical quality and decomposition of leaves of the dune grassland species *Calamagrostis epigeios*. *Plant Ecology* 128, 284–294.

Rozema, J., van de Staaij, J., Bjorn, L.O. and Caldwell, M. (1997b) UV-B as an environmental factor in plant life: stress and regulation. *Trends in Ecology and Evolution* 12, 22–28.

Rutledge, S., Campbell, D.I., Baldocchi, D. and Schipper, L.A. (2010) Photodegradation leads to increased CO_2 losses from terrestrial organic matter. *Global Change Biology* 16, 3065–3074.

Schade, G.W., Hormann, R.M. and Crutzen, P.J. (1999) CO emissions from degrading plant matter. *Tellus* 51B, 899–908.

Schimel, J.P. and Schaeffer, S.M. (2012) Microbial control over carbon cycling in soil. *Frontiers in Microbiology* 3, 348. DOI:10.3389/fmicb.2012.00348

Searles, P.S., Kropp, B.R., Flint, S.D. and Caldwell, M.M. (2001) Influence of solar UV-B radiation on peatland microbial communities of southern Argentina. *New Phytologist* 152, 213–221.

Setlow, R.B. (1974) The wavelengths in sunlight effective in producing skin cancer: a theoretical analysis. *Proceedings of the National Academy of Sciences of the United States of America* 71, 3363–3366.

Sinha, R.P. and Häder, D.P. (2002) UV-induced DNA damage and repair: a review. *Photochemical and Photobiological Sciences* 1, 225–236.

Smith, R., Prezelin, B., Baker, K., Bidigare, R., Boucher, N. *et al.* (1992) Ozone depletion: ultraviolet radiation and phytoplankton biology in antarctic waters. *Science* 255, 952–959.

Smith, W.K., Gao, W., Steltzer, H., Wallenstein, M.D. and Tree, R. (2010) Moisture availability influences the effect of ultraviolet-B radiation on leaf litter decomposition. *Global Change Biology* 16, 484–495.

Solomon, S. (1990) Progress towards a quantitative understanding of Antarctic ozone depletion. *Nature* 347, 347–354.

Solomon, S. (1999) Stratospheric ozone depletion: a review of concepts and history. *Reviews of Geophysics* 37, 275–316.

Solomon, S., Ivy, D.J., Kinnison, D., Mills, M.J., Neely III, R.R. and Schmidt, A. (2016) Emergence of healing in the Antarctic ozone layer. *Science* 353, 269–274.

Song, X., Peng, C., Jiang, H., Zhu, Q. and Wang, W. (2013) Direct and indirect effects of UV-B exposure on litter decomposition: a meta-analysis. *PLoS ONE* 8, e68858.

Stapleton, A.E. (1992) Ultraviolet radiation and plants: burning questions. *The Plant Cell* 4, 1353.

Stapleton, A.E., Thornber, C.S. and Walbot, V. (1997) UV-B component of sunlight causes measurable damage in field-grown maize (*Zea mays* L.): developmental and cellular heterogeneity of damage and repair. *Plant, Cell & Environment* 20, 279–290.

Stark, J.M. and Hart, S.C. (2003) UV-B radiation and soil microbial communities. *Nature* 423, 137–138.

Stratmann, J.W., Stelmach, B.A., Weiler, E.W. and Ryan, C.A. (2000) UVB/UVA radiation activates a 48 kDa myelin basic protein kinase and potentiates wound signaling in tomato leaves. *Photochemistry and Photobiology* 71, 116–123.

Strid, A., Chow, W.S. and Anderson, J.M. (1994) UV-B damage and protection at the molecular level in plants. *Photosynthesis Research* 39, 475–489.

Takeuchi, Y., Kubo, H., Kasahara, H. and Sakaki, T. (1996) Adaptive alterations in the activities of scavengers of active oxygen in cucumber cotyledons irradiated with UV-B. *Journal of Plant Physiology* 147, 589–592.

Tarr, M.A., Miller, W.L. and Zepp, R.G. (1995) Direct carbon monoxide production from plant matter. *Journal of Geophysical Research* 100, 11403–11413.

Teramura, A.H., and Sullivan, J.H. (1994) Effects of UV-B radiation on photosynthesis and growth of terrestrial plants. *Photosynthesis Research* 39, 463–473.

Tevini, M. and Teramura, A.H. (1989) UV-B effects on terrestrial plants. *Photochemistry and Photobiology* 50, 479–487.

Throop, H.L. and Archer, S.R. (2007) Interrelationships among shrub encroachment, land management, and litter decomposition in a semidesert grassland. *Ecological Applications* 17, 1809–1823.

UNEP (1994) *Environmental Effects of Ozone Depletion: 1994 Assessment.* United Nations Environment Programme, Nairobi.

Uselman, S.M., Snyder, K.A., Blank, R.R. and Jones T.J. (2011) UVB exposure does not accelerate rates of litter decomposition in a semi-arid riparian ecosystem. *Soil Biology and Biochemistry* 43, 1254–1265.

Vanderbilt, K., White, C., Hopkins, O. and Craig, J. (2008) Aboveground decomposition in arid environments: results of a long-term study in central New Mexico. *Journal of Arid Environments* 72, 696–709.

Veit, M., Bilger, W., Mühlbauer, T., Brummet, W. and Winter, K. (1996) Diurnal changes in flavonoids. *Journal of Plant Physiology* 148, 478–482.

Walker, G.C. (1984) Mutagenesis and inducible responses to deoxyribonucleic acid damage in *Escherichia coli*. *Microbiological Reviews* 48, 60–93.

Wang, J., Liu, L., Wang, X. and Chen, Y. (2015) The interaction between abiotic photodegradation and microbial decomposition under ultraviolet radiation. *Global Change Biology* 21, 2095–2104.

Whitford, W.G., Meentemeyer, V., Seastedt, T.R., Cromack Jr, K., Crossley Jr, D.A. *et al.* (1981) Exceptions to the AET model: deserts and clear-cut forests. *Ecology* 62, 275–277.

Willekens, H., Van Camp, W., Van Montagu, M., Inzé, D., Langebartels, C. and Sandermann Jr, H. (1994) Ozone, sulfur dioxide, and ultraviolet B have similar effects on mRNA accumulation of antioxidant genes in *Nicotiana plumbaginifolia* L. *Plant Physiology* 106, 1007–1014.

Williamson, C.E., Zepp, R.G., Lucas, R.M., Madronich, S., Austin, A.T. *et al.* (2014) Solar ultraviolet radiation in a changing climate. *Nature Climate Change* 4, 434–441.

Wu, D., Hu, Q., Yan, Z., Chen, W., Yan, C.Y. *et al.* (2012) Structural basis of ultraviolet-B perception by UVR8. *Nature* 484, 214–219.

Zavala, J.A., Mazza, C.A., Dillon, F.M., Chludil, H.D. and Ballaré, C.L. (2015) Soybean resistance to stink bugs (*Nezara viridula* and *Piezodorus guildinii*) increases with exposure to solar UV-B radiation and correlates with isoflavonoid content in pods under field conditions. *Plant, Cell and Environment* 38, 920–928.

4 UV-B-induced Changes in Secondary Plant Metabolites

Monika Schreiner*, Melanie Wiesner-Reinhold, Susanne Baldermann, Franziska S. Hanschen and Susanne Neugart
Leibniz Institute of Vegetable and Ornamental Crops, Großbeeren, Germany

Introduction – The New View on Secondary Plant Metabolites

In the current scientific literature, secondary plant metabolites are discussed in two key respects: (i) their relevance for the plant's fitness as regards its interactions with the environment, and (ii) their protective role for human health via plant-based nutrition. In this regard, several epidemiological studies have shown an inverse association between vegetable consumption and the incidence of chronic diseases such as different types of cancer, diabetes and cardiovascular disease. Moreover, secondary plant metabolites have been demonstrated to be the bioactive compounds accountable for this observed protective effect in several cellular and biochemical *in vitro* investigations as well as in *in vivo* experiments and human intervention studies (e.g. Watzl and Leitzmann, 2005; Verkerk *et al.*, 2009).

As regards plant interactions with the environment, secondary plant metabolites can act as pollination attractants, antioxidants or signalling molecules, as well as protective compounds against pathogens and herbivorous insects or various abiotic stresses such as radiation impact. In particular, they prevent solar UV-B stress, which has been shown to increase due to stratospheric ozone depletion (e.g. Rozema *et al.*, 1999). Recently, however, in contrast to former investigations that highlighted the impact of elevated UV-B radiation on plant physiological and metabolic responses, a paradigm shift has occurred. The regulatory properties of lower, but also ecologically relevant, UV-B levels have been demonstrated (Jansen, 2012), revealing the UV-B-induced triggering of distinct changes in plants' secondary metabolisms (e.g. summarized in the review by Schreiner *et al.*, 2012). Subsequently, secondary plant metabolites are accumulated *in planta* in an extremely structure-specific manner and lead to compositional changes that result in vegetables and fruit enriched with health-promoting secondary plant metabolites. Such vegetables and fruit can either serve as fresh products or be used as raw material for functional foods and supplements. Thus, this chapter focuses on pre-harvest UV-B effects on selected secondary plant metabolites in edible plants.

*Corresponding author: schreiner@igzev.de

Structure-specific Plant Responses to UV-B Exposure

UV-B-induced accumulation of certain secondary plant metabolites (Fig. 4.1) is not only greatly affected by the time of exposure and the dose of UV-B but also by other environmental factors such as photosynthetic photon flux density (PPFD), far-red radiation, atmospheric CO_2 concentration, and temperature, as well as by the genotype (Schreiner *et al.*, 2012). In addition, several biosynthetic pathways might be in competition with each other, thereby resulting in a trade-off between primary and secondary plant metabolites. This occurs when there is limited access to resources, e.g. water, nutrients, radiation or even metabolic precursors, and when a plant has to deal with various concurrent biotic and abiotic interactions in its environment (van Dam, 2009). With regard to UV-B exposure, a plant's short-term response is reflected by an increase of primary metabolites, which may act as precursors of UV-B-absorbing secondary metabolites that are synthesized later (Kusano *et al.*, 2011). Such a UV-B-induced response on the metabolite level *in planta* results in various temporary secondary metabolite profiles.

Phenolic Compounds

Phenolic compounds, especially flavonoids, in plants are diverse (Fig. 4.1) and naturally occur as flavonoid glycosides that, based on aglycones, can be divided into flavanols, flavanones, flavones, flavonols, anthocyanidines and isoflavonoids. In fact, more than 6500 flavonoid structures are known due to hydroxylation, glycosylation, acylation and/or methoxylation (Scalbert and Williamson, 2000; Heim *et al.*, 2002; Huang *et al.*, 2004; Edreva, 2005; Ferreres *et al.*, 2009; Calderon-Montano *et al.*, 2011; Mierziak *et al.*, 2014). Of note is that key flavonoid biosynthesis genes are regulated by UV-B exposure and dosage (Tilbrook *et al.*, 2013) and that flavonoids accumulate in a range of cellular compartments, including cell walls, vacuoles, chloroplasts and the nucleus, as well as in trichomes (Agati and Tattini, 2010). Moreover, intracellular accumulation at sites of reactive oxygen species (ROS) production (e.g. chloroplasts) highlights the important antioxidant properties of this class of metabolites (Hernandez *et al.*, 2009). Finally, flavonoid compounds have received considerable attention because of their potential health-promoting

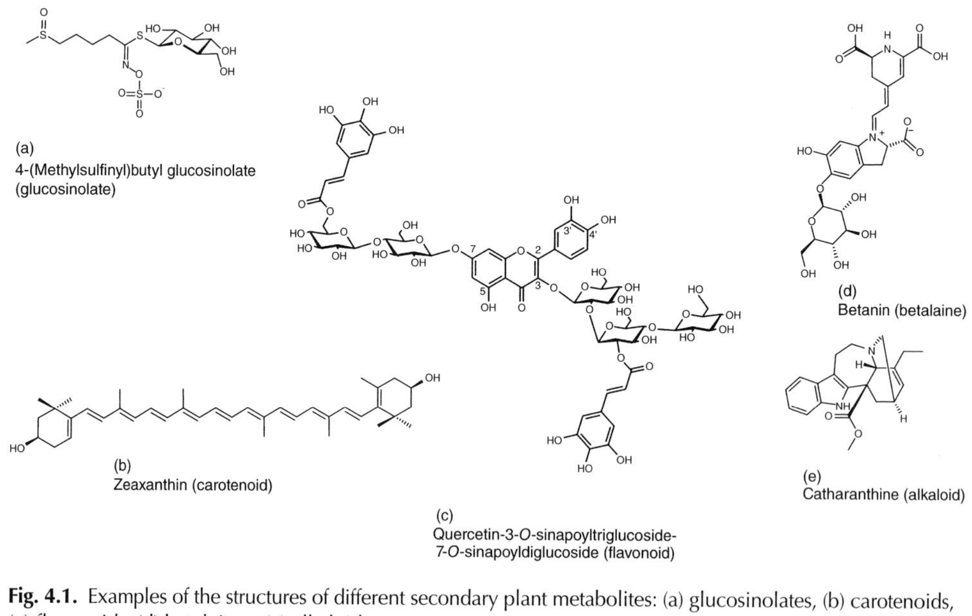

Fig. 4.1. Examples of the structures of different secondary plant metabolites: (a) glucosinolates, (b) carotenoids, (c) flavonoids, (d) betalaines, (e) alkaloids.

benefits to humans, e.g. anti-oxidative, anti-inflammatory and anti-cancerogenic properties (Pan *et al.*, 2010).

Ambient and increased UV-B radiation is able to modify the flavonoid profile of different plant species. The effect of UV-B is modified by UV-B dose, the structure of the flavonoids and other phenolics, as well as environmental factors such as PPFD and temperature. The exclusion of UV-B radiation is known to result in lower concentrations of UV-absorbing compounds in faba bean (*Vicia faba*) leaves (Barnes *et al.*, 2013) and a decrease in quercetin and kaempferol glycosides in vine leaves (*Vitis vinifera*) (Kolb *et al.*, 2001). In the model plant *Arabidopsis thaliana*, Morales *et al.* (2013) also found that mainly quercetin glycosides are decreased by the exclusion of UV-B radiation, but that kaempferol glycosides are less affected. However, the response of flavonoids is species specific. For example, mustard (*Sinapis alba*) decreases its flavonoid concentration as a response to lower UV-B doses, whereas nasturtium (*Nasturtium officinale*) shows no change in its flavonoid concentration even when UV-B is excluded (Reifenrath and Mueller, 2007). Experiments have also shown that the intensity, the duration and repetition of UV-B treatment can affect the total phenolic content. For example, in lettuce (*Lactuca sativa*), the total phenolic content increased after repeated doses of UV-B; this was linked to phenylalanine ammonia lyase activity (Lee *et al.*, 2014b). Moreover, UV-B exposure results in changes in the phenolic compound profile. Specifically, lettuce grown under higher UV-B doses had higher concentrations of total phenolics, anthocyanins and phenolic acids (Ordidge *et al.*, 2010). Furthermore, in Lollo Rosso lettuce (*Lactuca sativa* var. *crispa*), a strong negative correlation of the wavelengths (in UV-A and UV-B range) and the concentration of total phenolics and anthocyanins was demonstrated (Tsormpatsidis *et al.*, 2008). Finally, in Lollo Rosso lettuce grown under UV-transmitting plastic films, increased concentrations of flavonoids and a higher antioxidant activity were found compared to greenhouse-grown plants (Garcia-Macias *et al.*, 2007). Recent studies (see Table 4.1) underline the structure-dependent response of flavonoids and other phenolics to UV-B radiation. For flavonoid glycosides, there are structural characteristics regarding the aglycone, the glycosylated sugars as well as the acylated phenolic and organic acids.

The type of aglycone or the flavonoid class has a great impact on the response to environmental conditions including UV-B. Levels of quercetin and *ortho*-dihydroxylated flavonoids are often enhanced, while kaempferol and *ortho*-monohydroxylated flavonoids remain unaffected by UV-B radiation or higher intensities of solar radiation (Reifenrath and Mueller, 2007; Winter and Rostas, 2008). UV-B radiation is also known to lead to an enhanced quercetin to kaempferol ratio in the model plant *Arabidopsis thaliana* as well as in vegetables (Zhang *et al.*, 2003; Reifenrath and Mueller, 2007; Winter and Rostas, 2008; Goetz *et al.*, 2010). Quercetin glycosides have a higher antioxidative activity than their corresponding kaempferol glycosides (Zietz *et al.*, 2010) and are produced in greater quantities in response to UV-B radiation in various edible species and *A. thaliana* (Jansen *et al.*, 2008; Goetz *et al.*, 2010; Hectors *et al.*, 2012; Hectors *et al.*, 2014). In addition, levels of anthocyanins and catechins increased with higher UV-B doses corresponding to the gene expression in grapes and *A. thaliana* (Cominelli *et al.*, 2008; Zhang *et al.*, 2013). In broccoli (*Brassica oleracea* var. *italica*) and canola (*Brassica napus*), quercetin and its glycosides as well as kaempferol and most kaempferol glycosides were also shown to be present in greater quantities after additional UV-B radiation (Olsson *et al.*, 1998; Kuhlmann and Mueller, 2009). In kale (*Brassica oleracea* var. *sabellica*), after a single dose of UV-B radiation, quercetin glycoside quantities decreased, whereas the formation of kaempferol glycosides in response to UV-B was influenced by the number of sugar moieties and acylated hydroxycinnamic acid residues (Neugart *et al.*, 2012).

In a second experiment investigating the interaction of moderate UV-B radiation and temperature on structurally different phenolic compounds in kale, levels of monoacylated

Table 4.1. Examples of studies investigating the effect of UV-B on phenolic compounds.

Species	UV-B level/range	UV-B-induced changes in phenolic compounds	Reference
Canola (*Brassica napus*)	13 kJ m^{-2} d^{-1} UV-B_{BE}	quercetin glycosides ↓ kaempferol glycosides ↑ ↓ (depending on glycosylation and acylation pattern)	Olsson *et al.* (1998)
Grape (*Vitis vinifera*)	UV-B cut off filters (polyester film)	caffeic acid derivative ↑ coumaric acid derivative - quercetin glycosides ↓ kaempferol glycosides ↓	Kolb *et al.* (2001)
Kale (*Brassica oleracea* var. *sabellica*)	24 W m^{-2}	quercetin ↑ kaempferol ↑ quercetin/kaempferol ratio ↑	Zhang *et al.* (2003)
Nasturtium (*Nasturtium officinale*), mustard (*Sinapis alba*)	1.12 W m^{-2} (UV-B 280-315 nm)	quercetin glycosides ↑ kaempferol glycosides ↑ hydroxycinnamic acids -	Reifenrath and Mueller (2007)
Lollo Rosso lettuce (*Lactuca sativa* var. *crispa*)	UV-B cut off filters (polythene film)	total phenolics ↓ anthocyanins ↓ luteolin ↓ quercetin ↓	Garcia-Macias *et al.* (2007)
Lollo Rosso lettuce (*Lactuca sativa* var. *crispa*)	UV-B cut off filters (polythene film)	total phenolics ↓ total flavonoids ↓ total anthocyanins ↓	Tsormpatsidis *et al.* (2008)
Soybean (*Glycine max*)	UV-B cut off filters (polyester film)	quercetin glycosides ↓ kaempferol glycosides - isorhamnetin glycosides 1 ↓ isorhamnetin glycosides 3 -	Winter and Rostas (2008)
Various herbs, e.g. basil, rosemary	various UV-B doses	flavonoids ↑ poly-hydroxylated increase higher than mono-hydroxylated	Jansen *et al.* (2008)
Arabidopsis thaliana	3 μmol m^{-2} s^{-1}	*CHS* ↑ *F3H* ↑ *DFR* hardly detectable *LDOX* hardly detectable anthocynins ↑	Cominelli *et al.* (2008)
Arabidopsis thaliana	0-12 μmol m^{-2} s^{-1}	*HY5* ↑ (dose-dependent) *HYH* ↑ (dose-dependent) *CHS* ↑ (dose-dependent)	Brown and Jenkins (2008)
Tomato (*Solanum lycopersicum*)	UV-B cut off filters (polythene film)	naringenin-chalcone ↓ quercetin ↑ - (cultivar-dependent) rutin ↓ ↑ (cultivar-dependent) quercetin-3-pentosyl-rutinoside – ↑ (cultivar-dependent) *CHS* ↑ - (cultivar-dependent) *CHI* ↓ ↑ (cultivar-dependent) *F3H* ↑ ↑ (cultivar-dependent) *F3′H* ↑ ↑ (cultivar-dependent) *FLS* ↓ (cultivar-dependent)	Giuntini *et al.* (2008)
Broccoli (*Brassica oleracea* var. *italica*)	UV-B cut off filters (teflon film)	total flavonoids and hydroxycinnamic acids ↓	Kuhlmann and Mueller (2009)

(Continued)

Table 4.1. Continued.

Species	UV-B level/range	UV-B-induced changes in phenolic compounds	Reference
Lettuce (*Lactuca sativa*)	UV-B cut off filters (polythen film)	total phenolics ↓ anthocyanins ↓ luteolin ↓ quercetin ↓	Ordidge *et al.* (2010)
Arabidopsis thaliana	UV-B cut off filters (glass film)	quercetin glycosides ↓	Goetz *et al.* (2010)
Pak choi (*Brassica rapa* ssp. *chinensis*)	0.35 W m^{-2}	epidermal absorption 375 ↑ kaempferol glycosides ↑ malates of hydroxycinnamic acids	Harbaum-Piayda *et al.* (2010)
Peppermint (*Mentha* x *piperita*)	7.1 kJ m^{-2} d^{-1} $UV\text{-}B_{BE}$	eriocitrin ↑ hesperidin ↑ kaempferol 7-*O*-rutinoside ↑ narirutin ↓ 4′-metOkaempferol 7-*O*-rutinoside ↓	Dolzhenko *et al.* (2010)
Arabidopsis thaliana	0.56 kJ m^{-2} d^{-1} 280-315 nm	quercetin glycosides ↑ kaempferol glycosides ↑ (depending on glycosylation pattern)	Hectors *et al.* (2012)
Kale (*Brassica oleracea* var. *sabellica*)	0.22-0.88 kJ m^{-2} d^{-1} $UV\text{-}B_{BE}$ (single dose)	quercetin glycosides ↓ kaempferol glycosides ↑ ↓ (depending on glycosylation and acylation pattern)	Neugart *et al.* (2012)
Faba bean (*Vicia faba*)	UV-B cut off filters (polyester film)	UV-absorbing compounds ↓	Barnes *et al.* (2013)
Arabidopsis thaliana	UV-B cut off filters (polyester film)	quercetin glycosides ↓ kaempferol glycosides -	Morales *et al.* (2013)
Grape (*Vitis vinifera*)	1.8 kJ m^{-2} d^{-1}	*LAR1* ↑ in 3-week old berries *LAR2* - in 3-week old berries *ANR* ↑ in 3-week old berries flavan-3-ols ↑ in 3-week old berries after 1 h and then decrease again	Zhang *et al.* (2013)
Lettuce (*Lactuca sativa*)	4.2 W m^{-2} (UV-B 306 nm)	total phenolics ↑ after 3 d *PAL* ↑ after 4 d	Lee *et al.* (2014b)
Arabidopsis thaliana	0.59 kJ m^{-2} d^{-1} 280-315 nm	quercetin glycosides ↑ kaempferol glycosides ↑ (depending on glycosylation pattern)	Hectors *et al.* (2014)
Kale (*Brassica oleracea* var. *sabellica*)	0.25-1.25 kJ m^{-2} d^{-1} $UV\text{-}B_{BE}$	quercetin glycosides ↑ kaempferol glycosides ↑↓ (depending on glycosylation and acylation pattern)	Neugart *et al.* (2014)

ANR, anthocyanidin reductase; *CHI*, chalcone isomerase; *CHS*, chalcone synthase; *DFR*, dihydroflavonol 4-reductase; *FLS*, flavonol synthase; *F3H*, flavanone 3-hydroxylase; *F3′H*, flavanol 3-hydroxylase; *HYH*, HY5-homologue; *HY5*, elongated hypocotyl 5; *LAR*, leucoanthocyanidin reductase; *LDOX*, leucoanthocyanidin dioxygenase; also called anthocyanidin synthase/(*ANS*); *PAL*, Phenylalanine ammonia-lyase; $UV\text{-}B_{BE}$ biologically effective UV-B

quercetin glycosides were increased by higher UV-B doses and temperature (Neugart *et al.*, 2014). Concomitantly, enhanced mRNA expression of flavonol 3'-hydroxylase was found under the same conditions. The response of acylated kaempferol glycosides was more diverse and dependent on the hydroxycinnamic acid residue and the number of glucose moieties in the 7-*O* position. In peppermint (*Mentha* x *piperita*), the concentration of narirutin was decreased by higher UV-B doses, whereas quantities of the related compounds eriocitrin and hesperidin were increased (Dolzhenko *et al.*, 2010).

Finally, naringenin chalcone concentration decreased in the peel of tomato (*Solanum lycopersicum*) fruits exposed to higher UV-B doses (Giuntini *et al.*, 2008).

Recent data have also started to reveal UV-B-specific effects on flavonoid glycosylation patterns. The flavonoid glycosylation pattern is markedly influenced by higher UV-B doses as shown for kaempferol glycosides in kale (Neugart *et al.*, 2012). While monoacylated kaempferol tetraglucosides decreased following exposure to a single dose of UV-B, the monoacylated kaempferol diglucoside levels increased. Subsequent UV-B doses on kale resulted in different responses of acylated kaempferol glycosides that depended on the number of glucose moieties in the 7-*O* position (Neugart *et al.*, 2014). Moreover, the monoacylated kaempferol diglucosides were shown to be affected by an interaction of UV-B and temperature; this effect was not found for kaempferol tetraglycosides independent of the acylated hydroxycinnamic acid. Furthermore, after higher UV-B doses in *A. thaliana*, di- and triglycosides accumulate with a preponderance of 7-rhamnosylated flavonols (Hectors *et al.*, 2014). The accumulation of specific flavonoid glycosides appears to be an intrinsic part of the UV-B response, with the gene expression of several UDP-glucosyltransferases being directly controlled by UV-B (Brown and Jenkins, 2008). Rather paradoxically, glycosylation decreases the antioxidant activity of flavonoids as well as affecting their accumulation, stability and solubility (Gachon *et al.*, 2005; Bowles *et al.*, 2006).

The response to UV-B of flavonoid glycosides is dependent on the type of phenolic acid that is acylated to the flavonol glycoside (mainly hydroxycinnamic acids). In pak choi (*Brassica rapa* ssp. *chinensis*), total flavonoid levels increased with exposure to additional UV-B, but kaempferol glycosides acylated with ferulic, hydroxyferulic or sinapic acid did not respond to UV-B exposure at 22°C (Harbaum-Piayda *et al.*, 2010). In kale, the structures of the hydroxycinnamic acids themselves have an impact on the response to UV-B (Neugart *et al.*, 2014). While the levels of caffeic-acid- and hydroxyferulic-acid-monoacylated kaempferol triglycosides (containing a catechol structure) were increased with exposure to higher UV-B radiation, the ferulic- and sinapic-acid-monoacylated kaempferol triglycosides (no catechol structure) were not affected. In canola (*Brassica napus*), the levels of quercetin glycosides were enhanced when exposed to higher UV-B levels, while the kaempferol glycosides displayed additional changes dependent on their acylation pattern (Olsson *et al.*, 1998). Furthermore, in response to additional exposure to higher UV-B doses, the levels of non-acylated kaempferol-3-*O*-sophoroside-7-*O*-D-glucoside increased, while the sinapic-acid-monoacylated kaempferol glycoside did not respond (Olsson *et al.*, 1998).

Phenolic acids are also known to respond to UV-B radiation in a structure-dependent manner. In tomato fruit, the levels of caffeic acid, ferulic acid and *p*-coumaric acid were higher in plants exposed to higher UV-B doses compared to plants that were not (Giuntini *et al.*, 2008). However, in pak choi, Harbaum-Piayda *et al.* (2010) demonstrated that caffeoylmalate, hydroxyferuloylmalate, coumaroylmalate, feruloylmalate and sinapoylmalate were not affected by higher UV-B radiation at higher temperatures (22°C). The hydroxycinnamic acid derivatives of kale (caffeoylquinic acid, disinapoyl-gentiobiose and sinapoyl-feruloyl-gentiobiose) were hardly affected by subsequent doses of UV-B radiation (Neugart *et al.*, 2014). In contrast, a single UV-B dose led to a slight decrease of caffeoylquinic acid as well as an increase of disinapoyl-gentiobiose and sinapoyl-feruloyl-gentiobiose (Neugart *et al.*, 2012). Of note is that hydroxycinnamic acids were previously shown to act as scavengers to reactive oxygen species (ROS) induced by UV-B radiation (Edreva, 2005).

(Apo)Carotenoids, Abscisic Acid and Other Terpenoids

Plants contain various isoprenoid compounds that are composed of C5 units of 'active isoprene' (Fig. 4.1). These include the important classes of monoterpenes, sesquiterpenes, diterpenoids and tetraterpenes, containing C10, C15, C20 and C40 carbons, respectively.

Carotenoids are tetraterpenoids, which are photosynthetic pigments and participate in light harvesting and photoprotection. Moreover, they are potent scavengers of free radicals including ROS. Carotenoid-derived compounds, such as abscisic acid (ABA), are involved in the biotic stress response as well as growth regulation. The phytohormone ABA influences almost all aspects of plant growth and development and also affects plant stress responses (Zeevaart and Creelman, 1988; Seo and Koshiba, 2002; North *et al.*, 2007). Recently, short-chain carotenoid breakdown products have also been shown to be involved in plant stress responses (Havaux, 2014). However, whether these pathways are activated in response to UV radiation still needs to be investigated.

The interest in carotenoids and human health goes back to the time when the link between carotenoids and vitamin A was elucidated (summarized in Semba, 2012). Since then, carotenoids have had an increasingly important role in the human diet. Today, it is clear that carotenoids and derived compounds have other important functions, such as antioxidants, pigments in the human eye or immunomodulatory compounds (summarized in Britton *et al.*, 2009).

In higher plants, carotenoids can be synthesized under dark conditions, but their synthesis is controlled by blue light and UV receptors (Solovchenko and Merzlyak, 2008). UV-B receptors have mainly been characterized in model plants (Rizzini *et al.*, 2011). However, their function still remains to be elucidated in horticultural crops. In addition, the impact of UV-B on carotenoid metabolism has been poorly investigated to date, especially on a mechanistic level. However, it appears that ABA-responsive pathways are upregulated by UV-B exposure (Tossi *et al.*, 2009; Berli *et al.*, 2010).

UV-B radiation can trigger increases in the levels of carotenes and xanthophylls (Table 4.2). The response depends strongly on species and/or cultivar and is also time dependent. The carotenoid biosynthesis pathway is upregulated under UV exposure, but downregulated under chronic UV-B exposure in *A. thaliana* (Jansen *et al.*, 2008). In bunching onion (*Allium fistulosum*), only a few cultivars were sensitive to UV-B treatment (Abney *et al.*, 2013); whereas in tomato fruit, UV-B exposure led to an increase in carotenoids (Becatti *et al.*, 2009; Calvenzani *et al.*, 2010; Lazzeri *et al.*, 2012). Similarly, in tomato fruit after UV-B exposure, the level of lycopene, the most efficient oxygen quencher (Di Mascio *et al.*, 1989), as well as the levels of colourless precursors phytoene and phytofluene increased. It is interesting to note that changes seem to be affected by the presence or absence of other UV-shielding compounds such as anthocyanins. In green lettuce, an increase in total carotenoids in general and lutein and zeaxanthin in particular was observed after UV-B exposure; however, in the red lettuce variety, an opposite trend was found (Caldwell and Britz, 2006).

Finally, mono-, di- and sesquiterpenoids also belong to the class of terpenoids. Volatile terpenoids are emitted in response to biotic and abiotic stress and changes are also observed after UV-B treatment (Table 4.2). For example, after UV-B exposure, there are increases in carnosic acid (diterpenoid) in rosemary (*Rosmarinus officinalis*) (Luis *et al.*, 2007), mono- and sesquiterpenes in peppermint (Dolzhenko *et al.*, 2010), and artemisinin in sweet wormwood (*Artemisia annua*).

Glucosinolates

To date, about 130 glucosinolates are known (Clarke, 2010; Agerbirk and Olsen, 2012) and almost all are found in the order Brassicales. Glucosinolates are β-D-thioglucoside-*N*-hydroxysulfates and have a variable side chain (Fig. 4.1). Due to the positive biofunctional effects of their breakdown products in the plant–insect interaction (Mewis *et al.*, 2005; Halkier and Gershenzon, 2006) as well as in humans (Mithen, 2001; Shapiro *et al.*, 2006; Verkerk *et al.*, 2009), this class of secondary plant metabolites has been the subject of much intensive research.

However, with respect to the effects of UV-B radiation, glucosinolates have not been

Table 4.2. Examples of studies investigating the effect of UV-B on (apo)carotenoids, ABA and other terpenoids.

Species	UV-B level/range	UV-B-induced changes in (apo)carotenoids, ABA and other terpenoids	Reference
Grape (*Vitis vinifera* cv. Riesling)	UV-B cut off filters	more rapid degradation of zeaxanthin ↓	Schultz *et al.* (1998)
Leaf lettuce (*Lactuca sativa*)	not specified	response cultivar-dependent: green lettuce: carotenoids ↑, e.g., neoxanthin ↑, lutein ↑ red lettuce: carotenoids ↓, e.g., neoxanthin ↓, lutein ↓	Caldwell and Britz (2006)
Rosemary (*Rosmarinus officinalis* cv. Sissinghurst English)	5.4 and 31 kJ m^{-2} d^{-1}	carnosic acid ↑	Luis *et al.* (2007)
Tomato (*Solanum lycopersicum* cv. Ailsa Craig)	UV-B absorber benzophenone	total carotenoids ↑, phytoene ↑, phytofluene ↑, lutein ↑, lycopene ↑	Becatti *et al.* (2009)
Maize (*Zea mays*)	3.3 W m^{-2} UV-B for 3 h	ABA↑	Tossi *et al.* (2009)
Grape (*Vitis vinifera* cv. Malbec)	solar UV-B radiation was supplemented by 15 μW cm^{-2} over 5 h, arrow spectrum of 310-315 nm and a maximum at 311 nm	ABA↑	Berli *et al.* (2010)
Sweet wormwood (*Artemisia annua*)	4.2 kJ m^{-2} d^{-1}	artemisinin ↑	
Canola (*Brassica napus*)	18, 25, 40 μW cm^{-2} d^{-1}	carotenoids ↑	Tohidi-Moghadam *et al.* (2012)
Tomato (*Solanum lycopersicum* cv. Money Maker)	UV-B absorber benzophenone	flesh: phytoene ↑, phytofluene ↑, lycopene ↑ peel: lycopene ↑	Lazzeri *et al.* (2012), Calvenzani *et al.* (2010)
Peppermint (*Mentha X piperita*)	7.1 kJ m^{-2} d^{-1}	e.g., limonene ↑, linalool ↑, β-caryophyllene ↑, germacrene D ↑	Dolzhenko *et al.* (2010)
Sweet basil (*Ocimum basilicum* cv. cinnamon)	2 and 4 kJ UV-B m^{-2} d^{-1} for 7 d	young basil, no changes in carotenoids flowering stage, carotenoids ↓	Sakalauskaitė *et al.* (2013)
Bunching onion (*Allium fistulosum*)	7.0 μmol m^{-2} s^{-1} (2.68 W m^{-2}); 313 nm	response cultivar-dependent: sensitive cv Feast: lutein ↑, zeaxanthin ↑, neoxanthin ↑	Abney *et al.* (2013)
Winter cherry (*Withania somnifera*)	3.6 kJ m^{-2} d^{-1}	leaf: lycopene ↑, β-carotene ↑	Takshak and Agrawal (2014)
Grape (*Vitis vinifera* cv. Malbec)	UV-B cut off filters	α-pinene ↑, 3-carene ↑, terpinolene↑, nerolidol ↑, phytol ↑, squalene ↑	Alonso *et al.* (2015)

a significant focus of research because they do not have UV-shielding or ROS scavenging effects like phenolic compounds (see above). Thus, literature on the effects of UV-B on glucosinolate content and composition in the context of a pre-harvest treatment is currently scarce (see Table 4.3).

The UV-B-induced glucosinolate content has been reported to have species-specific differences – although the studies in the

Table 4.3. Examples of studies investigating the effect of UV-B on glucosinolates (GS).

Species	UV-B level/range	UV-B-induced changes in glucosinolates	Reference
Ethiopian kale (*Brassica carinata*)	3.0 J m^{-2} s^{-1} for 30 sec to 40 min	total GS with strong variability: reduction by 53% or increase up to 123%	Barro *et al.* (2003)
Nasturtium (*Nasturtium officinale*), mustard (*Sinapis alba*)	1.12 W m^{-2} (UV-B 280-315 nm)	total GS ↑, 2-phenylethyl GS ↑, 4-methoxyindole-3-ylmethyl GS ↑ (*Nasturtium officinale*) aromatic GS ↑ (*Sinapis alba*)	Reifenrath and Muller (2007)
Saltwater cresses (*Thellungiella halophila, T. salsuginea*)	2 lamps each with 13 W (UV 368 nm) for 60 min; or 1 lamp 30 W (254 nm) for 60 min	4-methylsulphinylbutyl GS ↑	Pedras and Zheng (2010)
Arabidopsis thaliana	1.55 W m^{-2}	total GS ↓, indole GS ↓	Wang *et al.* (2011)
Broccoli sprouts (*Brassica oleracea* var. *italica*)	0.6-1.0 kJ m^{-2} d^{-1} (240 min)	4-methylsulfinylbutyl GS ↑, 4-methoxyindole-3-ylmethyl GS ↑	Mewis *et al.* (2012)
Canola (*Brassica napus*)	18, 25, 40 μW cm^{-2} d^{-1}	total GS ↑	Tohidi-Moghadam *et al.* (2012)

current literature do not report consistent results. For example, Wang *et al.* (2011) published a decrease of total glucosinolate content in *A. thaliana*, whereas Tohidi-Moghadam *et al.* (2012) as well as Reifenrath and Mueller (2007) found increased total glucosinolate contents in canola as well as nasturtium and mustard, respectively. However, these differences might be due to different doses of UV-B radiation. For example, in Ethiopian kale (*Brassica carinata*), Barro *et al.* (2003) found a range of responses, from a decrease of total glucosinolate content by 53% up to an increase of 123% depending on the duration of exposure to UV-B radiation (from 30 s to 40 min).

However, some structure-specific similarities of individual glucosinolates have been reported. In particular, an increase of 4-methoxyindole-3-ylmethyl glucosinolate concentration was determined in broccoli and nasturtium (Reifenrath and Mueller, 2007; Mewis *et al.*, 2012). These results were supported by corresponding gene expression studies. Immediately after UV-B application and consecutive stimulation of a signal transduction network by upregulation of genes associated with salicylate and jasmonic acid signalling (*PR-1, PR-2, PR-4, BG3*), a transient expression of the transcription factor *MYB51* as a regulator of genes of the indole glucosinolate biosynthesis is induced. Moreover, temporal delayed genes encoding the CYP monooxygenases (*CYP81F2/F3*) were upregulated resulting in higher levels of 4-hydroxyindole-3-ylmethyl glucosinolate as precursor of the UV-B-induced 4-methoxyindole-3-ylmethyl glucosinolate (Mewis *et al.*, 2012). Interestingly, another gene expression study found that *CYP79B2* is downregulated (Hectors *et al.*, 2007). *CYP79B2* encodes an enzyme involved in the early steps of indole glucosinolate biosynthesis promoting the conversion of tryptophan to indole-3-acetaldoxime, which could be used for indole glucosinolate or tryptophan-dependent auxin synthesis. The same study also reveals the downregulated expression of other genes involved in glucosinolate biosynthesis, e.g. 2-oxoglutarate-dependent dioxygenase, which is involved in the biosynthesis of aliphatic glucosinolates (Hectors *et al.*, 2007). In addition, the expression of *FMO GS-OX5* was seen to increase remarkably. The translated product of *FMO GS-OX*$_5$ is involved specifically in the conversion of methylthioalkyl glucosinolate into methylsulfinylalkyl glucosinolate. This is accompanied with an upregulation of *MAM1* (methyl-thioalkylmalatsynthase), a gene which is responsible for the formation of short-chain C3 and C4 glucosinolates, such as 3-methylsulfinylpropyl and 4-methylsulfinylbutyl glucosinolate. Comparing the different UV studies (Table 4.3), a short-term application of UV-B results in an induction of transcription factors for aliphatic (*MYB76*) and indole (*MYB51*) glucosinolate biosynthesis as well as increased metabolite levels (higher 4-methylsulfinyl, indole-3-ylmethyl and 4-methoxyindole-3-ylmethyl glucosinolate). The highest 4-methoxyindole-3-ylmethyl glucosinolate levels were determined six days after treatment with UV (254 nm) for 60 min (Pedras and Zheng, 2010).

Glucosinolate Breakdown Products

Upon hydrolysis of glucosinolates, bioactive and volatile degradation products, such as nitriles, epithionitriles or isothiocyanates, can be released. As isothiocyanates are electrophilic compounds, they have several properties beneficial to health, such as antimicrobial (Fahey *et al.*, 2002), anti-inflammatory (Cho *et al.*, 2013), anti-thrombotic (Ku and Bae, 2014), as well as chemo-preventive (Singh and Singh, 2012; Veeranki *et al.*, 2015) effects. Therefore, consumption of *Brassica* vegetables is linked with a decreased risk for several types of cancer (Voorrips *et al.*, 2000; Terry *et al.*, 2001; Tang *et al.*, 2010; Wu *et al.*, 2012).

The hydrolysis process of glucosinolates is quite complex and many factors affect the degradation. When plant cells are disrupted, myrosinase, a β-D-thioglucosidase, and glucosinolates come into contact. D-glucose is released and an instable aglucon (thiohydroximate-*O*-sulfonate) is formed. The rate-limiting step, the release of the glucose molecule from the active site of myrosinase, can be promoted by L-ascorbic acid (vitamin C), which is a cofactor for myrosinase (Burmeister *et al.*, 2000), and the aglucon then rearranges spontaneously via a Lossen rearrangement to form the isothiocyanate or nitriles (Uda *et al.*, 1986; Wittstock and Burow, 2010). Some species of the order Brassicales contain additional proteins, e.g. the Fe^{2+} dependent epithiospecifier proteins (ESPs). These proteins modify the degradation of the aglucon, and if an alkenyl glucosinolate is present, epithionitriles are released. Moreover, ESPs also favour the formation of nitriles from other (non-alkenyl) glucosinolates (Wittstock and Burow, 2010; Kissen *et al.*, 2012). Furthermore, in *A. thaliana*, in addition to the ESPs, there are also nitrile specifier proteins (NSPs) that catalyse the release of nitriles from the aglucon; these NSPs are also Fe^{2+} dependent. Also in *A. thaliana*, the epithiospecifier modifier protein (ESM) blocks epithionitrile and nitrile formation and promotes isothiocyanate formation (Zhang *et al.*, 2006; Burow *et al.*, 2009; Kissen and Bones, 2009). Many *Brassica* vegetables, among them cabbage and broccoli, contain ESPs. Therefore, epithionitriles and nitriles, instead of health-promoting isothiocyanates, can be the main hydrolysis products (Matusheski *et al.*, 2006; Hanschen *et al.*, 2015, Kupke *et al.*, 2016).

To our knowledge, no studies have been published on the effect of UV-B radiation on the breakdown of glucosinolates. Recently, however, in an experiment on UV exclusion during growth of pak choi, our group found that UV-B treatment affects glucosinolate breakdown products (Hanschen *et al.*, unpublished data). Moreover, while reduction of UV-B had no effect on isothiocyanates or epithionitriles, nitrile concentrations increased under reduced (0.017 kJ m^{-2} d^{-1}) and low (0.002 kJ m^{-2} d^{-1}) UV-B treatments compared to normal UV-B treatment (0.059 kJ m^{-2} d^{-1}) (Heinze *et al.*, unpublished results; see Table 4.4).

In the current scientific literature (for examples, see Table 4.4), there are some reports that indicate that the hydrolysis of glucosinolates is affected by UV-B treatment. For example, ESP transcript levels were shown to decrease about twofold in broccoli sprouts treated with moderate UV-B doses (0.6 kJ m^{-2} d^{-1}) (Mewis *et al.*, 2012) and that this decrease in ESP activity would be followed with an increase in isothiocyanate formation. Further, effects of

Table 4.4. Examples of studies investigating the effect of UV-B on glucosinolate breakdown products.

Species	UV-B level/range	UV-B-induced changes in glucosinolate breakdown products	Reference
Arabidopsis thaliana Col-0	1.152 kJ m^{-2} (480 min)	L-ascorbic acid ↑	Gao and Zhang (2008)
Broccoli sprouts (*Brassica oleracea* var. *italica*)	0.6 kJ m^{-2} d^{-1} (240 min, single dose)	ESP transcripts ↓	Mewis *et al.* (2012)
Broccoli (*Brassica oleracea* var. *italica*)	2.2-16.4 kJ m^{-2} d^{-1}	L-ascorbic acid ↑	Topcu *et al.* (2015)
Pak choi (*Brassica rapa* ssp. *chinensis*)	0.002-0.059 kJ m^{-2} d^{-1}	nitriles ↓	Heinze *et al.* (unpublished results)

UV-B radiation on antioxidative L-ascorbic acid have also been reported. In *A. thaliana* wild-type Col-0, a more than 2.6-fold increase of L-ascorbic acid was found after exposure to 0.04 J m^{-2} for 8h (1.152 kJ m^{-2} d^{-1}) (Gao and Zhang, 2008). Moreover, in broccoli, after repeated applications of moderate to high doses (2.2, 8.8 and 16.4 kJ m^{-2} d^{-1}) of UV-B, a dose-dependent increase in L-ascorbic acid was observed (Topcu *et al.*, 2015). Finally, as vitamin C is a cofactor for the myrosinase, its activity can increase with higher L-ascorbic acid concentration. If ESPs or NSPs are therefore present, the higher myrosinase activity should favour isothiocyanate formation instead of epithionitrile or nitrile release (Burow *et al.*, 2006).

Sulfides

Allium vegetables, such as onions (*Allium cepa*), leeks (*Allium ampeloprasum*) or garlic (*Allium sativum*), contain S-alk(en)yl-L-cysteine sulfoxides, which are precursors to volatile flavour compounds that deter herbivores. If plant cells are disrupted, the *Allium* alliinase or the cystathionine β-lyase of *Brassica* species (Tocmo *et al.*, 2015) hydrolyses the sulfoxides, releasing reactive sulfenic acids that condense to thiosulfinates. Moreover, in onions, the 1-propenylsulfenic acid is enzymatically transformed to 1-propanethial-*S*-oxide, which confers the lachrymatory nature of onions. Thiosulfinates themselves are unstable and decompose to form a variety of dialk(en)ylsulfides, -disulfides, -trisulfides, cyclic sulfides or mercaptanes (Block, 1992). These organosulfurs are responsible for the typical flavour of these vegetables. Further, the unsaturated organopolysulfides, such as diallyl- and diallyldisulfide, derived from the alliin-rich garlic, are linked with antimicrobial, cardioprotective as well as cancer preventive effects (Rose *et al.*, 2005). However, high *S*-methyl-L-cysteine sulfoxide content as found in brassicas is associated with decreased growth rates and haemolytic anaemia due to livestock feeding on *Brassica* species (Stoewsand, 1995). To the best of our knowledge, there are currently no reports on the effect of UV-B radiation on the formation of these compounds.

Alkaloids

Alkaloids are a group of low molecular weight nitrogen-containing compounds (Fig. 4.1) that are of interest due to their physiological and medicinal properties. This group of secondary metabolites is classified by either their origin or their molecular precursor, e.g. pyridine (nicotine), tropane (cocaine), isoquinone (codeine) and purine (caffeine). Alkaloids are mostly linked to poisonous properties, but some species, e.g. Madagascar periwinkle (*Catharanthus roseus*), were cultivated due to the medicinal properties of their alkaloids such as vinblastine and vincristine. Such alkaloids act as chemotherapeutic agents in leukaemia and lymphoma patients (Binder *et al.*, 2009).

Alkaloids such as caffeine are known better as UV-B-absorbing compounds and

Table 4.5. Examples of studies investigating the effect of UV-B on alkaloids.

Species	UV-B level/range	UV-B-induced changes in alkaloids	Reference
Madagascar periwinkle cell culture (*Catharanthus roseus*)	Supplementary UV-B dose	catharanthine ↑	Ramani and Chelliah (2007)
Madagascar periwinkle root culture (*Catharanthus roseus*)	0–1.25 W m^{-2} d^{-1} UV-B dose (0–20 min)	lochnericine, serpentine, ajmalicine ↑; hörhammericine ↓	Binder *et al.* (2009)

are able to inhibit UV-B-induced damage (Ahsan *et al.*, 2007; Koo *et al.*, 2007; Kerzendorfer and O'Driscoll, 2009). Unfortunately, only a few studies (see Table 4.5) have investigated the pre-harvest response of alkaloids in edible plants under UV-B treatment. As an example, one study reported that UV-B radiation induced the signalling pathway of alkaloid biosynthesis genes, thereby leading to increased synthesis of catharanthine in cell cultures of Madagascar periwinkle (Ramani and Chelliah, 2007). In a further example, Binder *et al.* (2009) reported an increase in lochnericine, serpentine, ajmalicine and decreased levels of hörhammericine in the hairy roots of that plant after treatment with high levels of UV-B (90 W m^{-2}, up to 20 min) (Binder *et al.*, 2009). It is noteworthy that the biosynthesis of indole alkaloids could be induced by UV-B treatment. Tryptophan decarboxylase, a key enzyme catalysing the first step of the biosynthesis, is induced by UV-B (Ouwerkerk *et al.*, 1999), as is strictosidine synthase (Ramani and Chelliah, 2007). Both enzymes are required for the synthesis of indole alkaloids.

Solanaceous vegetables like potato (*Solanum tuberosum*), aubergine (*Solanum melongena*) and tomato (*Solanum lycopersicum*) also contain alkaloids that are light sensitive. To the best of our knowledge, no research has been performed to investigate the response of these alkaloids to UV-B exposure.

Betalains

Betalains are nitrogen-containing tyrosine-derived compounds (Jain and Gould, 2015; Fig. 4.1) and can be divided into red betacyanins and yellow/orange betaxanthins (Tanaka *et al.*, 2008; Jain and Gould, 2015). They are present in species of the order Caryophyllales such as swiss chard (*Beta vulgaris* var. *flavescens*), beetroot (*Beta vulgaris* ssp. *vulgaris*), amaranth (*Amaranthus*) and spinach (*Spinacia oleracea*) (Stintzing and Carle, 2004). Exposure to light increases the biosynthesis of betalains in amaranth and portulaca (*Portulaca grandifolia*) (Woodhead and Swain, 1974; Kishima *et al.*, 1991). Betacyanins and betaxanthins concentrations in beetroot are higher in field-grown cultivars (+UV-B) compared to greenhouse-grown cultivars (–UV-B) (Lee *et al.*, 2014a). However, the response of betalains to UV-B exposure is structure-specific. For example, vulgaxanthine was increased in both the cvs. Bulls Blood and Burpee's Golden Globe, whereas betanin was only increased in cv. Bulls Blood and not in cv. Burpee's Golden Globe (Lee *et al.*, 2014a; Table 4.6). To the best of our knowledge, no specific treatments with UV-B have been investigated in relation to betalain biosynthesis.

Future View

A plant's exposure to different UV-B dosages will directly affect which UV-B signalling pathways will be activated (e.g. Jenkins, 2009, Mewis *et al.*, 2012) as well as which genes will be upregulated (Lang-Mladek *et al.*, 2012). These UV-B dosage-mediated variations in UV-B signalling pathways and gene expression suggest that the exposure to different UV-B spectra and UV-B wavelengths might induce a differentiated plant

Table 4.6. Examples of studies investigating the effect of UV-B on betalains.

Species	UV-B level/range	UV-B-induced changes in betalains	Reference
Beetroot (*Beta vulgaris* ssp. *vulgaris*)	greenhouse (–UV-B) vs. field (+UV-B)	total betacyanins ↑ total betaxanthins ↑ vulgaxanthine ↑ betanin ↑ (cultivar-dependent)	Lee *et al.* (2014a)

response in its secondary metabolism. To date, mainly UV-B radiation sources with a broad emission spectrum peaking between 280 and 360 nm have been used within the UV-B studies. However, the use of a broad emission spectrum does not allow us to obtain a differentiated plant response to UV-B radiation of a certain wavelength. Furthermore, undesired crosstalk or even harmful stress from UV radiation of different wavelengths on certain secondary plant metabolites or other plant properties is possible (Schreiner *et al.*, 2014).

Semiconductor-based UV-B light-emitting diodes (LEDs) are an interesting alternative to the fluorescent lamps that are commonly used as a UV-B light source. The emission spectrum of UV-B LEDs is fairly narrow with full width at half maximum of less than 10 nm. Furthermore, the peak emission wavelength can be tailored to ideally match the effective spectrum for triggering the secondary plant metabolism. Moreover, UV-B LEDs do not exhibit any additional and unwanted side emission peaks that may damage the plants (Schreiner *et al.*, 2016). Thus, UV-B LEDs will open up a new field of UV-B applications in which a more tailor-made and accurate radiation treatment for targeted triggering of specific secondary plant metabolites can be applied.

It will also be of great importance to elucidate time-dependent changes, e.g. the carotenoid biosynthesis pathway is known to be upregulated under UV-B exposure, but downregulated under chronic UV-B exposure (Jansen *et al.*, 2008). Furthermore, a better understanding of rapid and adaptive response reactions of plants to UV-B exposure can be obtained by metabolic studies. For example, in *A. thaliana*, a short-term response was obtained at the level of primary metabolites meaning that cells were subsequently better equipped to produce UV-B-absorbing secondary metabolites (Kusano *et al.*, 2011). In maize, a broader systemic approach using transcriptome, proteome and metabolome data has been used to elucidate short-term effects of UV-B signalling (Casati *et al.*, 2011). However, there is currently a clear lack of a systems biology approach for horticultural species after UV-B treatment. The future application of different omics technologies should therefore make a major contribution to improving our basic knowledge of UV-B induced responses.

References

Abney, K.R., Kopsell, D.A., Sams, C.E., Zivanovic, S. and Kopsell, D.E. (2013) UV-B radiation impacts shoot tissue pigment composition in *Allium fistulosum* L. cultigens. *Scientific World Journal* 2013, 513867.

Agati, G. and Tattini, M. (2010) Multiple functional roles of flavonoids in photoprotection. *New Phytologist* 186, 786–793.

Agerbirk, N. and Olsen, C.E. (2012) Glucosinolate structures in evolution. *Phytochemistry* 77, 16–45.

Ahsan, H., Reagan-Shaw, S., Eggert, D.M., Tan, T.C., Afaq, F., Mukhtar, H. and Ahmad, N. (2007) Protective effect of sanguinarine on ultraviolet B-mediated damages in SKH-1 hairless mouse skin: implications for prevention of skin cancer. *Photochemistry and Photobiology* 83, 986–993.

Alonso, R., Berli, F.J., Bottini, R. and Piccoli, P. (2015) Acclimation mechanisms elicited by sprayed abscisic acid, solar UV-B and water deficit in leaf tissues of field-grown grapevines. *Plant Physiology and Biochemistry* 91, 56–60.

Barnes, P.W., Kerating, A.R., Flint, S.D., Beyschlag, W. and Ryel, R.J. (2013) Adjustments in epidermal UV-transmittance of leaves in sun-shade transitions. *Physiologia Plantarum* 149, 200–213.

Barro, F., Fernandez-Escobar, J., De la Vega, M. and Martin, A. (2003) Modification of glucosinolate and erucic acid contents in doubled haploid lines of *Brassica carinata* by UV treatment of isolated microspores. *Euphytica* 129, 1–6.

Becatti, E., Petroni, K., Giuntini, D., Castagna, A., Calvenzani, V. *et al.* (2009) Solar UV-B radiation influences carotenoid accumulation of tomato fruit through both ethylene-dependent and -independent mechanisms. *Journal of Agricultural and Food Chemistry* 57, 10979–10989.

Berli, F.J., Moreno, D., Piccoli, P., Hespanhol-Viana, L., Silva, M.F. *et al.* (2010) Abscisic acid is involved in the response of grape (*Vitis vinifera* L.) cv. Malbec leaf tissues to ultraviolet-B radiation by enhancing ultraviolet-absorbing compounds, antioxidant enzymes and membrane sterols. *Plant, Cell & Environment* 33, 1–10.

Binder, B.Y.K., Peebles, C.A.M., Shanks, J.V. and San, K.-Y. (2009) The effects of UV-B stress on the production of terpenoid indole alkaloids in *Catharanthus roseus* hairy roots. *Biotechnology Progress* 25, 861–865.

Block, E. (1992) The organosulfur chemistry of the genus *Allium*: Implications for the organic chemistry of sulfur. *Angewandte Chemie International Edition in English* 31, 1135–1178.

Bowles, D., Lim, E.K., Poppenberger, B. and Vaistij, F.E. (2006) Glycosyltransferases of lipophilic small molecules. *Annual Review of Plant Biology* 57, 567–597.

Britton, G., Liaaen-Jensen, S. and Pfander, H. (eds) (2009) *Carotenoids. Volume 5: Nutrition and Health.* Birkhäuser, Basel.

Brown, B.A. and Jenkins, G.I. (2008) UV-B signaling pathways with different fluence-rate response profiles are distinguished in mature Arabidopsis leaf tissue by requirement for UVR8, HY5, and HYH. *Plant Physiology* 146, 576–588.

Burmeister, W.P., Cottaz, S., Rollin, P., Vasella, A. and Henrissat, B. (2000) High resolution x-ray crystallography shows that scorbate is a cofactor for myrosinase and substitutes for the function of the catalytic case. *Journal of Biological Chemistry* 275, 39385–39393.

Burow, M., Losansky, A., Müller, R., Plock, A., Kliebenstein, D.J. and Wittstock, U. (2009) The genetic basis of constitutive and herbivore-induced ESP-independent nitrile formation in Arabidopsis. *Plant Physiology* 149, 561–574.

Burow, M., Markert, J., Gershenzon, J. and Wittstock, U. (2006) Comparative biochemical characterization of nitrile-forming proteins from plants and insects that alter myrosinase-catalysed hydrolysis of glucosinolates. *FEBS Journal* 273, 2432–2446.

Calderon-Montano, J.M., Burgos-Moron, E., Perez-Guerrero, C. and Lopez-Lazaro, M. (2011) A review on the dietary flavonoid kaempferol. *Mini-Reviews in Medicinal Chemistry* 11, 298–344.

Caldwell, C.R. and Britz, S.J. (2006) Effect of supplemental ultraviolet radiation on the carotenoid and chlorophyll composition of green house-grown leaf lettuce (*Lactuca sativa* L.) cultivars. *Journal of Food Composition and Analysis* 19, 637–644.

Calvenzani, V., Martinelli, M., Lazzeri, V., Giuntini, D., Dall'Asta, C. *et al.* (2010) Response of wild-type and high pigment-1 tomato fruit to UV-B depletion: flavonoid profiling and gene expression. *Planta* 231, 755–765.

Casati, P., Campi, M., Morrow, D.J., Fernandes, J.F. and Walbot, V. (2011) Transcriptomic, proteomic and metabolomic analysis of UV-B signaling in maize. *BMC Genomics* 12, 321. DOI:10.1186/1471-2164-12-321

Cho, H., Lee, K. and Park, J. (2013) Erucin exerts anti-inflammatory properties in murine macrophages and mouse skin: Possible mediation through the inhibition of NFκB signaling. *International Journal of Molecular Sciences* 14, 20564.

Clarke, D.B. (2010) Glucosinolates, structures and analysis in food. *Analytical Methods* 2, 310–325.

Cominelli, E., Gusmaroli, G., Allegra, D., Galbiati, M., Wade, H.K., Jenkins, G.I. and Tonelli, C. (2008) Expression analysis of anthocyanin regulatory genes in response to different light qualities in *Arabidopsis thaliana*. *Journal of Plant Physiology* 165, 886–894.

Di Mascio, P., Kaiser, S. and Sies, H. (1989) Lycopene as the most efficient biological carotenoid singlet oxygen quencher. *Archives of Biochemistry and Biophysics* 274, 532–538.

Dolzhenko, Y., Bertea, C.M., Occhipinti, A., Bossi, S. and Maffei, M.E. (2010) UV-B modulates the interplay between terpenoids and flavonoids in peppermint (*Mentha x piperita* L.). *Journal of Photochemistry and Photobiology B-Biology* 100, 67–75.

Edreva, A. (2005) The importance of non-photosynthetic pigments and cinnamic acid derivatives in photoprotection. *Agriculture Ecosystems & Environment* 106, 135–146.

Fahey, J.W., Haristoy, X., Dolan, P.M., Kensler, T.W., Scholtus, I. *et al.* (2002) Sulforaphane inhibits extracellular, intracellular, and antibiotic-resistant strains of *Helicobacter pylori* and prevents benzo[a]pyrene-induced stomach tumors. *Proceedings of the National Academy of Sciences of the United States of America* 99, 7610–7615.

Ferreres, F., Fernandes, F., Oliveira, J.M.A., Valentão, P., Pereira, J.A. and Andrade, P.B. (2009) Metabolic profiling and biological capacity of *Pieris brassicae* fed with kale (*Brassica oleracea* L. var. *acephala*). *Food and Chemical Toxicology* 47, 1209–1220.

Gachon, C.M.M., Langlois-Meurinne, M. and Saindrenan, P. (2005) Plant secondary metabolism glycosyltransferases: the emerging functional analysis. *Trends in Plant Science* 10, 542–549.

Gao, Q. and Zhang, L. (2008) Ultraviolet-B-induced oxidative stress and antioxidant defense system responses in ascorbate-deficient vtc1 mutants of *Arabidopsis thaliana*. *Journal of Plant Physiology* 165, 138–148.

Garcia-Macias, P., Ordidge, M., Vysini, E., Waroonphan, S., Battey, N.H. *et al.* (2007) Changes in the flavonoid and phenolic acid contents and antioxidant activity of red leaf lettuce (Lollo Rosso) due to cultivation under plastic films varying in ultraviolet transparency. *Journal of Agricultural and Food Chemistry* 55, 10168–10172.

Giuntini, D., Lazzeri, V., Calvenzani, V., Dall'asta, C., Galaverna, G. *et al.* (2008) Flavonoid profiling and biosynthetic gene expression in flesh and peel of two tomato genotypes grown under UV-B-depleted conditions during ripening. *Journal of Agricultural and Food Chemistry* 56, 5905–5915.

Goetz, M., Albert, A., Stich, S., Heller, W., Scherb, H. *et al.* (2010) PAR modulation of the UV-dependent levels of flavonoid metabolites in *Arabidopsis thaliana* (L.) Heynh. leaf rosettes: cumulative effects after a whole vegetative growth period. *Protoplasma* 243, 95–103.

Halkier, B.A. and Gershenzon, J. (2006) Biology and biochemistry of glucosinolates. *Annual Review of Plant Biology* 57, 303–333.

Hanschen, F.S., Herz, C., Schlotz, N., Kupke, F., Bartolomé Rodríguez, M.M. *et al.* (2015) The *Brassica* epithionitrile 1-cyano-2,3-epithiopropane triggers cell death in human liver cancer cells in vitro. *Molecular Nutrition & Food Research* 59, 2178–2189.

Harbaum-Piayda, B., Walter, B., Bengtsson, G.B., Hubbermann, E.M., Bilger, W. and Schwarz, K. (2010) Influence of pre-harvest UV-B irradiation and normal or controlled atmosphere storage on flavonoid and hydroxycinnamic acid contents of pak choi (*Brassica campestris* L. ssp *chinensis* var. communis). *Postharvest Biology and Technology* 56, 202–208.

Havaux, M. (2014) Carotenoid oxidation products as stress signals in plants. *The Plant Journal* 79, 597–606.

Hectors, K., Prinsen, E., De Coen, W., Jansen, M.A. and Guisez, Y. (2007) *Arabidopsis thaliana* plants acclimated to low dose rates of ultraviolet b radiation show specific changes in morphology and gene expression in the absence of stress symptoms. *New Phytologist* 175, 255–270.

Hectors, K., Van Oevelen, S., Guisez, Y., Prinsen, E. and Jansen, M.A.K. (2012) The phytohormone auxin is a component of the regulatory system that controls UV-mediated accumulation of flavonoids and UV-induced morphogenesis. *Physiologia Plantarum* 145, 594–603.

Hectors, K., Van Oevelen, S., Geuns, J., Guisez, Y., Jansen, M.A.K. and Prinsen, E. (2014) Dynamic changes in plant secondary metabolites during UV acclimation in *Arabidopsis thaliana*. *Physiologia Plantarum* 152, 219–230.

Heim, K.E., Tagliaferro, A.R. and Bobilya, D.J. (2002) Flavonoid antioxidants: chemistry, metabolism and structure-activity relationships. *Journal of Nutritional Biochemistry* 13, 572–584.

Hernandez, I., Alegre, L., Van Breusegem, F. and Munne-Bosch, S. (2009) How relevant are flavonoids as antioxidants in plants? *Trends in Plant Science* 14, 125–132.

Huang, T.S., Anzellotti, D., Dedaldechamp, F. and Ibrahim, R.K. (2004) Partial purification, kinetic analysis, and amino acid sequence information of a flavonol 3-O-methyltransferase from *Serratula tinctoria*. *Plant Physiology* 134, 1366–1376.

Jain, G. and Gould, K.S. (2015) Are betalain pigments the functional homologues of anthocyanins in plants? *Environmental and Experimental Botany* 119, 48–53.

Jansen, M.A.K. (2012) Ultraviolet-B radiation: from stressor to regulatory signal. In: Shabala, S. (ed.) *Plant Stress Physiology*, CABI, Wallingford, UK, pp. 266–290.

Jansen, M.A.K., Hectors, K., O'Brien, N.M., Guisez, Y. and Potters, G. (2008) Plant stress and human health: Do human consumers benefit from UV-B acclimated crops? *Plant Science* 175, 449–458.

Jenkins, G.I. (2009) Signal transduction in responses to UV-B radiation. *Annual Review of Plant Biology* 60, 407–431.

Kerzendorfer, C. and O'Driscoll, M. (2009) UVB and caffeine: inhibiting the DNA damage response to protect against the adverse effects of UVB. *Journal of Investigative Dermatology* 129, 1611–1613.
Kishima, Y., Nozaki, K., Akashi, R. and Adachi, T. (1991) Light-inducible pigmentation in portulaca callus: selection of a high betalain producing cell-line. *Plant Cell Reports* 10, 304–307.
Kissen, R. and Bones, A.M. (2009) Nitrile-specifier proteins involved in glucosinolate hydrolysis in *Arabidopsis thaliana*. *Journal of Biological Chemistry* 284, 12057–12070.
Kissen, R., Hyldbakk, E., Wang, C.W.V., Sormo, C.G., Rossiter, J.T. and Bones, A.M. (2012) Ecotype dependent expression and alternative splicing of epithiospecifier protein (ESP) in *Arabidopsis thaliana*. *Plant Molecular Biology* 78, 361–375.
Kolb, C.A., Kaser, M.A., Kopecky, J., Zotz, G., Riederer, M. and Pfundel, E.E. (2001) Effects of natural intensities of visible and ultraviolet radiation on epidermal ultraviolet screening and photosynthesis in grape leaves. *Plant Physiology* 127, 863–875.
Koo, S.W., Hirakawa, S., Fujii, S., Kawasumi, M. and Nghiem, P. (2007) Protection from photodamage by topical application of caffeine after ultraviolet irradiation. *British Journal of Dermatology* 156, 957–964.
Ku, S.-K. and Bae, J.-S. (2014) Antithrombotic activities of sulforaphane via inhibiting platelet aggregation and FIIa/FXa. *Archives of Pharmacalogical Research* 37, 1454–1463.
Kuhlmann, F. and Mueller, C. (2009) Development-dependent effects of UV radiation exposure on broccoli plants and interactions with herbivorous insects. *Environmental and Experimental Botany* 66, 61–68.
Kupke, F., Herz, C., Hanschen, F.S., Platz, S., Odongo, G. *et al.* (2016) Cytotoxic and genotoxic potential of food-borne nitriles in a liver in vitro model. *Scientific Reports* 6, 37631.
Kusano, M., Tohge, T., Fukushima, A., Kobayashi, M., Hayashi, N. *et al.* (2011) Metabolomics reveals comprehensive reprogramming involving two independent metabolic responses of Arabidopsis to UV-B light. *Plant Journal* 67, 354–369.
Lang-Mladek, C., Xie, L.S., Nigam, N., Chumak, N., Binkert, M., Neubert, S. and Hauser, M.T. (2012) UV-B signaling pathways and fluence rate dependent transcriptional regulation of ARIADNE12. *Physiologia Plantarum* 145, 527–539.
Lazzeri, V., Calvenzani, V., Petroni, K., Tonelli, C., Castagna, A. and Ranieri, A. (2012) Carotenoid profiling and biosynthetic gene expression in flesh and peel of wild-type and hp-1 tomato fruit under UV-B depletion. *Journal of Agricultural and Food Chemistry* 60, 4960–4969.
Lee, E.J., An, D., Nguyen, C.T.T., Patil, B.S., Kim, J. and Yoo, K.S. (2014a) Betalain and betaine composition of greenhouse- or field-produced beetroot (*Beta vulgaris* L.) and inhibition of HepG2 cell proliferation. *Journal of Agricultural and Food Chemistry* 62, 1324–1331.
Lee, M.J., Son, J.E. and Oh, M.M. (2014b) Growth and phenolic compounds of *Lactuca sativa* L. grown in a closed-type plant production system with UV-A, -B, or -C lamp. *Journal of the Science of Food and Agriculture* 94, 197–204.
Luis, J.C., Pérez, R.M. and González, F.V. (2007) UV-B radiation effects on foliar concentrations of rosmarinic and carnosic acids in rosemary plants. *Food Chemistry* 101, 1211–1215.
Matusheski, N.V., Swarup, R., Juvik, J.A., Mithen, R., Bennett, M. and Jeffery, E.H. (2006) Epithiospecifier protein from broccoli (*Brassica oleracea* L. ssp. *italica*) inhibits formation of the anticancer agent sulforaphane. *Journal of Agricultural and Food Chemistry* 54, 2069–2076.
Mewis, I., Appel, H.M., Hom, A., Raina, R. and Schultz, J.C. (2005) Major signaling pathways modulate arabidopsis glucosinolate accumulation and response to both phloem-feeding and chewing insects. *Plant Physiology* 138, 1149–1162.
Mewis, I., Schreiner, M., Nguyen, C.N., Krumbein, A., Ulrichs, C., Lohse, M. and Zrenner, R. (2012) UV-B irradiation changes specifically the secondary metabolite profile in broccoli sprouts: Induced signaling overlaps with defense response to biotic stressors. *Plant and Cell Physiology* 53, 1546–1560.
Mierziak, J., Kostyn, K. and Kulma, A. (2014) Flavonoids as important molecules of plant interactions with the environment. *Molecules* 19, 16240–16265.
Mithen, R.F. (2001) Glucosinolates and their degradation product. *Advances in Botanical Research* 35, 213–232.
Morales, L.O., Brosche, M., Vainonen, J., Jenkins, G.I., Wargent, J.J. *et al.* (2013) Multiple roles for UV RESISTANCE LOCUS8 in regulating gene expression and metabolite accumulation in Arabidopsis under solar ultraviolet radiation. *Plant Physiology* 161, 744–759.

Neugart, S., Fiol, M., Schreiner, M., Rohn, S., Zrenner, R., Kroh, L.W. and Krumbein, A. (2014) Interaction of moderate UV-B exposure and temperature on the formation of structurally different flavonol glycosides and hydroxycinnamic acid derivatives in kale (*Brassica oleracea* var. *sabellica*). *Journal of Agricultural and Food Chemistry* 62, 4054–4062.

Neugart, S., Zietz, M., Schreiner, M., Rohn, S., Kroh, L.W. and Krumbein, A. (2012) Structurally different flavonol glycosides and hydroxycinnamic acid derivatives respond differently to moderate UV-B radiation exposure. *Physiologia Plantarum* 145, 582–593.

North, H.M., Almeida, A.D., Boutin, J.-P., Frey, A., To, A. *et al.* (2007) The Arabidopsis ABA-deficient mutant aba4 demonstrates that the major route for stress-induced ABA accumulation is via neoxanthin isomers. *The Plant Journal* 50, 810–824.

Olsson, L.C., Veit, M., Weissenbock, G. and Bornman, J.F. (1998) Differential flavonoid response to enhanced UV-B radiation in *Brassica napus*. *Phytochemistry* 49, 1021–1028.

Ordidge, M., Garcia-Macias, P., Battey, N.H., Gordon, M.H., Hadley, P. *et al.* (2010) Phenolic contents of lettuce, strawberry, raspberry, and blueberry crops cultivated under plastic films varying in ultraviolet transparency. *Food Chemistry* 119, 1224–1227.

Ouwerkerk, P.B., Hallard, D., Verpoorte, R. and Memelink, J. (1999) Identification of UV-B light-responsive regions in the promoter of the tryptophan decarboxylase gene from *Catharanthus roseus*. *Plant Molecular Biology* 41, 491–503.

Pan, M.H., Lai, C.S. and Ho, C.T. (2010) Anti-inflammatory activity of natural dietary flavonoids. *Food & Function* 1, 15–31.

Pedras, M.S.C. and Zheng, Q.A. (2010) Metabolic responses of *Thellungiella halophila/salsuginea* to biotic and abiotic stresses: metabolite profiles and quantitative analyses. *Phytochemistry* 71, 581–589.

Ramani, S. and Chelliah, J. (2007) UV-B-induced signaling events leading to enhanced-production of catharanthine in *Catharanthus roseus* cell suspension cultures. *BMC Plant Biology* 7, 61.

Reifenrath, K. and Mueller, C. (2007) Species-specific and leaf-age dependent effects of ultraviolet radiation on two Brassicaceae. *Phytochemistry* 68, 875–885.

Rizzini, L., Favory, J.-J., Cloix, C., Faggionato, D., O'Hara, A. *et al.* (2011) Perception of UV-B by the Arabidopsis UVR8 protein. *Science* 332, 103–106.

Rose, P., Whiteman, M., Moore, P.K. and Zhu, Y.Z. (2005) Bioactive S-alk(en)yl cysteine sulfoxide metabolites in the genus *Allium*: the chemistry of potential therapeutic agents. *Natural Product Reports* 22, 351–368.

Rozema, J., Van De Staaij, V., Oudejans, A., Meijkamp, B. and De Bakker, N. (1999) UV-B effects in natural ecosystems. *Photochemistry and Photobiology* 69, 60S.

Sakalauskaitė, J., Viskelis, P., Dambrauskienė, E., Sakalauskienė, S., Samuolienė, G. *et al.* (2013) The effects of different UV-B radiation intensities on morphological and biochemical characteristics in *Ocimum basilicum* L. *Journal of the Science of Food and Agriculture* 93, 1266–1271.

Scalbert, A. and Williamson, G. (2000) Dietary intake and bioavailability of polyphenols. *Journal of Nutrition* 130, 2073S–2085S.

Schreiner, M., Mewis, I., Huyskens-Keil, S., Jansen, M.A.K., Zrenner, R. *et al.* (2012) UV-B-induced secondary plant metabolites – potential benefits for plant and human health. *Critical Reviews in Plant Sciences* 31, 229–240.

Schreiner, M., Martínez-Abaigar, J., Glaab, J. and Jansen, M. (2014) UV-B induced secondary plant metabolites. *Optik & Photonik* 9, 34–37.

Schreiner, M., Mewis, I., Neugart, S., Zrenner, R., Glaab, J., Wiesner, M. and Jansen, M.A. (2016) UV-B elicitation of secondary plant metabolites. In: Kneissl, M. and Raß, J. (eds) *III-Nitride Ultraviolet Emitters – Technology & Applications*. Springer, Heidelberg, Germany, pp. 387–414.

Schultz, H.R., Löhnertz, O., Bettner, W., Bálo, B., Linsenmeier, A. *et al.* (1998) Is grape composition affected by current levels of UV-B radiation? *Vitis* 37, 191–192.

Semba, R.D. (2012) On the 'discovery' of vitamin A. *Annals of Nutrition and Metabolism* 61, 192–198.

Seo, M. and Koshiba, T. (2002) Complex regulation of ABA biosynthesis in plants. *Trends in Plant Science* 7, 41–48.

Shapiro, T.A., Fahey, J.W., Dinkova-Kostova, A.T., Holtzclaw, W., Stephenson, K.K. *et al.* (2006) Safety, tolerance, and metabolism of broccoli sprout glucosinolates and isothiocyanates: a clinical phase I study. *Nutrition and Cancer – An International Journal* 55, 53–62.

Singh, S.V. and Singh, K. (2012) Cancer chemoprevention with dietary isothiocyanates mature for clinical translational research. *Carcinogenesis* 33, 1833–1842.

Solovchenko, A.E. and Merzlyak, M.N. (2008) Screening of visible and UV radiation as a photoprotective mechanism in plants. *Russian Journal of Plant Physiology* 55, 719–737.
Stintzing, F.C. and Carle, R. (2004) Functional properties of anthocyanins and betalains in plants, food, and in human nutrition. *Trends in Food Science & Technology* 15, 19–38.
Stoewsand, G.S. (1995) Bioactive organosulfur phytochemicals in *Brassica oleracea* vegetables – a review. *Food and Chemical Toxicology* 33, 537–543.
Takshak, S. and Agrawal, S.B. (2014) Secondary metabolites and phenylpropanoid pathway enzymes as influenced under supplemental ultraviolet-B radiation in *Withania somnifera* Dunal, an indigenous medicinal plant. *Journal of Photochemistry and Photobiology B: Biology* 140, 332–343.
Tanaka, Y., Sasaki, N. and Ohmiya, A. (2008) Biosynthesis of plant pigments: anthocyanins, betalains and carotenoids. *Plant Journal* 54, 733–749.
Tang, L., Zirpoli, G.R., Jayaprakash, V., Reid, M.E., McCann, S.E. *et al.* (2010) Cruciferous vegetable intake is inversely associated with lung cancer risk among smokers: a case-control study. *BMC Cancer* 10, 162.
Terry, P., Wolk, A., Persson, I. and Magnusson, C. (2001) Brassica vegetables and breast cancer risk. *Journal of the American Medical Association* 285, 2975–2977.
Tilbrook, K., Arongaus, A.B., Binkert, M., Heijde, M., Yin, R. and Ulm, R. (2013) The UVR8 UV-B photoreceptor: perception, signaling and response. *Arabidopsis Book* 11, e0164.
Tocmo, R., Liang, D., Lin, Y. and Huang, D. (2015) Chemical and biochemical mechanisms underlying the cardioprotective roles of dietary organopolysulfides. *Frontiers in Nutrition* 2, 1–18.
Tohidi-Moghadam, H.R., Ghooshchi, F., Jamshidpour, F. and Zahedi, H. (2012) Effect of UV radiation and elevated CO_2 on physiological attributes of canola (*Brassica napus* L.) grown under water deficit stress. *Polish Journal of Environmental Studies* 21, 1417–1427.
Topcu, Y., Dogan, A., Kasimoglu, Z., Sahin-Nadeem, H., Polat, E. and Erkan, M. (2015) The effects of UV radiation during the vegetative period on antioxidant compounds and postharvest quality of broccoli (*Brassica oleracea* L.). *Plant Physiology and Biochemistry* 93, 56–65.
Tossi, V., Lamattina, L. and Cassia, R. (2009) An increase in the concentration of abscisic acid is critical for nitric oxide-mediated plant adaptive responses to UV-B irradiation. *New Phytologist* 181, 871–879.
Tsormpatsidis, E., Henbest, R.G.C., Davis, F.J., Battey, N.H., Hadley, P. and Wagstaffe, A. (2008) UV irradiance as a major influence on growth, development and secondary products of commercial importance in Lollo Rosso lettuce 'Revolution' grown under polyethylene films. *Environmental and Experimental Botany* 63, 232–239.
Uda, Y., Kurata, T. and Arakawa, N. (1986) Effects of pH and ferrous ion on the degradation of glucosinolates by myrosinase. *Agricultural and Biological Chemistry* 50, 2735–2740.
Van Dam, N.M. (2009) How plants cope with biotic interactions. *Plant Biology* 11, 1–5.
Veeranki, O., Bhattacharya, A., Tang, L., Marshall, J. and Zhang, Y. (2015) Cruciferous vegetables, isothiocyanates, and prevention of bladder cancer. *Current Pharmacology Reports* 1, 272–282.
Verkerk, R., Schreiner, M., Krumbein, A., Ciska, E., Holst, B. *et al.* (2009) Glucosinolates in brassica vegetables: The influence of the food supply chain on intake, bioavailability and human health. *Molecular Nutrition & Food Research* 53, S219.
Voorrips, L.E., Goldbohm, R.A., Van Poppel, G., Sturmans, F., Hermus, R.J.J. and Van Den Brandt, P.A. (2000) Vegetable and fruit consumption and risks of colon and rectal cancer in a prospective cohort study. The Netherlands cohort study on diet and cancer. *American Journal of Epidemiology* 152, 1081–1092.
Wang, Y., Xu, W.J., Yan, X.F. and Wang, Y. (2011) Glucosinolate content and related gene expression in response to enhanced UV-B radiation in *Arabidopsis*. *African Journal of Biotechnology* 10, 6481–6491.
Watzl, B. and Leitzmann, C. (2005) *Bioaktive Substanzen in Lebensmitteln*. Georg Thieme, Stuttgart, Germany.
Winter, T.R. and Rostas, M. (2008) Ambient ultraviolet radiation induces protective responses in soybean but does not attenuate indirect defense. *Environmental Pollution* 155, 290–297.
Wittstock, U. and Burow, M. (2010) Glucosinolate breakdown in arabidopsis: mechanism, regulation and biological significance. *The Arabidopsis Book* 8, e0134.
Woodhead, S. and Swain, T. (1974) Effect of light on betalain and cinnamic acid biosynthesis in *Amaranthus caudatus*. *Phytochemistry* 13, 953–956.
Wu, Q.J., Yang, Y., Vogtmann, E., Wang, J., Han, L.H., Li, H.L. and Xiang, Y.B. (2012) Cruciferous vegetables intake and the risk of colorectal cancer: a meta-analysis of observational studies. *Annals of Oncology* 24, 1079–1087.

Zeevaart, J. and Creelman, R. (1988) Metabolism and physiology of abscisic acid. *Annual Review of Plant Physiology and Plant Molecular Biology* 39, 439–473.

Zhang, J.M., Satterfield, M.B., Brodbelt, J.S., Britz, S.J., Clevidence, B. and Novotny, J.A. (2003) Structural characterization and detection of kale flavonoids by electrospray ionization mass spectrometry. *Analytical Chemistry* 75, 6401–6407.

Zhang, Z., Ober, J.A. and Kliebenstein, D.J. (2006) The gene controlling the quantitative trait locus EPITHIOSPECIFIER MODIFIER1 alters glucosinolate hydrolysis and insect resistance in Arabidopsis. *The Plant Cell* 18, 1524–1536.

Zhang, Z.Z., Che, X.N., Pan, Q.H., Li, X.X. and Duan, C.Q. (2013) Transcriptional activation of flavan-3-ols biosynthesis in grape berries by UV irradiation depending on developmental stage. *Plant Science* 208, 64–74.

Zietz, M., Weckmuller, A., Schmidt, S., Rohn, S., Schreiner, M., Krumbein, A. and Kroh, L.W. (2010) Genotypic and climatic influence on the antioxidant activity of flavonoids in kale (*Brassica oleracea* var. *sabellica*). *Journal of Agricultural and Food Chemistry* 58, 2123–2130.

5 UV-B-induced Morphological Changes – an Enigma

Marcel A.K. Jansen[1]*, Karel Klem[2], T. Matthew Robson[3], and Otmar Urban[2]

[1]*University College Cork, School of Biological, Earth and Environmental Science, and Environmental Research Institute, North Mall, Cork, Ireland;* [2]*Global Change Research Institute CAS, v.v.i., Bělidla 4a, CZ 60300, Brno, Czech Republic;* [3]*Department of Biosciences, Viikki Plant Science Center, University of Helsinki, Helsinki 00014, Finland*

No two trees are the same to Raven.
No two branches are the same to Wren.
If what a tree of a bush does is lost on you,
You are surely lost. Stand still. The forest knows.

(David Wagoner, 'Lost', 1999)

Introduction

David Wagoner (1999) wrote in his poem 'Lost' about the variation in architecture that is so characteristic of plants. The poem also refers to 'knowledge' – information that is shared between organisms present in the forest environment, information that is important to all. Notwithstanding the poetic interpretation, these lines are in many ways an accurate statement on the high degree of variation in plant architecture and the important ecological consequences of variation for the plant as well as the entire ecosystem. The intraspecific plasticity in plant architecture is controlled by endogenous growth processes and external environmental influences (Barthélémy and Caraglio, 2007). Morphological processes that determine plant architecture include primary growth (organogenesis and elongation), branching, morphological differentiation of axes, and positioning of reproductive structures (Barthélémy and Caraglio, 2007). Thus, plant architecture is dependent on the arrangement of what are, in essence, modular structures in a particular pattern.

Environmental parameters can impact on plant architecture by altering the arrangement of organs in a 3D structure, the identity of organs formed, and/or the morphology of organs. These responses to environmental cues are vital for optimizing growth under different conditions. Temperature, radiation and nutrient supply are known to modulate organ identity, branching and phenology (Costes *et al.*, 2013). The role of solar radiation is complex as light constitutes both energy and information. Optimal intensities of Photosynthetically Active Radiation (PAR) alter growth and overall plant architecture through the improved supply of photosynthates, while specific wavelengths control architecture via photoreceptors that perceive the informational content of light. Photoreceptors can trigger responses to minor

*Corresponding author: m.jansen@ucc.ie

changes in the direction, duration, dose and wavelength of light, and this underlies processes such as photoperiodicity, phototropisms and photomorphogenesis. The best-documented examples of light-mediated changes in plant architecture are those facilitated by phytochrome (red/far-red responses including shade-avoidance), cryptochrome (blue light responses including hypocotyl elongation) and phototropin (blue light responses including effects on tropisms and leaf architecture) (Möglich *et al.*, 2010; Galvão and Fankhauser, 2015). In recent years, effects of ultraviolet-B (UV-B; 280–315 nm) radiation on plant architecture have also drawn the attention of the scientific community (Robson *et al.*, 2015b) with research focused on mechanistic, ecological and commercial aspects. Here we review the concept of the UV-B phenotype, describing UV-B-induced morphological changes, analysing underlying regulatory pathways and exploring the functional importance.

The UV-B Phenotype

Reports on UV-mediated changes in plant architecture have been around for a considerable period. More than 60 years ago, Brodführer (1955) reported that solar UV radiation altered the architecture of the *Arabidopsis thaliana* inflorescence. Teramura (1983) concluded that 'Ultraviolet-B radiation has been shown to affect anatomical and morphological plant characteristics'. Since the publication of these early reports, many studies have shown that UV-B radiation can alter plant architecture (reviewed by Jansen, 2002; Robson *et al.*, 2015b). Generally, the term 'UV-B phenotype' refers to a more compact plant. At the organismal level, the most common UV-B responses are decreases in leaf area and/or increases in thickness, together with changes in leaf shape, shorter petioles and, in some cases, leaf curling (Wargent *et al.*, 2009; Hectors *et al.*, 2010; Klem *et al.*, 2012, Robson and Aphalo, 2012). A few studies have also reported UV effects on root development, and especially an increase in root–shoot ratio (Robson *et al.*, 2015b). In parallel with UV-B-induced decreases in leaf size, leaf venation also changes, with a notable decrease in the width of the mid-rib of soybean (*Glycine max*) leaves (Fatima *et al.*, 2016). Typically, stems will remain shorter as detailed for various species (Barnes *et al.*, 1990; Hofmann and Campbell, 2011; Germ *et al.*, 2013). Although the length of the main stem may decrease in UV-B-acclimated plants, overall stem length does not necessarily decrease due to enhanced axillary branching and/or tillering (cf. Jansen, 2002). For example, *Taxus chinensis* exposed to supplemental UV-B under growth room conditions displays an almost sixfold increase in the number of secondary branches (Zu *et al.*, 2010). Yet, caution is required when analysing published data on the UV-B phenotype. UV-B exposure conditions vary dramatically between studies, and involve exposure to low or high UV-B doses, to filtered UV-B radiation or mixtures of UV-A, UV-B and UV-C radiation, and to various UV-B:PAR ratios. Moreover, experiments are performed under indoor or outdoor conditions, and using different red:far-red ratios. Given such variation in experimental conditions, it is not surprising that there is considerable variation in observed UV-B phenotype, and that some studies fail to report the 'prototype' UV-B phenotype of a 'compact' plant.

Despite experimental variations, the existence of a UV-B phenotype has been firmly established. Studies with UV-B photoreceptor (UVR8) mutants have unambiguously shown the role of UV-B, and that of UVR8, in controlling plant architecture (Favory *et al.*, 2009; Heijde and Ulm, 2012). Indeed, UVR8 was discovered in a screen for UV-B-induced hypocotyl shortening (Favory *et al.*, 2009). The failure of UVR8 mutants to undergo UV-induced shortening of the hypocotyl was the first evidence linking UVR8 to control of plant architecture. UVR8-deficient mutants do not just fail to display a shorter hypocotyl after UV-B exposure, but their petiole length and rosette diameter also remain relatively large despite UV-B exposure (Hayes *et al.*, 2014). Yet, UVR8-deficient mutants still display 'dwarfing' when exposed to high UV doses. Therefore, not all UV-B-mediated effects on plant architecture are mediated by UVR8, and it must be concluded that there is more than one UV-B-induced phenotype.

Existential Doubts

The UV-B phenotype in the natural environment

The UV-B phenotype is routinely observed in plants raised under supplemental UV-B in controlled conditions. Barnes *et al.* (1990) observed reductions in leaf length, leaf area, and shoot height, as well as increases in leaf and axillary shoot production across a collection of 12 dicot and monocot species kept in a glasshouse. Cooley *et al.* (2001) showed UV-B-induced reductions in leaf area, petiole length, and leaf number in a range (but not all) of *Arabidopsis thaliana* accessions exposed for 21 days to supplemental UV-B under outdoor conditions. Yet, long-term outdoor studies have yielded results that are more variable. For example, Indian cress (*Tropaeolum majus*) grown outdoors under supplemental UV-B for three months, displayed no UV-induced alterations in specific leaf area, internode length, and petiole length (Germ *et al.*, 2016). In contrast, work by the same group on common and tartary buckwheat (*Fagopyrum esculentum* and *F. tataricum*, respectively) grown outdoors under supplemental UV-B revealed strong UV-B-induced decreases in leaf area, and plant height as well as increases in leaf thickness (Breznik *et al.*, 2005). It is not clear how many studies have explored UV-B effects on morphology under natural growth conditions, as it is likely that studies that did not observe a significant effect have not been published. Sun *et al.* (2016) reported how leaf morphological traits of *Quercus guyavifolia* change along an altitudinal gradient on the Qinghai–Tibet plateau. With increasing UV dose, leaf length, leaf length–width ratio, and petiole length all decreased. Although these data appear to suggest that a UV phenotype does occur in the natural environment, this is not necessarily the case, as other altitude dependant factors such as temperature and rainfall are similarly associated with leaf architecture, thus complicating the interpretation of the data. A more extensive experiment was done by Roro *et al.* (2016) who combined an altitudinal gradient with the use of UV filters. This revealed that UV radiation decreases total leaf area, but increases stem branching and specific leaf area in pea plants (*Pisum sativum*). Effects on branching and specific leaf area were particularly pronounced during the dry season, emphasising that other environmental factors moderate UV-B effects on morphology. Nevertheless, differences in PAR transmission between the filters may also have affected the results. Perhaps the most ecologically relevant data on UV-induced morphological change are those generated at Abisko Research station in Sweden where outdoor UV-supplementation studies lasted decades. In an early study, leaf thickness of *Vaccinium vitis-idaea* increased following two years of UV supplementation, although co-existing *Vaccinium myrtillus* and *V. uliginosum* both developed thinner leaves in the same exposure experiment (Johanson *et al.*, 1995). Tellingly, the year-on-year variation in leaf thickness of non-UV control plants was greater than the actual UV effect in each particular year. After seven years of UV-B treatment there were no discernible effects of UV-B on leaf thickness (Semerdjieva *et al.*, 2003). These data underline that the UV-B phenotype is not reliably observed under natural conditions. Indeed, UV filtration experiments in peatlands showed effects of UV-B on height growth and morphology of the shrub *Empetrum rubrum*, and the tree *Nothofagus antarctica* in some years, but not in others (Robson *et al.*, 2003). It is likely that in some years the UV-B effects on plant architecture are masked by other environmental factors, such as light, temperature, and water availability, which are known to exert strong effects on plant architecture. Apart from environmental factors, there also appears to be a strong effect of plant genotype on the UV-B phenotype. Different Arabidopsis accessions display distinct morphological responses to the same UV-B treatment (Cooley *et al.*, 2001). Moreover, Klem *et al.*

(2012) demonstrated the importance of leaf ontogeny for UV-B responses. Thus, rather than a simple on/off scenario, the induction of the UV-B phenotype is a specific phenomenon that can be observed under specific environmental conditions in specific species and/or ecotypes.

The UV-B phenotype as a transient phenomenon

Plant organs display determinate or indeterminate growth. Leaves typically have a final form and size, depending on environmental conditions. In contrast, stems often exhibit indeterminate growth. Awareness of growth patterns is essential when assessing the impact of an environmental factor on organ size. Unfortunately, single time-point studies constitute the bulk of knowledge about the UV-B phenotype, and these studies fail to clarify whether UV-B exposure leads to a permanently more dwarfed phenotype or slows down the expansion rate to yield a transiently smaller organism. Few studies have investigated this question, but it appears that both scenarios do occur. In silver birch (*Betula pendula*), leaf elongation is delayed by supplemental UV-B, but as elongation growth continues slightly longer in the UV-B-exposed leaves, only a transient effect on leaf size is observed (Robson and Aphalo, 2012). In contrast, in downy birch (*Betula pubescens*) UV-B decreases the size of the fully developed leaf (Robson and Aphalo 2012). Effects on fully developed leaves were also described by Johanson *et al.* (1995) who reported UV-induced changes in leaf thickness in three *Vaccinium* species grown outdoors, under supplemental UV-B. Transient effects of UV-B on leaf morphology have been studied in some detail in *Arabidopsis thaliana*. Hectors *et al.* (2010) showed that supplemental UV-B initially mostly impeded longitudinal growth. However, in leaves exposed for longer periods to UV-B, the length–width ratio was restored as a result of a stronger impediment of elongation along the transverse axis of the leaf. Thus, not only are some UV-B effects transient, it also appears that plants are capable of compensatory responses that restore the geometric balance of the leaf. Lake *et al.* (2009) reported a transient effect of supplemental UV-B on leaf elongation in Arabidopsis. Following an initial (acute) phase of decreased expansion, plants exposed to chronic UV-B exposure recovered growth. Interestingly, a permanent phenotypic effect was observed for the Arabidopsis *fah-1* mutant. This mutant is UV-sensitive as it lacks sinapic acid due to a mutation in the enzyme ferulate-5-hydroxylase. This observation implies that permanent, morphological UV-B effects are associated with stress, while transient UV effects are associated with lower UV-B doses. Given the mixture of transient and permanent UV-B effects, a key message is that single time-point studies are inadequate for analysing UV-B induced morphological changes. Indeed, the possibility cannot be excluded that the failure of some studies to detect a UV-B effect on plant morphology is due to the transient character of the UV-B phenotype, in combination with an unfortunate choice of time-point for analysis.

The dose response for induction of the UV-B phenotype

Nearly all reports on the UV-B phenotype are based on single-dose studies, and therefore fail to elucidate any dose response relationship. The few studies that investigated the effects of different doses of UV-B on plant architecture show that the relationship is not necessarily linear. Brodführer (1955) revealed that increasing the UV-B dose from 2% to 33% of ambient solar UV-B resulted in an increase in the length of the main stem of the Arabidopsis inflorescence. Increasing the UV-B dose from 33% to 100% of solar UV-B did not cause a further increase in stem length, but rather a substantial decrease in stem length. Similarly, low UV doses increased inflorescence branching, while high doses inhibited the same process. Van de Staaij *et al.* (1997) observed a similar (but inverse) bell-shaped UV-B dose-response. Low doses of UV-B decreased flower formation in *Silene vulgaris*, whilst higher UV doses stimulated this process. An inverse, bell-shaped dose response was also found by Qaderi *et al.* (2008), who reported that low doses of UV-B decreased the number of leaves in *Silene noctiflora*, although higher UV doses increased leaf numbers. At present there are not enough dose response curves of UV-B mediated plant morphology to draw firm conclusions. However, the three examples of bell-shaped dose response curves imply the possibility that distinct UV-B response pathways are triggered by low as opposed to high UV-B doses. Consistently, UVR8-mutants fail to display a shorter hypocotyl length when exposed to low doses of UV-B, but display a 'dwarfing' response to high doses (Favory *et al.*, 2009).

The UV-B-induced phenotype exists, and some of its architectural characteristics are mediated by the UV-B photoreceptor, UVR8. Nevertheless, reported dose-response curves, and mixtures of transient and permanent UV-B effects, strongly suggest the existence of at least two different UV-B phenotypes. Furthermore, there is evidence that UVR8-mediated responses are modulated by wavelengths other than UV-B, for example due to photoreceptor interactions (Morales *et al.*, 2013).

An Anatomical Perspective of the UV-B Phenotype

A cellular perspective

The size of plant organs is determined by interactions between genotype, physiology and environment, through effects on cell proliferation and expansion. During the proliferation phase, the size of densely cytoplasmic cells is relatively constant, while in the post-mitotic organ, cells start to enlarge and this is often accompanied by increases in ploidy (Hepworth and Lenhard, 2014). Environmental factors can alter organ size through impacts on cell proliferation and/or cell expansion. However, this view is overly simplistic, as 'compensatory' cell expansion can mask decreases in cell proliferation. Indeed, organ size is co-modulated by the identity of the organ itself, i.e. a top-down control function (Hepworth and Lenward, 2014). UV-B has been shown to decrease cell proliferation and/or cell expansion. UV-B can impede cell division through the accumulation of DNA damage (primarily cyclobutane pyrimidine dimers and pyrimidine (6-4) pyrimidone dimers) which slows down the G1-to-S step in the cell cycle (Jiang *et al.*, 2011). Oxidative stress can also impede the cell cycle, through interactions with oxidative stress checkpoints (Tsukagoshi, 2012). The cell cycle block can facilitate DNA repair before further replication occurs (Jiang *et al.*, 2011), but does not necessarily result in smaller numbers of cells in a particular organ, as plants can delay the transition from cell proliferation to expansion (Hepworth and Lenhard, 2014). Compensatory effects of UV-B radiation on cell expansion have been related to increases in ploidy. UV-B can enhance endoreduplication resulting in increased ploidy which, in turn, has been associated with cellular expansion (Radziejwoski *et al.*, 2011).

UV-B exposure can inhibit cell proliferation (Wargent *et al.*, 2009), expansion (Hectors *et al.*, 2010), or have a complex effect on both processes. Both cell numbers and cell size decreased when a UV-sensitive *Arabidopsis thaliana fah-1* mutant was exposed to UV-B. The observed tenfold decrease in leaf area was probably associated with abiotic stress (Lake *et al.*, 2009). In comparison, larger cells were reported on the abaxial (but not adaxial) leaf surface when wildtype Arabidopsis was exposed to the same UV-B dose (Lake *et al.*, 2009). This may represent a compensatory response as described by Hepworth and Lenward (2014). Wargent *et al.* (2009) also reported an increase in cell size in UV-B-exposed Arabidopsis, although this was offset by a decrease in cell number. Hectors *et al.* (2010) found that UV-B had no measurable effect on the numbers of cells in Arabidopsis, but cell expansion was decreased by UV-B along a developmentally controlled pattern. Thus, effects on cell size became apparent first for the distal zone, and only later for the middle and proximal zones of the leaf. Interpretation of these data needs to consider differences in the UV-B dose and exposure protocols used. Nevertheless, the data show the variation in

UV-induced cellular responses that can occur in UV-B-exposed leaves.

A tissue perspective

There is a substantial knowledge gap between UV-B effects on epidermal cells, and on plant organs. In fact, up-scaling is complicated because tissues within a leaf respond differently to UV-B exposure. Leaf thickness increased substantially in blueberry (*Vaccinium corymbosum*) cultivar Legacy exposed for 40 days to supplemental UV-B, due to increased thickness of the mesophyll (Reyes-Díaz *et al.*, 2016). This observation is consistent with data by Robson and Aphalo (2012) who reported UV-B-induced increases in palisade thickness in birch leaves, and by Nagel *et al.* (1998), who reported increases in hypodermal thickness of pine (*Pinus ponderosa*) needles. Other studies have reported increased epidermal thickness of UV-acclimated leaves. In lemon (*Citrus limon*) fruits, UV induces cell wall thickening in the epidermis, as well as in the underlying parenchyma and collenchyma (Ruiz *et al.*, 2016). Although Reyes-Díaz *et al.* (2016) reported increased mesophyll thickness in UV-B-exposed blueberry cultivar Legacy, this was not the case for cultivar Bluegold. In the latter cultivar, leaf thickening was associated with disorganisation of the mesophyll cells, and the formation of substantial intercellular cavities. Thus, under the same exposure conditions one blueberry cultivar appears to display a form of acclimation, whilst another cultivar displays stress, reinforcing the message that there is more than one UV-B-mediated process that mediates alterations in plant architecture.

Underpinning Regulatory Mechanisms

UVR8-mediated control of architecture

Understanding of UVR8-mediated changes in plant architecture has increased in recent years. Interactions with hormonal pathways are a key feature of UVR8 activity. Hayes *et al.*, (2014) demonstrated that UVR8 slows elongation growth through interactions with gibberellic acid (GA) and auxin metabolism. GA homeostasis is affected through a UV-B-mediated increase in GA2-oxidase transcript levels. Evidence for a drop in GA concentrations is indirect, through an increase in (elongation inhibiting) DELLA proteins. Consistently, several other studies have reported induction of genes encoding GA oxidases (cf. Vanhaelewyn *et al.*, 2016). Peng and Zhou (2009) reported a decrease in actual GA levels in soybean (*Glycine max*). In contrast, Yang *et al.* (2004) showed that GA levels in tomato leaves doubled following UV-B exposure. Thus, measurements of GA levels in UV-B-exposed plants do not yet yield a coherent story.

There is good evidence for a role of auxin in UV-B-mediated morphological changes. Auxin is a key regulator of elongation, axillary branching, leaf development, and root growth. Initially, auxins were associated with the UV-B phenotype based on architectural similarities between the UV-B phenotype and auxin mutants (Jansen, 2002). Hectors *et al.*, (2012) demonstrated a UV-B-mediated decrease in free auxin levels in young leaves of Arabidopsis, while Yang *et al.* (2004) reported an overall decrease in auxin levels in UV-B-exposed tomato (*Solanum lycopersicum*). Hayes *et al.* (2014) showed UVR8-mediated effects on auxin homeostasis using *pDR5:-GUS* reporter constructs. Consistently, UV-B acclimation involves the differential expression of a range of auxin-related genes (Favory *et al.*, 2009; Hectors *et al.*, 2010, 2012; Hayes *et al.*, 2014; Vandenbussche *et al.*, 2014). Furthermore, the Arabidopsis auxin influx mutant *axr4-1*, and auxin biosynthesis mutant nit1-3 display relatively strong morphological responses to UV-B exposure (Hectors *et al.*, 2012). Thus, there is diverse evidence for a central role of auxin in mediating UV-B-induced morphological acclimation.

Stress-mediated control of architecture

It is unlikely that UVR8-mediated responses comprise the only mechanism of UV-B-mediated changes in plant morphology. Favory *et al.* (2009) reported 'dwarfing' of Arabidopsis UVR8-deficient plants grown in a solar sunlight simulator. UVR8-deficient plants are hypersensitive to UV-B due to a lack of protective responses (Heijde and Ulm, 2012), and it is likely that UV-B-induced alterations in the architecture of these mutants are associated with stress. The notion of Stress Induced Morphogenic Responses (SIMR) is based on the similarities in phenotype following exposure and acclimation to different stressors (Potters *et al.*, 2007). SIMR comprises a redirection of growth, rather than a cessation. The resulting phenotype can be more dwarfed, with increasing leaf thickness and/or branching. SIMRs are thought to be associated with generic stress-related processes such as enhanced production of Reactive Oxygen Species (ROS) and changed metabolism of auxin (Potters *et al.*, 2007). Although UV-B-induced stress is considered to be rare in the natural environment, UV-B is potentially damaging to plants (Jansen and Bornman 2012). UV-B can trigger oxidative-stress responses (cf. Hideg *et al.*, 2013) including the activation of mitogen-activated protein kinase phosphatases (Besteiro and Ulm, 2013). UV-B-mediated ROS production has also been linked with nitric oxide (NO) signalling (Lytvyn *et al.*, 2016). UV-B-induced NO has been linked with changes in microtubuli organization (Krasylenko *et al.*, 2012), which in turn can affect morphology though regulation of cell division, cell elongation and initiation of lateral growth.

The generic SIMR is likely to play a key role under oxidative stress conditions caused by exposure to high doses of UV-B (for a discussion of high and low UV-B doses see Hideg *et al.*, 2013). In contrast, UVR8-mediated morphological responses can occur under very low UV-B fluences (Brown and Jenkins, 2008; Fig. 5.1). Yet, the two potential response pathways are not mutually exclusive, and it is likely that there is considerable overlap of the two responses under the fluctuating UV intensities that are characteristic of natural sunlight.

UV-B acclimation and morphology

UV-B induces a broad range of biochemical acclimation responses, some of which can interfere with the mechanism controlling plant growth, while others may affect growth through incurring a fitness cost (Fig. 5.1). UV-B-induced changes in plant architecture and in the concentration of protective flavonoids are typically co-occurring phenomena (Qi *et al.*, 2003; Klem *et al.*, 2012). Flavonoids play a central role in UV-B protection due to their antioxidant and UV-screening properties. However, flavonoid aglycones are also regulators of polar auxin transport (Peer and Murphy 2007). Consistently, exposure of tobacco seedlings to exogenous flavonoids resulted in reduced leaf expansion, increased root length, a decrease in lateral and adventitious roots and an increase in free auxin in the shoot (Mahajan *et al.*, 2011). Arabidopsis tt4 and ugt78d2 flavonoid mutants also display alterations in both auxin distribution and plant morphology (Peer and Murphy 2007; Yin *et al.*, 2014). Nevertheless, UV-B-induced alterations in flavonoid profile comprise specific glycosylated compounds, accumulating in specific subcellular domains. At present, it remains to be proven that such specific alterations in UV-induced flavonoids can, through their effect on auxin transport, 'fine-tune' the plant phenotype mediated by UVR8 and/or stress.

A Biological Function of UV-B Phenotype

Many reports on the UV-B phenotype refer to a potential role in protecting plants from UV-B stress. It has been hypothesised that thicker leaves contain 'UV-free' zones (Jansen 2002). Yet, in most plant species little (<10% of incident dose) UV-B reaches the mesophyll due to UV-screening by epidermal cells (Barnes *et al.*, 2008). Thus, the importance of leaf thickening for UV-B protection remains unproven, especially as UV-B transmission is patchy due to UV-B penetration via stomatal pores and anticlinal cell walls. It has also been argued that a lack of elongation growth increases self-shading, and therefore decreases UV-B exposure. Yet, despite the obvious attraction of such a concept, shading does not necessarily equate to decreased UV-B exposure. The diffuse fraction of global UV-B irradiance is larger (0.57 to 0.91) than

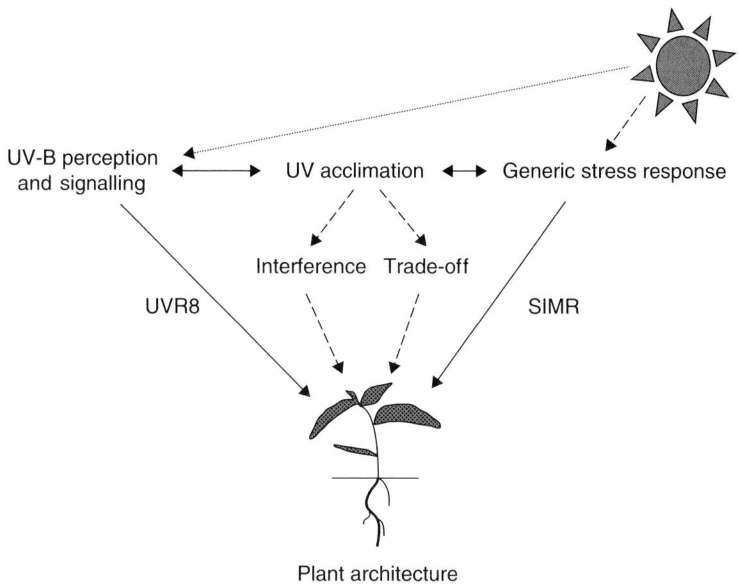

Fig. 5.1. UV-B and plant morphology. Low doses of UV-B can alter morphology directly via the UV-B photoreceptor, UVR8. Alternatively, metabolites formed during UV-B acclimation can interfere with morphology, as argued for flavonoids. A trade-off cost associated with UV acclimation has also been postulated, but not conclusively demonstrated. High UV-B doses can affect plant morphology through a generic Stress Induced Morphogenic Response (SIMR), as observed for many other stressors.

that of visible wavelengths (0.25 to 0.70) (Webb and Steven, 1984) which results in relatively strong penetration of UV-B into canopies (Fig. 5.2). Within a forest canopy the UV:PAR ratio in sunflecks (i.e. exposure to direct sunlight) is slightly enhanced compared to sunlight in open environments, while in the shaded understorey the UV:PAR ratio can reach at least five times that of sunlight in the open (Yang *et al.*, 1993; Brown *et al.*, 1994). Thus, a more dwarfed architecture does not necessarily reduce UV-B exposure, and may even increase the UV:PAR ratio which is, in general, considered a determinant of UV-B stress.

Thus, there is no conclusive evidence that UV-induced alterations in morphology contribute to UV-B protection. The observation that some UV-B effects on morphology are transient (Lake *et al.*, 2009; Robson and Aphalo, 2012) implies, at best, a temporary role in UV protection. Furthermore, the observation of bell shaped dose response curves (Brodführer, 1955; Van de Staaij *et al.*, 1997; Qaderi *et al.*, 2008) triggers the question: how can opposing morphological responses be linked with a single, functional role? Given the lack of an obvious association between morphology and UV-B tolerance, the possibility that (aspects of) the UV-B phenotype have a function other than UV-protection should be considered.

An exciting hypothesis on the role of UV-B-induced morphological changes was proposed by Hayes *et al.* (2014) who argued that UV-B, via the UVR8 photoreceptor, represses plant shade avoidance. Plants perceive shading through phytochrome which senses the decrease in red:far-red ratio. This triggers elongation growth involving, amongst others, PHYTOCHROME INTERACTING FACTORS (PIFs) and changes in auxin distribution. UV-B counters this response by triggering degradation of PIF4 and PIF5, while increasing DELLA stability (Hayes *et al.*, 2014). The antagonistic interaction between UVR8 and phytochrome responses creates a system of 'checks and balances' whereby elongation occurs under shaded conditions (low red to far-red ratio), while UV-B perception

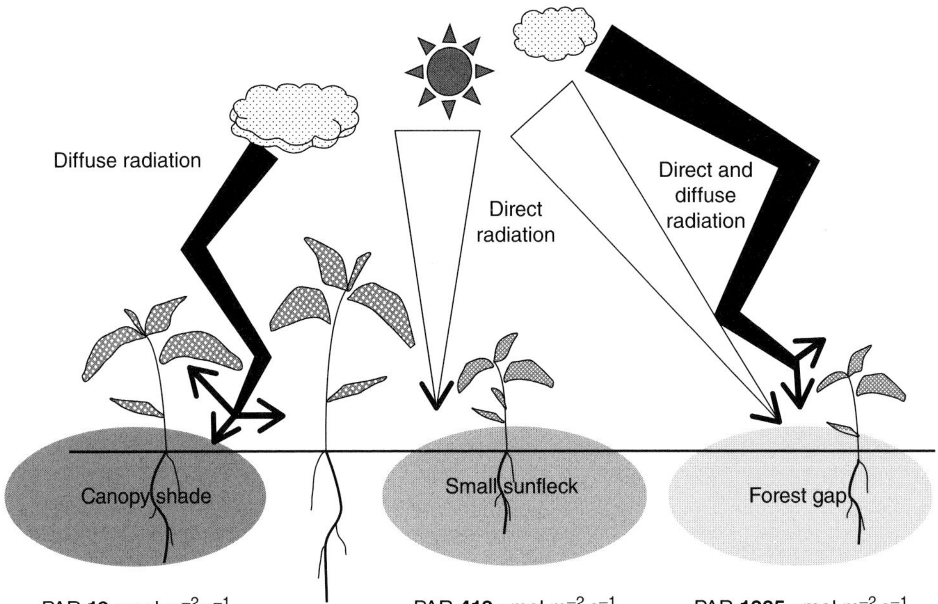

Fig. 5.2. UV-B within a canopy. Canopy transmittance of solar radiation depends on vegetation characteristics but is generally lower for direct (isotropic; open arrow) than for diffuse (anisotropic; filled arrow) light. UV-B and PAR intensities are low under a canopy, compared to incident radiation. Yet, in shaded areas, UV-B: PAR ratios may increase substantially due to the relatively large component of diffuse radiation enriched in UV-B. Values in the figure were calculated from spectral photon irradiance measurements with a diode array spectroradiometer (Ocean Optics Maya Pro2000+). Both PAR and UV-B:PAR $\times 10^4$ are given. Measurements represent canopy shade, a sunfleck, and a 10 m diameter gap on the floor of an old-growth *Fagus sylvatica* forest (el Hayedo de Montejo, central Spain) at solar noon (17/5/2014).

under exposed conditions impedes this process (Hayes *et al.*, 2014). However, this is not necessarily the case as the UV:PAR ratio can be strongly enriched in the understory (Yang *et al.*, 1993; Brown *et al.*, 1994) (Fig. 5.2) with the degree of enrichment depending on vegetation structure including species-specific leaf reflectance and absorbance (Robson *et al.*, 2015b). To understand the antagonism between phytochrome and UVR8 pathways in plant shade responses, there is a need for experimental approaches that cover the natural range of variation in the red/far-red and UV-B fluences (Mazza and Ballaré, 2015).

The idea that UV-B-induced morphology has a function different from increasing UV-B tolerance is intriguing. In the natural environment, exposure to high UV-B will normally be paralleled by exposure to high intensities of PAR, and therefore typically high temperatures, and possibly drought. Therefore, UV-B-induced morphological changes might play a role in acclimation to high levels of PAR, heat and/or drought. A reduction in leaf area in combination with increased leaf thickness is a typical characteristic of a sun-leaf (Lichtenthaler *et al.*, 2007; Niinemets, 2010). A smaller but thicker leaf is associated with a decrease in transpirational water loss (Anyia and Herzog, 2004). Consistently, recent work by Robson *et al.* (2015a) demonstrated that UV-B exposure induced drought tolerance in silver birch (*Betula pendula*). In contrast, Bandurska *et al.* (2013) argued that there is no direct association between UV acclimation and drought tolerance. Thus, while a role for the UV-B phenotype in acclimation to various solar and/or weather conditions is not proven, it is an attractive prospect that deserves study.

The UV-B Phenotype and Biomass Growth

Morphological traits are good indicators of plant performance and adaptation (Poorter and Bongers, 2006), through effects on light capture and photosynthetic performance. Alterations in leaf area and/or leaf thickness will alter light absorption, but also CO_2 availability, nitrogen use, heat load, transpirational water loss and self-shading (Nunes-Nesi *et al.*, 2016). Thus, UV-B-induced alterations in architecture are likely to have consequences for growth, but few studies have explored this. Some studies report UV-B-induced changes in plant architecture, and concomitant decreased biomass accumulation (Breznik *et al.*, 2005; Chen *et al.*, 2016). Yet, it is likely that negative effects on biomass are due to parallel, damaging impacts of UV-B on the cellular machinery, rather than as a fitness cost of the new phenotype per se. In some studies, UV-B-induced morphological changes are not accompanied by a loss in shoot biomass (Barnes *et al.*, 1990). For example, field exclusion studies showed how near-ambient UV-B caused reduced height growth but had no effect on biomass production by the moss *Sphagnum magellanicum* (Robson *et al.*, 2003). This may be interpreted as meaning that UV-B-induced morphological changes do not necessarily carry a yield penalty. However, this is far from proven, particularly as many studies are short, and therefore not suitable for visualising small incremental differences in biomass yield. Thus, the effect of UV-B-induced morphological changes on plant biomass production remains largely unknown.

Alterations in architecture can have indirect effects on growth. For example, the spatial distribution of leaves will determine the microclimate which may, in turn, affect susceptibility for pest and pathogen attack (Costes *et al.*, 2013; Ben-Yakir and Fereres, 2016). The best evidence for a potential yield penalty of the more dwarfed UV-B phenotype is generated by studies on plant–plant competition. UV-B-induced changes in morphology are large enough to affect competition for light capture in a canopy (Ryel *et al.*, 1990). Indeed, UV-B-induced alterations in the competitive balance between wheat (*Triticum aestivum*) and wild oat (*Avena fatua*) were linked to alterations in the relative position of leaves (Barnes *et al.*, 1988). Yet, it is important to be aware that UV-B radiation can also affect plant–plant interactions through other routes, such as a stimulation of production and release of allelochemicals. For example, Li *et al.* (2009) found that the allelopathic potential of *Zanthoxylum bungeanum* was stimulated under enhanced UV-B radiation.

What next?

UV-B-induced changes in plant morphology comprise a decrease in elongation growth, resulting in a more compact plant displaying decreases in petiole length, leaf area and/or enhanced leaf thickness together with shorter, but more branched stems. Here, we argue that there are at least two distinct UV-B phenotypes. One phenotype is mediated by the UV-B photoreceptor UVR8. The second UV-B-induced phenotype does not require functional UVR8 and is associated with plant stress. It is likely that both phenotypes do occur simultaneously in the natural environment. It is also likely that this mixture of two phenotypes is a cause of (i) contradictory information on UV-B-induced morphological changes; (ii) complex dose response curves; (iii) a mixture of transient and permanent morphological changes; and (iv) distinct effects on cell and organismal development. To distinguish the two UV-B phenotypes, detailed dose-response curves and action spectra need to be developed. In turn, these can be used to identify molecular, physiological and/or biochemical markers representative of distinct phenotypes. Only when this has been achieved, is there a realistic chance to explore the functional role of the UV-B phenotypes and to identify regulatory interactions with other environmental parameters which co-modulate plant morphology.

Acknowledgements

OU was supported by the Ministry of Education, Youth and Sports of the Czech Republic within NPU I project (LO1415), and TMR by the Academy of Finland decision number 266523. MAKJ acknowledges support by Science Foundation Ireland (11/RFP.1/EOB/3303) and WoB.

References

Anyia, A.O. and Herzog, H. (2004) Water-use efficiency, leaf area and leaf gas exchange of cowpeas under mid-season drought. *European Journal of Agronomy* 20, 327–339. DOI:10.1016/S1161-0301(03)00038-8

Bandurska, H., Niedziela, J. and Chadzinikolau, T. (2013) Separate and combined responses to water deficit and UV-B radiation. *Plant Science* 213, 98–105. DOI:10.1016/j.plantsci.2013.09.003

Barnes, P.W., Jordan, P.W., Gold, W.G. *et al.* (1988) Competition, morphology and canopy structure in wheat (*Triticum aestivum* L.) and wild oat (*Avena fatua* L.) exposed to enhanced ultraviolet-B radiation. *Functional Ecology* 2, 319–330. DOI:10.2307/2389404

Barnes, P.W., Flint, S.D. and Caldwell, M.M. (1990) Morphological responses of crop and weed species of different growth forms to ultraviolet-B radiation. *American Journal of Botany* 77, 1354–1360. DOI:10.2307/2444596

Barnes, P.W., Flint, S.D., Slusser, J.R. *et al.* (2008) Diurnal changes in epidermal UV transmittance of plants in naturally high UV environments. *Physiologia Plantarum* 133, 363–372. DOI:10.1111/j.1399-3054.2008.01084.x

Barthélémy, D. and Caraglio, Y. (2007) Plant architecture: a dynamic, multilevel and comprehensive approach to plant form, structure and ontogeny. *Annals of Botany* 99, 375–407. DOI:10.1093/aob/mcl260

Ben-Yakir, D. and Fereres, A. (2016) The effects of UV radiation on arthropods: a review of recent publications (2010–2015), *Acta Horticulturae* 1134, 335–342. DOI:10.17660/ActaHortic.2016.1134.44

Besteiro, M.A.G. and Ulm, R. (2013) Phosphorylation and stabilization of Arabidopsis MAP kinase phosphatase 1 in response to UV-B stress. *Journal of Biological Chemistry* 288, 480–486. DOI:10.1074/jbc.M112.434654

Breznik, B., Germ, M., Gaberscik, A. and Kreft, I. (2005) Combined effects of elevated UV-B radiation and the addition of selenium on common (*Fagopyrum esculentum* Moench) and tartary [*Fagopyrum tataricum* (L.) Gaertn.] buckwheat. *Photosynthetica* 43, 583–589. DOI:10.1007/s11099-005-0091-1

Brodführer, U. (1955) Der einfluss einer abgestuften dosierung von ultravioletter sonnenstrahlung auf das wachstum der pflanzen. *Planta* 45, 1–56.

Brown, B.A., and Jenkins, G.I. (2008) UV-B signaling pathways with different fluence-rate response profiles are distinguished in mature Arabidopsis leaf tissue by requirement for UVR8, HY5, and HYH. *Plant Physiology* 146, 576–588. DOI:10.1104/pp.107.108456

Brown, M.J., Parker, G.G. and Posner, N.E. (1994) A survey of ultraviolet-B radiation in forests. *Journal of Ecology* 82, 843–854.

Chen, M., Huang, Y. and Liu, G. (2016) Effects of enhanced UV-B radiation on morphology, physiology, biomass, leaf anatomy and ultrastructure in male and female mulberry (*Morus alba*) saplings. *Environmental and Experimental Botany* 129, 85–93. DOI:10.1016/j.envexpbot.2016.03.006

Cooley, N.M., Higgins, J.T., Holmes, M.G. and Attridge, T.H. (2001) Ecotypic differences in responses of *Arabidopsis thaliana* L. to elevated polychromatic UV-A and UV-B+A radiation in the natural environment: a positive correlation between UV-B+A inhibition and growth rate. *Journal of Photochemistry and Photobiology B: Biology* 60, 143–150. DOI:10.1016/S1011-1344(01)00140-3

Costes, E., Lauri, P.E., Simon, S., and Andrieu, B. (2013) Plant architecture, its diversity and manipulation in agronomic conditions, in relation with pest and pathogen attacks. *European Journal of Plant Pathology* 135, 455–470. DOI:10.1007/s10658-012-0158-3

Fatima, A., Kataria, S., Guruprasad, K.N. *et al.* (2016) Synchrotron X-ray phase contrast imaging of leaf venation in soybean (*Glycine max*) after exclusion of solar UV (280–400 nm) radiation. *Journal of Synchrotron Radiation* 23, 795–801. DOI:10.1107/S1600577516003507

Favory, J.J., Stec, A., Gruber, H. *et al.* (2009) Interaction of COP1 and UVR8 regulates UVB-induced photomorphogenesis and stress acclimation in Arabidopsis. *The EMBO Journal* 28, 591–601. DOI:10.1038/emboj.2009.4

Galvão, V.C. and Fankhauser, C. (2015) Sensing the light environment in plants: photoreceptors and early signaling steps. *Current Opinion in Neurobiology* 34, 46–53. DOI:10.1016/j.conb.2015.01.013

Germ, M., Breznik, B., Dolinar, N. *et al.* (2013) The combined effect of water limitation and UV-B radiation on common and tartary buckwheat. *Cereal Research Communications* 41, 97–105. DOI:10.1556/CRC.2012.0031

Germ, M., Spahić, I. and Gaberščik, A. (2016) Morphological, biochemical and physiological responses of Indian cress (*Tropaeolum majus*) to elevated UV-B radiation. *Periodicum Biologorum* 117, 357–364. DOI:10.18054/pb.v117i3.2858

Hayes, S., Velanis, C.N., Jenkins, G.I. and Franklin, K.A. (2014) UV-B detected by the UVR8 photoreceptor antagonizes auxin signaling and plant shade avoidance. *Proceedings of the National Academy of Sciences of the United States of America* 111, 11894–11899. DOI:10.1073/pnas.1403052111

Hectors, K., Jacques, E., Prinsen, E. *et al.* (2010) UV radiation reduces epidermal cell expansion in leaves of *Arabidopsis thaliana*. *Journal of Experimental Botany* 61, 4339–4349. DOI:10.4161/psb.6.1.14127

Hectors, K., van Oevelen, S., Guisez, Y. *et al.* (2012) The phytohormone auxin is a component of the regulatory system that controls UV-mediated accumulation of flavonoids and UV-induced morphogenesis. *Physiologia Plantarum* 145, 594–603. DOI:10.1111/j.1399-3054.2012.01590.x

Heijde, M. and Ulm, R. (2012) UV-B photoreceptor-mediated signalling in plants. *Trends in Plant Science* 17, 230–237. DOI:10.1016/j.tplants.2012.01.007

Hepworth, J. and Lenhard, M. (2014) Regulation of plant lateral-organ growth by modulating cell number and size. *Current Opinion in Plant Biology* 17, 36–42. DOI:10.1016/j.pbi.2013.11.005

Hideg, E., Jansen, M.A.K. and Strid, A. (2013) UV-B exposure, ROS, and stress: inseparable companions or loosely linked associates? *Trends in Plant Science* 18, 107–115. DOI:10.1016/j.tplants.2012.09.003

Hofmann, R.W. and Campbell, B.D. (2011) Response of *Trifolium repens* to UV-B radiation: morphological links to plant productivity and water availability. *Plant Biology* 13, 896–901. DOI:10.1111/j.1438-8677.2011.00458.x

Jansen, M.A.K. (2002) Ultraviolet-B radiation effects on plants: induction of morphogenic responses. *Physiologia Plantarum* 116, 423–429. DOI:10.1034/j.1399-3054.2002.1160319.x

Jansen, M.A.K. and Bornman, J.F. (2012) UV-B radiation: from generic stressor to specific regulator. *Physiologia Plantarum* 145, 501–504. DOI:10.1111/j.1399-3054.2012.01656.x

Jiang, L., Wang, Y., Björn, L.O and Li, S. (2011) UV-B-induced DNA damage mediates expression changes of cell cycle regulatory genes in Arabidopsis root tips. *Planta* 233, 831–841. DOI:10.1007/s00425-010-1340-5

Johanson, U., Gehrke, C., Bjorn, L.O. and Callaghan, T.V. (1995) The effects of enhanced UV-B radiation on the growth of dwarf shrubs in a subarctic heathland. *Functional Ecology* 713–719. DOI:10.1023/A:1009771125992

Klem, K., Ac, A., Holub, P. *et al.* (2012) Interactive effects of PAR and UV radiation on the physiology, morphology and leaf optical properties of two barley varieties. *Environmental and Experimental Botany* 75, 52–64. DOI:10.1016/j.envexpbot.2011.08.008

Krasylenko, Y.A., Yemets, A.I., Sheremet, Y.A. and Blume, Y.B. (2012) Nitric oxide as a critical factor for perception of UV-B irradiation by microtubules in Arabidopsis. *Physiologia Plantarum* 145, 505–515. DOI:10.1111/j.1399-3054.2011.01530.x

Lake, J.A., Field, K.J., Davey, M.P. *et al.* (2009) Metabolomic and physiological responses reveal multiphasic acclimation of *Arabidopsis thaliana* to chronic UV radiation. *Plant, Cell & Environment* 32, 1377–1389. DOI:10.1111/j.1365-3040.2009.02005.x

Li, H., Pan, K., Liu, Q. and Wang, J. (2009) Effect of enhanced ultraviolet-B on allelopathic potential of *Zanthoxylum bungeanum*. *Scientia Horticulturae* 119, 310–314. DOI:10.1016/j.scienta.2008.08.010

Lichtenthaler, H.K., Ac, A., Marek, M.V. *et al.* (2007) Differences in pigment composition, photosynthetic rates and chlorophyll fluorescence images of sun and shade leaves of four tree species. *Plant Physiology and Biochemistry* 45, 577–588. DOI:10.1016/j.plaphy.2007.04.006

Lytvyn, D.I., Raynaud, C., Yemets, A.I. *et al.* (2016) Involvement of inositol biosynthesis and nitric oxide in the mediation of UV-B induced oxidative stress. *Frontiers in Plant Science* 7, 430. DOI:10.3389/fpls.2016.00430

Mahajan, M., Kumar, V. and Yadav, S.K. (2011) Effect of flavonoid-mediated free IAA regulation on growth and development of in vitro-grown tobacco seedlings. *International Journal of Plant Developmental Biology* 5, 42–48.

Mazza, C.A. and Ballaré, C.L. (2015) Photoreceptors UVR8 and phytochrome B cooperate to optimize plant growth and defense in patchy canopies. *New Phytologist* 207, 4–9. DOI:10.1111/nph.13332

Möglich, A., Yang, X., Ayers, R.A., and Moffat, K. (2010) Structure and function of plant photoreceptors. *Annual Review of Plant Biology* 61, 21–47. DOI:10.1146/annurev-arplant-042809-112259

Morales, L.O., Brosché, M., Vainonen, J. *et al.* (2013) Multiple roles for UV RESISTANCE LOCUS8 in regulating gene expression and metabolite accumulation in Arabidopsis under solar ultraviolet radiation. *Plant Physiology* 161, 744–759. DOI:10.1104/pp.112.211375

Nagel, L.M., Bassman, J.H., Edwards, G.E. *et al.* (1998) Leaf anatomical changes in *Populus trichocarpa, Quercus rubra, Pseudotsuga menziesii* and *Pinus ponderosa* exposed to enhanced ultraviolet-B radiation. *Physiologia Plantarum* 104, 385–396. DOI:10.1034/j.1399-3054.1998.1040314.x

Niinemets, Ü (2010) A review of light interception in plant stands from leaf to canopy in different plant functional types and in species with varying shade tolerance. *Ecological Research* 25, 693–714. DOI:10.1007/s11284-010-0712-4

Nunes-Nesi, A., de Laia Nascimento, V., de Oliveira Silva, F.M. *et al.* (2016) Natural genetic variation for morphological and molecular determinants of plant growth and yield. *Journal of Experimental Botany* 67, 2989–3001. DOI:10.1093/jxb/erw124

Peer, W.A. and Murphy, A.S. (2007) Flavonoids and auxin transport: modulators or regulators? *Trends in Plant Science* 12, 556–563. DOI:10.1016/j.tplants.2007.10.003

Peng, Q. and Zhou, Q. (2009) Antioxidant capacity of flavonoid in soybean seedlings under the joint actions of rare earth element La (III) and ultraviolet-B stress. *Biological Trace Element Research* 127, 69–80. DOI:10.1007/s12011-008-8218-4

Poorter, L. and Bongers, F. (2006) Leaf traits are good predictors of plant performance across 53 rain forest species. *Ecology* 87, 1733–1743. DOI:10.1890/0012-9658(2006)87[1733:LTAGPO]2.0.CO;2

Potters, G., Pasternak, T.P., Guisez, Y. *et al.* (2007) Stress-induced morphogenic responses: growing out of trouble? *Trends in Plant Science* 12, 98–105. DOI:10.1016/j.tplants.2007.01.004

Qaderi, M.M., Yeung, E.C. and Reid, D.M. (2008) Growth and physiological responses of an invasive alien species, *Silene noctiflora*, during two developmental stages to four levels of ultraviolet-B radiation. *Ecoscience* 15, 150–159. DOI:10.2980/15-2-3052

Qi, Y., Bai, S. and Heisler, G.M. (2003) Changes in ultraviolet-B and visible optical properties and absorbing pigment concentrations in pecan leaves during a growing season. *Agricultural and Forest Meteorology* 120, 229–240. DOI:10.1016/j.agrformet.2003.08.018

Radziejwoski, A., Vlieghe, K., Lammens, T. *et al.* (2011) Atypical E2F activity coordinates PHR1 photolyase gene transcription with endoreduplication onset. *The EMBO Journal* 30, 355–363. DOI:10.1038/emboj.2010.313

Reyes-Díaz, M., Meriño-Gergichevich, C., Inostroza-Blancheteau, C. *et al.* (2016) Anatomical, physiological, and biochemical traits involved in the UV-B radiation response in highbush blueberry. *Biologia Plantarum* 60, 355–366. DOI:10.1007/s10535-015-0580-4

Robson, T.M. and Aphalo, P.J. (2012) Species-specific effect of UV-B radiation on the temporal pattern of leaf growth. *Physiologia Plantarum* 144, 146–160. DOI:10.1111/j.1399-3054.2011.01546.x

Robson, T.M., Pancotto, V.A., Flint, S.D. *et al.* (2003) Six years of solar UV-B manipulations affect growth of *Sphagnum* and vascular plants in a Tierra del Fuego peatland. *New Phytologist* 160, 379–389. DOI:10.1046/j.1469-8137.2003.00898.x

Robson, T., Hartikainen, S.M. and Aphalo, P.J. (2015a) How does solar ultraviolet-B radiation improve drought tolerance of silver birch (*Betula pendula* Roth.) seedlings? *Plant, Cell & Environment* 38, 953–967. DOI:10.1111/pce.12405

Robson, T.M., Klem, K., Urban, O. and Jansen, M.A.K. (2015b) Re-interpreting plant morphological responses to UV-B radiation. *Plant, Cell & Environment* 38, 856–866. DOI:10.1111/pce.12374

Roro, A.G., Terfa, M.T., Solhaug, K.A. *et al.* (2016) The impact of UV radiation at high altitudes close to the equator on morphology and productivity of pea (*Pisum sativum*) in different seasons. *South African Journal of Botany* 106, 119–128. DOI:10.1016/j.sajb.2016.05.011

Ruiz, V.E., Interdonato, R., Cerioni, L. *et al.* (2016) Short-term UV-B exposure induces metabolic and anatomical changes in peel of harvested lemons contributing in fruit protection against green mold. *Journal of Photochemistry and Photobiology B: Biology* 159, 59–65. DOI:10.1016/j.jphotobiol.2016.03.016

Ryel, R.J., Barnes, P.W., Beyschlag, W. *et al.* (1990) Plant competition for light analyzed with a multispecies canopy model. *Oecologia* 82, 304–310. DOI:10.1007/BF00317475

Semerdjieva, S.I., Phoenix, G.K., Hares, D. *et al.* (2003) Surface morphology, leaf and cuticle thickness of four dwarf shrubs from a sub-Arctic heath following long-term exposure to enhanced levels of UV-B. *Physiologia Plantarum* 117, 289–294. DOI:10.1034/j.1399-3054.2003.00006.x

Sun, M., Su, T., Zhang, S.B. *et al.* (2016) Variations in leaf morphological traits of *Quercus guyavifolia* (Fagaceae) were mainly influenced by water and ultraviolet irradiation at high elevations on the Qinghai-Tibet Plateau, China. *International Journal of Agriculture and Biology* 18, 266–273. DOI:10.17957/IJAB/15.0074

Teramura, A.H. (1983) Effects of ultraviolet-B radiation on the growth and yield of crop plants. *Physiologia Plantarum* 58, 415-427. DOI:10.1111/j.1399-3054.1983.tb04203.x

Tsukagoshi, H. (2012) Defective root growth triggered by oxidative stress is controlled through the expression of cell cycle-related genes. *Plant Science* 197, 30–39. DOI:10.1016/j.plantsci.2012.08.011

Vandenbussche, F., Tilbrook, K., Fierro, A.C. *et al.* (2014) Photoreceptor-mediated bending towards UV-B in Arabidopsis. *Molecular Plant* 7, 1041–1052. DOI:10.1093/mp/ssu039

Vanhaelewyn, L., Prinsen, E., Van Der Straeten, D. and Vandenbussche, F. (2016) Hormone-controlled UV-B responses in plants. *Journal of Experimental Botany* 67, 4469–4482. DOI:10.1093/jxb/erw261

Van de Staaij, J.W.M., Bolink, E., Rozema, J. and Ernst, W.H.O. (1997) The impact of elevated UV-B (280–320 nm) radiation levels on the reproduction of a Highland and a lowland population of *Silene vulgaris*. *Plant Ecology* 128, 173–179. DOI:10.1023/A:1009710907336

Wagoner, D. (1999) *Traveling Light: Collected and New Poems*, University of Illinois Press, Illinois.

Wargent, J.J., Gegas, V.C., Jenkins, G.I. *et al.* (2009) UVR8 in *Arabidopsis thaliana* regulates multiple aspects of cellular differentiation during leaf development in response to ultraviolet-B radiation. *New Phytologist* 183, 315–326. DOI:10.1111/j.1469-8137.2009.02855.x

Webb, A. and Steven, M.D. (1984) Measurement of solar UVB radiation in the English Midlands. *Archives for Meteorology, Geophysics, and Bioclimatology, Series B* 35, 221–231. DOI:10.1007/BF02263347

Yang, X.S., Miller, D.R. and Montgomery, M.E. (1993) Vertical distributions of canopy foliage and biologically-active radiation in a defoliated/refoliated hardwood forest. *Agricultural and Forest Meteorology* 67, 129–146. DOI:10.1016/0168-1923(93)90054-L

Yang, H., Zhao, Z., Qiang, W. *et al.* (2004) Effects of enhanced UV-B radiation on the hormonal content of vegetative and reproductive tissues of two tomato cultivars and their relationships with reproductive characteristics. *Plant Growth Regulation* 43, 251–258. DOI:10.1023/B:GROW.0000046002.39513.be

Yin, R., Han, K., Heller, W. *et al.* (2014) Kaempferol 3-O-rhamnoside-7-O-rhamnoside is an endogenous flavonol inhibitor of polar auxin transport in Arabidopsis shoots. *New Phytologist* 201, 466–475. DOI:10.1111/nph.12558

Zu, Y.G., Pang, H.H., Yu, J.H. *et al.* (2010) Responses in the morphology, physiology and biochemistry of *Taxus chinensis* var. *mairei* grown under supplementary UV-B radiation. *Journal of Photochemistry and Photobiology B: Biology* 98, 152–158. DOI:10.1016/j.jphotobiol.2009.12.001

6 Plant Responses to Fluctuating UV Environments

Paul W. Barnes[1]*, T. Matthew Robson[2], Mark A. Tobler[1], Isabelle N. Bottger[1] and Stephan D. Flint[3]

[1]Department of Biological Sciences and Environment Program, Loyola University New Orleans, 6363 St. Charles Avenue, New Orleans, Louisiana 70118, USA; [2]Department of Biosciences, Viikki Plant Science Centre, University of Helsinki, PO Box 65, 00014 Helsinki, Finland; [3]Department of Forest, Rangeland and Fire Sciences, UIPO 441135, University of Idaho, Moscow, Idaho 83844-1135, USA

Introduction

The terrestrial solar ultraviolet (UV: ~290–400 nm) radiation regime experienced by plants in nature varies across multiple timescales (interannual, seasonal and diurnal). Long-term (year-to-year) variability in UV irradiance at the Earth's surface in modern times is driven largely by changes in stratospheric ozone, which influences the attenuation of ultraviolet-B radiation (UV-B: 280–315 nm) and climate change, which can alter both UV-B and ultraviolet-A radiation (UV-A: 315–400 nm) via changes in cloud cover, aerosols and tropospheric ozone (Bais *et al.*, 2015; see also Chapter 1, this volume). Variation in incident UV over shorter timeframes (seasonal and diurnal) results primarily from the natural rhythms in prevailing solar angles that occur over a year or day, though intraseasonal fluctuations in ozone can play a significant role in certain regions (Madronich *et al.*, 2011; Bais *et al.*, 2015). Natural cycles in UV irradiance are routinely disrupted by changes in atmospheric conditions and vegetative cover (e.g. phenological patterns in canopy development). These cycles themselves can exhibit some degree of seasonal and diurnal periodicities, as shown in Fig. 6.1.

In detail, Fig. 6.1a shows integrated daily plant effective UV-B over three years at a Sonoran Desert location in southern Arizona, with the pronounced annual summer monsoon periods noted. Fig. 6.1b gives the daily plant effective UV-B at Pullman, Washington, over a period in early spring of heavy clouds followed by clear skies (April–early May 2014) when new leaves are emerging in many native plant species, while Fig. 6.1c shows diurnal patterns of plant effective UV-B (30 min means) under clear and cloudy sky conditions in the Arizona Sonoran Desert in July 2015. Thus, Figs. 6.1b and 6.1c demonstrate that rapid fluctuations in solar UV can also occur via changes in cloud cover, which can either increase or decrease UV irradiance depending on the position and type of clouds in relation to the solar disc (Thiel *et al.* 1997; Lopez *et al.* 2009). Fluctuations can also come from gaps in the canopy that create periodic sunflecks in the understorey (Fig. 6.1d; Flint and Caldwell, 1998; Heisler *et al.*, 2003; see also Chapter 2, this volume). Fig. 6.1d shows instantaneous

*Corresponding author: pwbarnes@loyno.edu

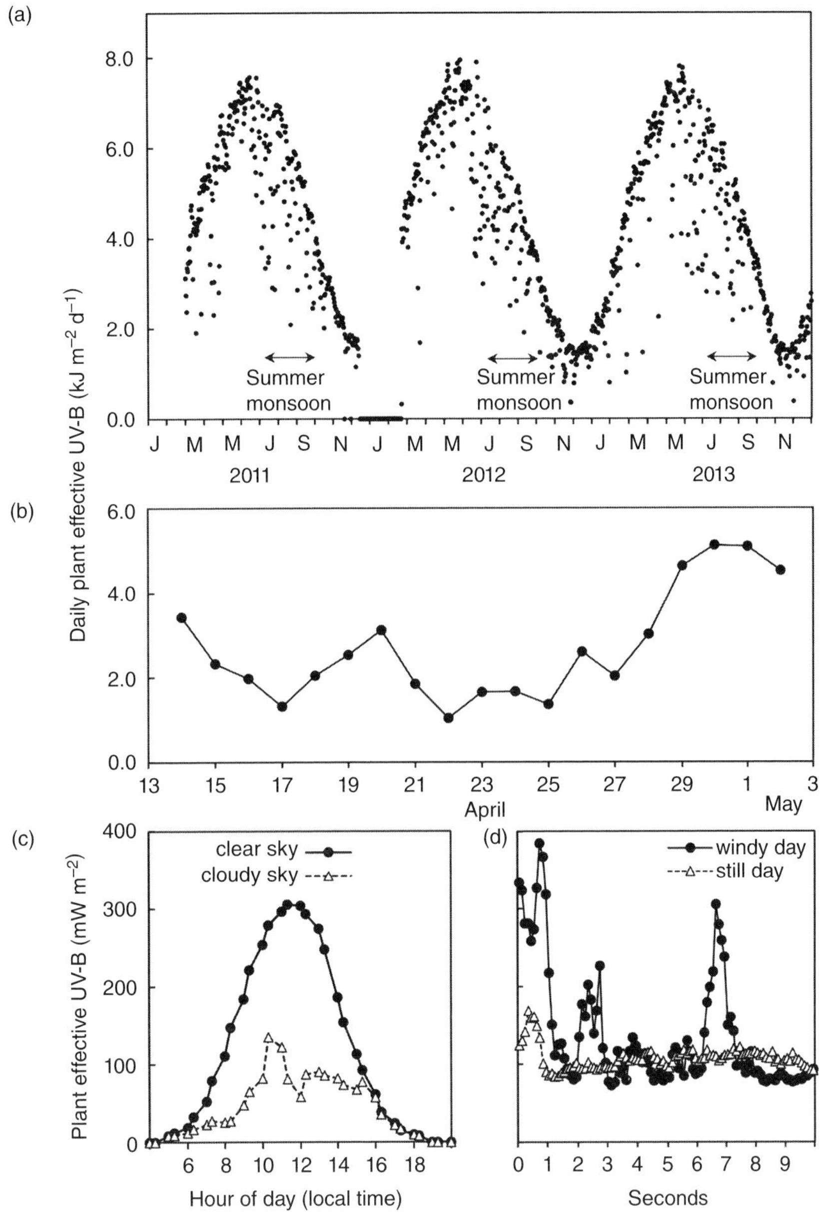

Fig. 6.1. Variability in ultraviolet-B irradiance at multiple time scales. See main text for description. All irradiance data are weighted according to the generalized plant action spectrum (Caldwell, 1971) and normalized to unity at 300 nm. Data in (a) and (c) were obtained from an on-site broadband UV-B sensor (YES UVB-1) calibrated against a double-monochromator spectroradiometer (Gooch & Housego 756). Data in (b) were obtained from the USDA UV monitoring network (http://uvb.nrel.colostate.edu/UVB/index.jsf (accessed 30 March 2017)) using data from narrow waveband spectrometers to produce synthetic spectra. Data in (d) were obtained from a diode array spectroradiometer (OceanOptics Maya Pro2000+) measuring spectral irradiance every 0.1 seconds for a 10-second period.

UV-B at the same location (to the nearest cm) in the understorey of a 20-year old *Betula pendula* stand in a provenance trial at Little Wittenham, Oxfordshire, UK, at the same time of day on a windy and relatively calm day. Consequently, plants and individual leaves can experience considerable change in UV exposure over their lifespans, from one day to the next, and within a given day, that is regular and, at times, erratic and unpredictable.

Over the past several decades, considerable attention has been given to understanding plant responses to changes in average UV conditions that occur as a result of stratospheric ozone depletion and to the interactive effects of enhanced UV-B and other climate change factors (e.g. elevated CO_2 and increased temperature) (Björn, 2015; Bornman *et al.* 2015; and references therein). By comparison, we know far less about plant responses to changes in solar UV fluxes that occur (i) naturally over the course of a given growing season or day; (ii) in response to changes in cloud cover; or (iii) as leaves alternate between shaded and sunlit conditions within canopies and in understorey environments.

Given that plants have evolved a photosensory system to sense ambient UV-B (Rizzini *et al.*, 2011; see also Chapter 8, this volume) and that this photoreceptor appears to interact with other photoreceptor systems (e.g. cryptochromes and phytochromes) to provide information to a plant about its light environment (Xie and Hauser, 2012; Tilbrook *et al.*, 2013; Mazza and Ballaré, 2015), it is relevant to ask how quickly and to what degree plants can respond to changes in their UV environment, and what these responses might mean for plant fitness. For instance, do rapid fluctuations in UV (seconds to minutes) in fact matter to plants or do they integrate and respond to UV over much longer time frames (hours, days or weeks)? Also, is the rate and magnitude of the response to changing UV conditions contingent upon developmental stage (e.g. young versus old leaves) and environmental factors that influence resource availability and levels of physiological stress (e.g. atmospheric CO_2 concentration, temperature and moisture availability)? Finally, what is the adaptive significance of variation among plant species in their abilities to respond to temporal fluctuations in UV, and what might these differences mean for the timing of defences against other abiotic (e.g. drought) and biotic (e.g. herbivory) factors that share commonalities in photosensory signalling (Demkura and Ballaré, 2012) and biosynthetic pathways (Izaguirre *et al.*, 2007) with UV responses?

In this chapter, we review research to date on what is known about the prevalence and mechanisms of plant responses to short-term fluctuations in UV exposure and explore the possible adaptive and ecological significance of these responses. We address only changes in total UV irradiances and not shifts in spectral quality, which can be associated with fluctuating UV conditions particularly in understorey environments (Parisi and Kimlin, 1999). Our emphasis is on physiological responses to UV, and we specifically focus on UV screening protection as this is a primary avenue of UV acclimation and one of the most frequently documented and readily observable responses to UV in plants (Searles *et al.*, 2001).

Overview of Plant Responses to UV and the Time Course of UV Acclimation

Abundant evidence from studies conducted in controlled environmental conditions indicates that exposure to UV can elicit specific photomorphogenic responses in plants as well as induce a variety of effects that are generally considered to be detrimental to plant function (Jansen and Bornman, 2012; Li *et al.*, 2013; Robson *et al.*, 2015). The highly energetic shorter wavelengths of UV (i.e. UV-B) are particularly efficient at producing a number of deleterious effects in plants, including disruption of the integrity and function of important macromolecules (DNA, proteins and lipids), oxidative damage, partial inhibition of photosynthesis and growth reduction (see Chapter 7, this volume). To combat these adverse effects, plants have developed a suite of biochemical, physiological and morphological mechanisms that collectively protect or repair sensitive targets from direct and indirect UV-induced injury (Britt, 1999; Favory *et al.*, 2009; Hectors *et al.*, 2009; Jacques *et al.*, 2009; see also Chapter 4, this volume). This acclimation to UV appears sufficient to largely minimize any detrimental effects of UV-B on plant growth and productivity when plants are grown under ambient or realistically enhanced UV-B in the field (Ballaré *et al.*, 2011) such that photomorphogenic (and often beneficial) effects of UV generally predominate under these conditions (Wargent and Jordan, 2013; see also Chapter 3, this volume).

One of the most important and widespread protective responses of plants to UV radiation involves the induction and synthesis of flavonoids, hydroxycinnamic acids (HCAs) and other related phenylpropanoid compounds that function as 'UV sunscreens' and antioxidants (Caldwell *et al.*, 1983; Agati *et al.*, 2012; Emiliani *et al.*, 2013). The accumulation of flavonoids and other UV-absorbing compounds in epidermal tissue decreases epidermal UV transmittance (T_{UV}) but has minimal effect on attenuating photosynthetically active radiation (PAR: 400–700 nm) needed for photosynthesis of underlying mesophyll tissue (Fig. 6.2; Day *et al.*, 1994; Mazza *et al.*, 2000; Bidel *et al.*, 2007). The epidermal screening of UV is a primary avenue by which plants acclimate to changing UV environments, including alterations resulting from stratospheric ozone depletion and climate change (Williamson *et al.*, 2014; Bornman *et al.*, 2015). This UV acclimation response entails a measurable energetic and fitness cost (Snell *et al.*, 2009; Guidi *et al.*, 2011; Hofmann and Jahufer, 2011), varies within and among species (Day *et al.*, 1992; Qi *et al.*, 2010; Randriamanana *et al.*, 2015) and is linked with cross-tolerance to other abiotic and biotic stresses (e.g. drought, herbivory and pathogen infection. See Mewis *et al.*, 2012; Bandurska *et al.*, 2013; Zavala *et al.*, 2015).

Flavonoid biosynthesis is influenced by UV-B as well as UV-A, PAR (Flint *et al.*, 2004; Siipola *et al.*, 2015) and other environmental factors, such as temperature (Bilger *et al.*, 2007). Orchestration of UV-B-induced flavonoid biosynthesis appears to involve the UV-B photoreceptor UV RESISTANCE LOCUS 8

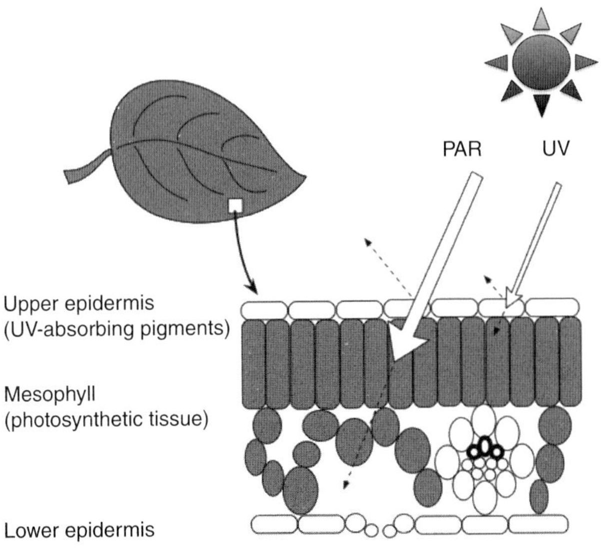

Fig. 6.2. Illustration of a typical dicot leaf showing the penetration of PAR (400–700 nm) and UV (280–400 nm) radiation through the epidermal tissue. The majority (c. 90%) of the incident UV is absorbed by flavonoids and other UV-absorbing compounds in the epidermis while most of the PAR is transmitted to the underlying mesophyll where photosynthesis occurs. The epidermal UV transmittance (T_{UV}) of the epidermis varies over time in relation to the stage of leaf development, UV and PAR exposure and, in some species, time of season, day and cloud cover.

(UVR8) (Jenkins, 2014) with UV-B exposure leading to the expression of UVR8-dependent gene transcripts involved in phenylpropanoid metabolism (Morales *et al.*, 2013). While exposure to UV-B has been shown to induce rapid (within minutes) activation of UVR8 (Kaiserli and Jenkins, 2007), the production and accumulation of UV-absorbing compounds and resultant decrease in T_{UV} typically occurs over much longer time frames (i.e. days. See Fig. 6.3a; Hectors *et al.*, 2014; Bidel *et al.*, 2015; Wargent *et al.*, 2015). During this period of acclimation, leaves may be susceptible to UV-induced injury (e.g. inhibition of photosynthesis. See Wargent *et al.*, 2015). Prolonged exposure to UV-B eventually leads to reduced sensitivity to UV and may even enhance photosynthetic performance and provide some protection against photoinhibition (Wargent *et al.*, 2011; Wargent *et al.*, 2015).

The epidermal UV transmittance of leaves is a highly plastic trait and within a given species can vary from <5% to near 100% depending upon the radiation environment (UV and PAR) experienced during leaf development (Fig. 6.3a,b). Thus, leaves that are produced in low UV and/or low PAR environments, such as those in deep shade in canopies or understorey environments, or in glasshouses that lack UV, typically display limited UV screening capabilities (Fig. 6.3a,b; Krause *et al.*, 2003a; Agati *et al.*, 2008; Pollastrini *et al.*, 2011; Barnes *et al.*, 2013). However, the relationship between UV shielding and the light (UV and PAR) regime is strongly non-linear such that leaves that develop under even rather moderate shade can still exhibit relatively high UV protection (low T_{UV}; Fig. 6.3a, inset; 6.3b). Leaves that do develop under deep shade or very low UV can be particularly susceptible to UV-induced injury if they suddenly encounter high UV environments. This would occur for certain vegetable crops that are propagated as seedlings in glasshouses or shade structures before being transplanted to the field (see Chapter 11, this volume), shade leaves within plant canopies that are exposed to full sunlight as a result of canopy gaps or perturbations due to herbivory, wind (Fig. 6.1d) or other factors (Kolb *et al.*, 2001, Barnes *et al.*, 2013), and understorey plants exposed to periodic sunflecks (Krause *et al.*, 2003b).

Responses to Sun–Shade Transitions

As indicated above, the ability of leaves to adjust their flavonoid levels and UV shielding in response to UV exposure during leaf development is well established. What is less clear, however, is how pliable these UV optical properties are once leaves have matured and developed under one set of conditions and are then confronted with rapid changes in these conditions. Some studies suggest that mature leaves or leaf segments are generally unresponsive in their UV-sunscreen capacities when exposed to sudden changes in environmental conditions such as temperature (Nybakken *et al.*, 2004b; Bilger *et al.*, 2007). Other studies however have shown that the UV-screening response can be more flexible in reaction to changes in the light environment (Krause *et al.*, 2007; Agati *et al.*, 2011; Morales *et al.*, 2011; Bidel *et al.*, 2015). For example, field studies with fully developed sun and shade leaves of *Populus tremuloides* (aspen) and *Vicia faba* (fava bean) have shown that there can be considerable flexibility in UV acclimation in mature leaves, at least for those produced under low light conditions (i.e. shade leaves, see Fig. 6.3c). Specifically, shade leaves were shown to decrease T_{UV} when transferred to sun environments, but sun leaves did not change their UV sunscreen protection when suddenly exposed to shade conditions (Barnes *et al.*, 2013). Although increases in UV shielding were detected 1–2 days following the transfer of plants from shade to full sun conditions, full transition of shade leaves to the equivalent protection of sun leaves required 4–10 days depending on species and conditions. These changes in leaf optical properties were generally associated with increases in UV-absorbing compounds. Similarly Krause *et al.* (2004) showed that mature leaves of tropical understorey plants can also respond to abrupt increases in solar UV radiation by increasing UV-absorbing compounds. In the case of *P. tremuloides*

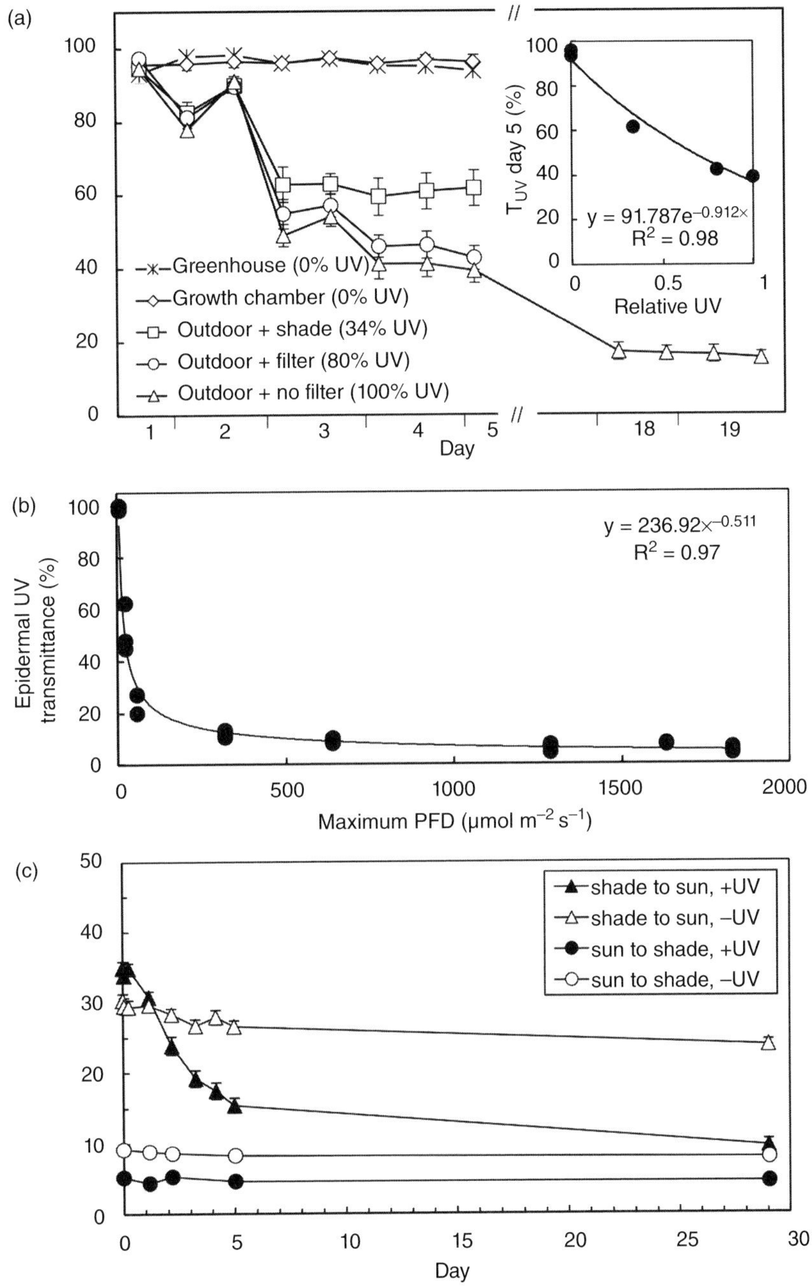

Fig. 6.3. The response of adaxial epidermal UV transmittance (T_{UV}; measured with a UVA-PAM) to different light environments in pot-grown *Arabidopsis thaliana* (a) and field-grown *Populus tremuloides* (aspen); (b), (c). (a) shows the time course of T_{UV} in *A. thaliana* grown in a growth chamber with no UV-B and then transferred to different conditions (indoor and outdoors) that differed in levels of UV-B (as measured with a calibrated broadband UV-B sensor). Inset shows the relationship between T_{UV} measured 5 days after transfer and relative UV irradiances. (b) shows the relationship between adaxial T_{UV} and midday photon flux density (PFD) of PAR (400–700 nm) in *P. tremuloides* leaves occupying different canopy environments. (c) shows the response of T_{UV} in sun and shade leaves of *P. tremuloides* subjected to different solar radiation treatments in the field. Data are means ± SE. (a) is from an unpublished study by Atunes, T., Tobler, M. and Barnes, P. (b) and (c) are reproduced from Barnes *et al.* (2013).

and *V. faba*, this adjustment in epidermal UV transmittance of shade leaves required both UV-B and UV-A, whereas PAR and UV were involved in the establishment of UV sunscreen protection during leaf development (Barnes *et al.*, 2013). Similarly, Bidel *et al.* (2015) showed that UV-B was required to induce additional UV shielding in mature leaves of *Centella asiatica*. Thus, relatively high levels of PAR appear to provide a high foundation of UV protection, which is further enhanced by subsequent UV exposure (Götz *et al.*, 2010; Wargent *et al.*, 2015).

Seasonal and Diurnal Changes in UV Sunscreen Protection

Occurrence and patterns among species and environments

The above studies indicate that mature leaves can, in some cases, respond, albeit relatively slowly, to sudden changes in their UV environment. Additionally, there is now growing evidence that fully developed leaves can modulate their levels of UV sunscreen protection on more rapid timescales such as one day to the next, or over the course of an individual day. The earliest report suggesting that plants may be capable of rapid adjustment in UV screening came from observations by Lautenschlager-Fleury (1955) who found that the UV-B transmittance of epidermal peels from *Vicia faba* leaves was low at midday on a sunny day but remained relatively high on a cloudy day. Subsequently, Sullivan *et al.* (2007) detected significant day-to-day changes in UV-absorbing compounds in field-grown *Hordeum vulgare* (barley) and *Glycine max* (soybean) and this variation was correlated with variation in ambient UV-B in both species and UV-A in soybean. This variability was dampened in plants grown under reduced (near 90% reduction) levels of UV-B and no significant relationship was found between variation in solar UV and UV-screening compounds in these plants. By comparison, Kotilainen *et al.* (2010) found no clear relationships between UV doses and seasonal variation in flavonoids and other phenolics in *Alnus incana* (alder) and *Betula pubescens* (birch) leaves. Other investigators have also shown that flavonoids and epidermal UV screening can vary over the course of a season (Fischbach *et al.*, 1999; Liakoura *et al.*, 2001; Louis *et al.*, 2009; Nenadis *et al.*, 2015) though it is sometimes difficult to assess whether this variation is the result of UV acclimation during leaf development as opposed to rapid adjustment in fully developed leaves, especially in species that produce long-lived leaves and multiple leaf cohorts/season.

The first evidence that plants could adjust levels of UV-screening compounds on a diurnal basis came from field studies conducted by Veit *et al.* (1996) who found that the concentration of flavonoids from whole-leaf extracts of *Anacardium excelsum*, a tropical tree, and *Cryptogramma crispa*, an alpine fern, increased progressively during the morning and then declined in a similar fashion in the afternoon in both species. More recently, Barnes *et al.* (2008) used chlorophyll fluorescence (UVA-PAM; Kolb *et al.*, 2005) to non-invasively measure epidermal UV-A shielding and reported small (c. 1–2% change in absolute T_{UV}; 13–16% change in relative T_{UV}), but statistically significant diurnal changes in epidermal UV transmittance in three plant species (*Vicia faba*, *Oenothera stricta* and *Verbascum thapsus*) growing in a high UV tropical alpine environment in Hawaii. Following this study, Barnes *et al.* (2016a) surveyed 37 species (63 taxa of wild and cultivated species with multiple cultivars/species for several cultivated species) growing in different locations (Hawaii, Utah, Idaho and Louisiana) and found that diurnal change in T_{UV} occurred in nearly half of the species examined. Diurnal changes in T_{UV} were found in plants at all locations, in monocots as well as dicots and in both herbaceous and woody growth forms. Species that did not exhibit diurnal change in T_{UV} included grasses and dicots of cultivated, wild and exotic species (e.g. *Symphoricarpus albus*, *Zea mays*, *Phalaris arundinacea* and *Malva parviflora*), and in some cultivated species (*Brassica rapa*, *Triticum aestivum*, and *Citrullus lanatus*) there was significant intraspecific variation in the prevalence of this phenomenon. Diurnal changes in flavonoid levels have also been observed in *Centaurea uniflora* and *Geum montanum*, two herbaceous alpine species growing in the French Alps, using similar non-invasive techniques (i.e. Dualex) (Goulas *et al.*, 2004; T.M. Robson and S.M. Hartikainen, unpublished data; Fig. 6.4). Thus, diurnal adjustment in UV sunscreen protection appears not limited to species in extreme UV environments but is in fact widespread among higher plants. However, at present there appears to be no apparent functional or ecological distinction between diurnally 'responsive' and 'unresponsive' species.

Results from the above multiple species survey (Barnes *et al.*, 2016a) further revealed that significant interspecific variation existed in the magnitude of these diurnal changes in epidermal UV transmittance. Certain species, such as *Abelmoschus esculentus* (okra) displayed large (62% relative change) decreases in T_{UV} from dawn to midday (absolute ΔT_{UV} = 16.1%; Fig. 6.5a) whereas others, such as *Typha latifolia*, showed minimal diurnal changes in T_{UV} over the day (ΔT_{UV} < 1%). Across taxa and locations, the largest diurnal changes in T_{UV} were found for plants growing in locations with warm nights (i.e. Louisiana). Low temperatures are known to induce the production of flavonoids (Neugart *et al.*, 2013) and decrease T_{UV} (Bilger *et al.*, 2007), such that many plants growing in cold, high-elevation or high-latitude sites often exhibit high constitutive levels of UV sunscreen protection (Barnes *et al.*, 2000; Nybakken *et al.*, 2004a; Albert *et al.*, 2009). The small diurnal changes in T_{UV} observed in plants from cooler locations (temperate latitudes and high elevations) may thus be a

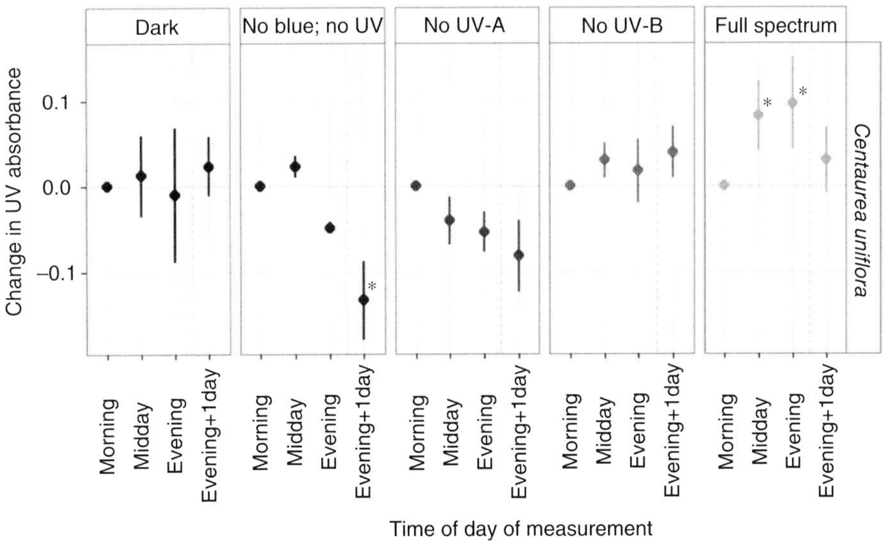

Fig. 6.4. The absorbance of UV at 375 nm by leaf adaxial epidermal flavonoids (absorbance units; AU) in *Centaurea uniflora* as measured using chlorophyll fluorescence (Dualex Scientific+, Force-A, Paris). Measurements were taken on plants growing in a subalpine meadow (2130 m asl) in the French Alps during July 2015. Cylindrical filters (220 mm diameter × 300 mm height) were made from special plastic films that completely attenuated blue, UV-A and UV-B radiation (Rosco 313 Canary Yellow; no blue no UV treatment); UV-A and UV-B radiation (Rosco #226 supergel, Westlighting, Helsinki, Finland; no UV-A; no UV-B treatment); and UV-B radiation (polyester, 0.125 mm thick, Autostat CT5; Thermoplast, Helsinki, Finland; no UV-B treatment). Plants filtered with polyethene (0.05 mm thick, 04 PE-LD; Etola, Jyväskylä, Finland) served as full-spectrum controls whereas plants in complete darkness (polyethene, solid white outside, solid black inside) served as dark controls. Data are means ± SE change in absorbance of each leaf from its initial value (n=5 per filter type). * = means within a filter treatment significantly different at P<0.05 based on ANOVA (T.M. Robson, and S.M. Hartikainen, unpublished data).

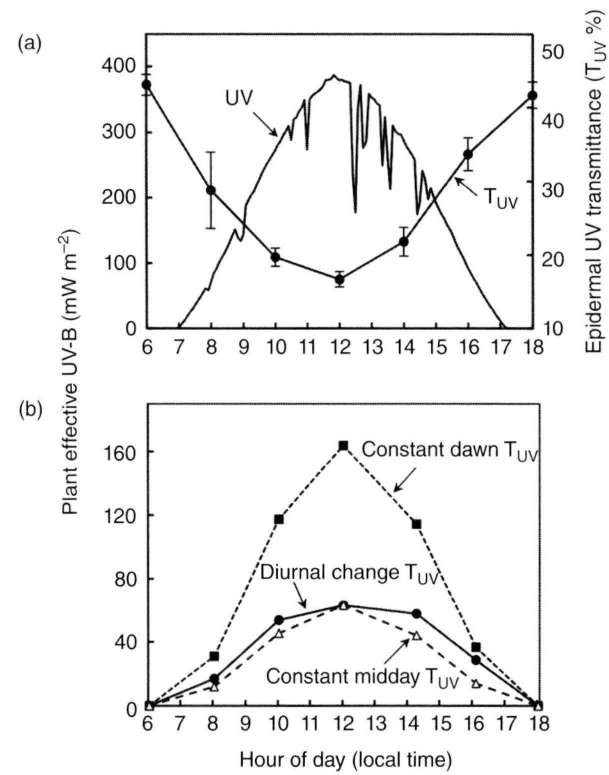

Fig. 6.5. Diurnal change in adaxial epidermal UV transmittance (T_{UV}; measured with a UVA-PAM) in *Abelmoschus esculentus* (okra) (a) and calculated plant effective UV-B irradiance directly beneath the epidermis (Panel (b) under near-clear summer sky conditions in New Orleans, Louisiana. T_{UV} data in (a) are means ± SE. (b) shows the diurnal course of calculated UV-B irradiance just beneath the adaxial (upper) epidermis for three scenarios: measured diurnal change in T_{UV} (circles), assuming constant dawn epidermal T_{UV} (squares) and assuming constant midday T_{UV} (triangles). Calculations assumed a horizontal leaf with a uniform epidermis and no qualitative change in the UV-absorbance spectrum. (Reproduced from Barnes *et al.*, 2015.)

consequence of low temperature effects on UV-absorbing compounds, which then restricts the range of diurnal transmittances possible in these species. Higher night-time temperatures may also allow for higher dark respiration rates, increased conversion rates of flavonoid structures, more rapid intercellular transport of flavonoids and/or other metabolic processes involving these compounds. Interestingly, across all taxa and locations the magnitude of ΔT_{UV} and the daily minimum value of T_{UV} (midday values) were not correlated with daily UV fluxes or the amplitude of diurnal changes in solar UV-B irradiance. The magnitude of the diurnal changes in epidermal UV transmittance therefore appears to be governed more by night-time values of T_{UV} than those at midday. Collectively, these findings suggest that the largest diurnal changes in T_{UV} will probably occur in plants inhabiting warm subtropical and tropical environments.

Mechanisms and drivers

Although diurnal adjustment in UV shielding appears common among plant species, the underlying mechanisms and drivers of these rapid responses remain unclear. Direct transmittance measurements made on epidermal peels of *V. faba* have confirmed that the rapid changes in T_{UV} measured using chlorophyll fluorescence are indeed the result of changes in the optical properties of the epidermis (Barnes *et al.*, 2016b). These diurnal alterations in T_{UV} are further associated with changes in the levels of whole-leaf UV-absorbing compounds in responsive species (e.g. *A. esculentus* and *Solanum lycopersicum*).

How plants actually achieve these rapid and reversible changes in UV screening and flavonoids is not known. As indicated above, the induction and accumulation of UV-absorbing compounds and resultant decreases in T_{UV} typically occur over timeframes (days) considerably longer than these rapid changes (Hectors *et al.*, 2014; Wargent *et al.*, 2015). The relocation of flavonoids among different pools in leaf tissues (Schnitzler *et al.*, 1996; Burchard *et al.*, 2000), the rapid alterations in absorptive properties of individual compounds (e.g. Dean *et al.*, 2014) and/or the UV-induced conversion of phenylpropanoid structures (e.g. Boulton, 2001) would seem more likely mechanisms to account for the rapid changes in these UV sunscreens than *de novo* synthesis and degradation of these compounds. Indeed, relatively rapid changes (<1 day) in the flavonoid composition of leaves have been observed in several plant species. For example, after one day of moderate UV-B exposure, Neugart *et al.* (2012) found that leaves of juvenile plants of *Brassica oleracea* showed a number of structurally dependant changes in flavonol (quercetin and kaempferol) glycosides, with some compounds increasing and others declining. Similarly, Barnes *et al.* (2016b) found significant diurnal shifts in the foliar composition of quercetin glycosides in *A. esculentus* with low molecular weight compounds changing to a greater degree than higher weight compounds. Because of the nature of these studies, there was no way to ascertain the precise cellular location of these compounds and thus it is unclear whether these compounds function primarily as UV sunscreens or as antioxidants.

It is conceivable that this diurnal change in UV shielding represents some aspect of an endogenous circadian rhythm in plants (McClung, 2001). Indeed, diurnal rhythms in gene expression, metabolites and the activities of key enzymes involved in phenylpropanoid biosynthesis (e.g. chalcone synthase) are known to occur (Peter *et al.*, 1991; Kim *et al.*, 2011), and the circadian clock in *Arabidopsis thaliana* appears to interact with UVR8-controlled UV-B signalling (Feher *et al.*, 2011). Whether these rapid UV-shielding responses are mediated, all or in part, by the UV-B photoreceptor UVR8 is unknown at present. Several studies have, however, demonstrated that manipulation of the light regime can significantly influence these diurnal changes. For example, Barnes *et al.* (2008) showed that the imposition of dense shade at different times of the day could effectively eliminate diurnal changes in T_{UV} in *Verbascum thapsus*, and conversely, the removal of shade caused T_{UV} to revert to 'normal' conditions within a matter of minutes. Subsequent studies with *A. esculentus* showed that the diurnal adjustment in T_{UV} in this species could be reduced by c. 50% when plants were placed under plastic film that attenuated much of the ambient solar UV (Barnes *et al.*, 2016a). In addition, the diurnal changes in UV shielding in this species and others are usually less pronounced on cloudy than sunny days, and for *A. esculentus* there is a strong negative correlation between T_{UV} during the day and incident UV-B and PAR when data from clear and cloudy days are combined (Barnes *et al.*, 2016b). Field filter experiments by Robson and Hartikainen (unpublished) have also shown that diurnal change in epidermal flavonoids in *Centaurea uniflora* was not detectable when plants were kept in continuous darkness or when UV-B or UV-B + UV-A was filtered out of sunlight (Fig. 6.4). Removing blue and UV radiation produced a diurnal pattern somewhat different from that of plants experiencing full sunlight. These findings are consistent with those of Veit *et al.* (1996) who reported that removal of solar UV-B eliminated the diurnal changes in flavonoids in their study species and with the studies of Sullivan *et al.* (2007) who found a significant positive correlation between day-to-day variation in UV-absorbing compounds and ambient solar UV-B. Thus, it appears that these rapid changes in leaf optical properties represent a specific response to sunlight (UV and perhaps visible) though some involvement with endogenous rhythms has not been ruled out (e.g. Atamian *et al.*, 2016).

Costs, benefits and ecological significance

Irrespective of mechanisms, these diurnal changes in UV shielding probably provide clear benefits to plants in UV protection, at least when compared to hypothetical situations where low predawn levels of UV shielding remain unchanged throughout the day (Fig. 6.5). However, the benefits of diurnal adjustment in UV protection relative to that of plants maintaining consistently high (i.e. midday equivalent) levels of UV protection over the day are less clear. In comparison to these kinds of plants (e.g. *Z. mays*), calculations (assuming a uniform epidermis) suggest that plants that exhibit diurnal adjustment in UV shielding (e.g. *A. esculentus*) may experience increased penetration of UV to the underlying mesophyll both in the morning and afternoon but not at midday. This is probably due to the fact that there is a time lag in T_{UV} responses to solar UV (i.e. T_{UV} does not change instantaneously in response to changes in UV). It is conceivable that increased penetration of UV to photosynthetic tissue at these times of day may protect leaves from photoinhibition (Wargent *et al.*, 2015) that can occur under high irradiances in the middle of the day. However, studies on *V. thapsus* have shown no correlation between T_{UV} and photochemical efficiency (F_v/F_m) when leaves encounter sudden changes in light regime (Barnes *et al.*, 2008). There is evidence that UV-A can drive photosynthesis (Turnbull *et al.*, 2013) and increased penetration of UV-A might thereby increase photosynthesis at times of the day when leaves are light-limited (mornings and afternoons). If this is the case, elimination or reduction in diurnal change in T_{UV} could potentially reduce plant carbon gain and growth. Finally, there is the possibility that maintaining constant high levels of flavonoids might interfere with plant growth during the night. Several of the flavonoids induced by UV (e.g. quercetin and kaempferol) are known to interfere with auxin metabolism and transport (Ringli *et al.*, 2008; Kuhn *et al.*, 2011), which in turn could influence plant growth and morphology (Hectors *et al.*, 2012). If this is the case, the benefits of diurnal cycling in UV shielding may lie not in UV protection but rather in other functional roles of flavonoids.

As an initial test to assess the costs and benefits of diurnal changes in UV sunscreen protection, we compared the effects of attenuating ambient solar UV on the growth and morphology of two species (*A. esculentus* and *Capsicum annuum*) that differed in their UV protection 'strategies'. Although statistically significant diurnal adjustments in T_{UV} were evident in both species, this change was much greater in *A. esculentus* than in *C. annuum* (Fig. 6.6a). In addition, overall levels of UV screening protection were higher and less affected by UV exclusion in *C. annuum* than *A. esculentus* (Fig. 6.6b). When comparing shoot growth and morphology, it appeared that *C. annuum* was influenced by UV exclusion to a greater degree than *A. esculentus* (Table 6.1). Thus, the species that was more flexible in its UV screening (*A. esculentus* in this case), was less affected in its growth by UV. Whether these differences in morphological sensitivity to UV are due to these differences in UV screening is difficult to assess, as these species probably differ in other aspects of UV protection (e.g. DNA damage repair, canopy architecture and others), which were not examined in this study. Nonetheless, these findings do suggest that, at the very least, there are no clear negative consequences in terms of UV effects on growth and production for species exhibiting diurnal changes in UV shielding. Additional study is needed to determine if there is any general association between flexibility in UV screening protection and sensitivity to UV-induced alterations in growth and morphology.

Diurnal fluctuations in UV sunscreen protection may also have consequences for the timing of plant responses to other abiotic and biotic stresses (e.g. drought and herbivory) that can vary in severity over the course of a day (e.g. Goodspeed *et al.*, 2012) and

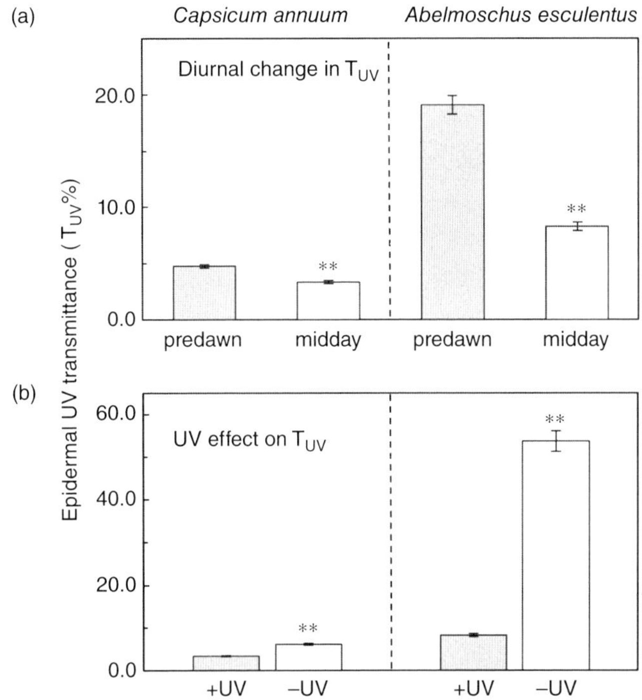

Fig. 6.6. Diurnal change (predawn to midday) in adaxial epidermal UV transmittance (T_{UV}; measured with a UVA-PAM) in pot-grown *Capsicum annuum* (pepper) and *Abelmoschus esculentus* (okra) (a) and midday T_{UV} in these species grown under ambient (+UV) and attenuated (-UV) solar UV (280–400 nm) in New Orleans, Louisiana (b). Data are means ± SE (n=24). ** = means within a species significantly different at $P<0.01$ based on ANOVA (I. Bottger, M. Tobler, and P. Barnes, unpublished data).

Table 6.1. Effects of ambient solar UV on the growth, morphology and midday epidermal UV transmittance (T_{UV}) in *Capsicum annuum* (pepper) and *Ablemoschus esculentus* (okra). Data are expressed as % difference of mean ambient – mean attenuated UV with ambient UV values as the base. Plants in the ambient UV treatment were grown under UV-transparent film (aclar) whereas plants in the attenuated UV treatment were grown under clear film that did not transmit UV-B or UV-A radiation (llumar). Significant treatment differences at $P<0.05$ (*), $P<0.01$ (**), and $P<0.001$(***) as determined by ANOVA; ns = not significant at $P>0.05$; n=24 per species and treatment (I. Bottger, M. Tobler, and P. Barnes, unpublished data).

	Capsicum annuum		*Abelmoschus esculentus*	
Parameter	% difference	*P*	% difference	*P*
Shoot height	–10.2	*	–4.4	ns
Internode length	–38.3	**	–16.8	*
Leaf area	–22.6	*	2.4	ns
Leaf dry mass	–17.4	ns	2.8	ns
Leaf mass/area	5.7	ns	1.3	ns
Shoot dry mass	–11.2	ns	5.0	ns
Midday T_{UV}	–19.2	***	–71.9	***

which, in some cases, employ similar suites of secondary compounds for both defence and UV protection (Kuhlmann and Müller, 2010; König *et al.*, 2014). For example, are plants that exhibit diurnal changes in flavonoids more susceptible to attack by herbivores or pathogens early and late in the day than in the middle of the day? No studies to date have explored this possibility, but understanding how UV protection interacts with these, and other, physiological and ecological functions may be required to fully evaluate the costs and benefits associated with 'static' vs. 'dynamic' UV protection strategies in plants.

Practical Considerations

The existence of a temporally dynamic UV sunscreen protection system in plants has a number of important practical implications for how plant UV research is conducted, and for culturing plants in controlled environments when UV-B (and UV-A) is employed to enhance food plant quality and vigour (Schreiner *et al.*, 2012; see also Chapter 11, this volume). For research aimed at quantifying the effects of UV radiation on flavonoids and UV protection it is important that the time of day (and time of year) when samples are collected or leaves are measured, is standardized and recorded. For field-grown plants, determination of the seasonality of flavonoid levels should be assessed and sample collections and measurements should then be done at midday during the seasonal peak to provide the best measure of maximal levels of UV protection. As noted by Kotilainen *et al.* (2010), this seasonal maximum may well differ for different species that are sympatric. Care should also be taken to collect samples/data under similar sky conditions.

When plants are grown in controlled environments (growth chambers or greenhouses) artificial sources of UV are typically employed (e.g. UV fluorescent bulbs or LEDs). While it is not yet clear whether plants grown in these environments exhibit 'typical' diurnal patterns in UV shielding, it is well known that plants are more susceptible to UV injury in these environments than those grown in the field (Caldwell and Flint, 1994). It is conceivable that this heightened sensitivity to UV, at least for some species, may well be the result of diminished UV sunscreen protection when natural diurnal cycles in UV radiation are muted or absent in controlled environments. In these environments it may be necessary to develop artificial lighting and UV exposure systems that promote natural diurnal (and seasonal) adjustments in UV screening so as to promote the beneficial effects of UV-B while avoiding excessive UV injury.

Summary and Conclusions

The leaf epidermis is a selective filter of solar radiation – absorbing much of the potentially detrimental solar UV while transmitting visible wavelengths (PAR) that drive photosynthesis in the underlying mesophyll. The epidermis is also a variable UV filter and the increase in the concentration of UV-absorbing compounds (flavonoids and related phenolics) and resultant decrease in epidermal UV transmittance represents a primary mechanism by which plants acclimate to changing UV environments. Understanding the nature and limitations of this acclimation response is fundamental to evaluating the ecologic and agronomic significance of variation in solar UV that occurs over multiple temporal scales (interannual, seasonal and diurnal).

Results from a number of relatively recent studies conducted on diverse cultivated and wild plant species reveal that this UV protective mechanism is much more dynamic and flexible than previously thought. While the mechanistic underpinnings of this temporal variation are relatively clear in some cases (e.g. the adjustments of UV shielding and flavonoid levels over ontogenetic time), the basis of other responses (e.g. diurnal changes in UV shielding), are less understood. To date, most studies on rapid responses to UV have focused on herbaceous plants that are adapted to high light environments (i.e. heliophytes); less attention has been given to exploring the temporal dynamics of UV protection in forest understorey species (i.e. sciophytes) that are adapted to survive in low light environments but which routinely encounter brief periods of intense UV and PAR in sunflecks. Additional studies are also needed to evaluate the adaptive significance of rapid modulation in UV shielding in plants and the potential implications of a dynamic UV protection system for other abiotic (drought, nutrient limitation and temperature extremes) and biotic factors (herbivore and pathogen protection) that are cross-linked to this UV protection response.

Acknowledgements

The research described here was supported, in part, by funds from the Louisiana Board of Regents (SURE/NSF grant LEQSF-EPS (2013)-SURE-85 to Loyola University), the US National Science Foundation (NSF Ecosystems Grant DEB-0815897 to Loyola), the US Department of Agriculture UV-Monitoring Program (USDA-CSREES No. 2004-34263-14270 to Utah State University via subcontract with Colorado State University), the Utah Agricultural Experiment Station, the Loyola University J.H. Mullahy Endowment in Environmental Biology, and the Academy of Finland decision # 266523 to TMR. We thank T. Atunes, A. Barkley, B. Burnet, B. Carneiro Alves, S. Gergen, M. Grabner, D. Hackenburg, S. Hartikainen, G. LeBleu and P. Pouta for their assistance. This chapter is dedicated to the memory of Ronald J. Ryel who died from complications arising from pancreatic cancer in October 2015 and who was instrumental in providing direction and expertise for a substantial amount of the research described here.

References

Agati, G., Cerovic, Z.G., Dalla Marta, A., Di Stefano, V., Pinelli, P., Traversi, M.L. and Orlandini, S. (2008) Optically-assessed preformed flavonoids and susceptibility of grapevine to *Plasmopara viticola* under different light regimes. *Functional Plant Biology* 35, 77–84.

Agati, G., Cerovic, Z.G., Pinelli, P. and Tattini, M. (2011) Light-induced accumulation of ortho-dihydroxylated flavonoids as non-destructively monitored by chlorophyll fluorescence excitation techniques. *Environmental and Experimental Botany* 73, 3–9.

Agati, G., Azzarello, E., Pollastri, S. and Tattini, M. (2012) Flavonoids as antioxidants in plants: location and functional significance. *Plant Science* 196, 67–76.

Albert, A., Sareedenchai, V., Heller, W., Seidlitz, H.K. and Zidorn, C. (2009) Temperature is the key to altitudinal variation of phenolics in *Arnica montana* L. cv. ARBO. *Oecologia* 160, 1–8.

Atamian, H.S., Creux, N.M., Brown, E.A., Garner, A.G., Blackman, B.K. and Harmer, S.L. (2016) Circadian regulation of sunflower heliotropism, floral orientation, and pollinator visits. *Science* 353, 587–590.

Bais, A.F., McKenzie, R.L., Bernhard, G., Aucamp, P.J., Ilyas, M., Madronich, S. and Tourpali, K. (2015) Ozone depletion and climate change: impacts on UV radiation. *Photochemical & Photobiological Sciences* 14, 19–52.

Ballaré, C.L., Caldwell, M.M., Flint, S.D., Robinson, S.A. and Bornman, J.F. (2011) Effects of solar ultraviolet radiation on terrestrial ecosystems: patterns, mechanisms, and interactions with climate change. *Photochemical & Photobiological Sciences* 10, 226–241.

Bandurska, H., Niedziela, J. and Chadzinikolau, T. (2013) Separate and combined responses to water deficit and UV-B radiation. *Plant Science* 213, 98–105.

Barnes, P.W., Searles, P.S., Ballaré, C.L., Ryel, R.J. and Caldwell, M.M. (2000) Non-invasive measurements of leaf epidermal transmittance of UV radiation using chlorophyll fluorescence: field and laboratory studies. *Physiologia Plantarum* 109, 274–283.

Barnes, P.W., Flint, S.D., Slusser, J.R., Gao, W. and Ryel, R.J. (2008) Diurnal changes in epidermal UV transmittance of plants in naturally high UV environments. *Physiologia Plantarum* 133, 363–372.

Barnes, P.W., Kersting, A.R., Flint, S.D., Beyschlag, W. and Ryel, R.J. (2013) Adjustments in epidermal UV-transmittance of leaves in sun-shade transitions. *Physiologia Plantarum* 149, 200–213.

Barnes, P.W., Flint, S.D., Ryel, R.J., Tobler, M.A., Barkley, A.E. and Wargent, J.J. (2015) Rediscovering leaf optical properties: new insights into plant acclimation to solar UV radiation. *Plant Physiology and Biochemistry* 93, 94–100.

Barnes, P.W., Flint, S.D., Tobler, M.A. and Ryel, R.J. (2016a) Diurnal adjustment in UV-sunscreen protection is widespread among higher plants. *Oecologia* 181, 55–63.

Barnes, P.W., Tobler, M.A., Keefover-Ring, K., Flint, S.D., Barkley, A.E., Ryel, R.J. and Lindroth, R.L. (2016b) Rapid modulation of ultraviolet shielding in plants is influenced by solar ultraviolet radiation and linked to alterations in flavonoids. *Plant, Cell & Environment* 39, 222–230.

Bidel, L.P.R., Meyer, S., Goulas, Y., Cadot, Y. and Cerovic, Z.G. (2007) Responses of epidermal phenolic compounds to light acclimation: In vivo qualitative and quantitative assessment using chlorophyll fluorescence excitation spectra in leaves of three woody species. *Journal of Photochemistry and Photobiology B:Biology* 88, 163–179.

Bidel, L.P.R., Chomicki, G., Bonini, F., Mondolot, L., Soulé, J. *et al.* (2015) Dynamics of flavonol accumulation in leaf tissues under different UV-B regimes in *Centella asiatica* (Apiaceae). *Planta* 242, 545–559.

Bilger, W., Rolland, M. and Nybakken, L. (2007) UV screening in higher plants induced by low temperature in the absence of UV-B radiation. *Photochemical & Photobiological Sciences* 6, 190–195.

Björn, L.O. (2015) On the history of phyto-photo UV science (not to be left in skoto toto and silence). *Plant Physiology and Biochemistry* 93, 3–8.

Bornman, J.F., Barnes, P.W., Robinson, S.A., Ballaré, C.L., Flint, S.D. and Caldwell, M.M. (2015) Solar ultraviolet radiation and ozone depletion-driven climate change: effects on terrestrial ecosystems. *Photochemical & Photobiological Sciences* 14, 88–107.

Boulton, R. (2001) The copigmentation of anthocyanins and its role in the color of red wine: a critical review. *American Journal of Enology and Viticulture* 52, 67–87.

Britt, A.B. (1999) Molecular genetics of DNA repair in higher plants. *Trends in Plant Science* 4, 20–25.

Burchard, P., Bilger, W. and Weissenböck, G. (2000) Contribution of hydroxycinnamates and flavonoids to epidermal shielding of UV-A and UV-B radiation in developing rye primary leaves as assessed by ultraviolet-induced chlorophyll fluorescence measurements. *Plant, Cell & Environment* 23, 1373–1380.

Caldwell, M.M. (1971) Solar UV irradiation and the growth and development of higher plants. *Photophysiology* 6, 131–177.

Caldwell, M.M. and Flint, S.D. (1994) Stratospheric ozone reduction, solar UV-B radiation and terrestrial ecosystems. *Climatic Change* 28, 375–394.

Caldwell, M.M., Robberecht, R. and Flint, S.D. (1983) Internal filters: prospects for UV-acclimation in higher plants. *Physiologia Plantarum* 58, 445–450.

Day, T.A., Vogelmann, T.C. and Delucia, E.H. (1992) Are some plant life forms more effective than others in screening out ultraviolet-B radiation? *Oecologia* 92, 513–519.

Day, T.A., Howells, B.W. and Rice, W.J. (1994) Ultraviolet absorption and epidermal-transmittance spectra in foliage. *Physiologia Plantarium* 92, 207–218.

Dean, J.C., Kusaka, R., Walsh, P.S., Allais, F. and Zwier, T.S. (2014) Plant sunscreens in the UV-B: ultraviolet spectroscopy of jet-cooled sinapoyl malate, sinapic acid, and sinapate ester derivatives. *Journal of the American Chemical Society* 136, 14780–14795.

Demkura, P.V. and Ballaré, C.L. (2012) UVR8 mediates UV-B-induced Arabidopsis defense responses against *Botrytis cinerea* by controlling sinapate accumulation. *Molecular Plant* 5, 642–652.

Emiliani, J., Grotewold, E., Falcone Ferreyra, M.L. and Casati, P. (2013) Flavonols protect Arabidopsis plants against UV-B deleterious effects. *Molecular Plant* 6, 1376–1379.

Favory, J.J., Stec, A., Gruber, H., Rizzini, L., Oravecz, A. *et al.* (2009) Interaction of COP1 and UVR8 regulates UV-B-induced photomorphogenesis and stress acclimation in Arabidopsis. *EMBO Journal* 28, 591–601.

Feher, B., Kozma-Bognar, L., Kevei, E., Hajdu, A., Binkert, M. *et al.* (2011) Functional interaction of the circadian clock and UV RESISTANCE LOCUS 8-controlled UV-B signaling pathways in *Arabidopsis thaliana. Plant Journal* 67, 37–48.

Fischbach, R.J., Kossmann, B., Panten, H., Steinbrecher, R., Heller, W. *et al.* (1999) Seasonal accumulation of ultraviolet-B screening pigments in needles of Norway spruce (*Picea abies* (L.) Karst.). *Plant, Cell & Environment* 22, 27–37.

Flint, S.D. and Caldwell, M.M. (1998) Solar UV-B and visible radiation in tropical forest gaps: measurements partitioning direct and diffuse radiation. *Global Change Biology* 4, 863–870.

Flint, S.D., Searles, P.S. and Caldwell, M.M. (2004) Field testing of biological spectral weighting functions for induction of UV-absorbing compounds in higher plants. *Photochemistry and Photobiology* 79, 399–403.

Goodspeed, D., Chehab, E.W., Min-Venditti, A., Braam, J. and Covington, M.F. (2012) *Arabidopsis* synchronizes jasmonate-mediated defense with insect circadian behavior. *Proceedings of the National Academy of Sciences of the United States of America* 109, 4674–4677.

Götz, M., Albert, A., Stich, S., Heller, W., Scherb, H. *et al.* (2010) PAR modulation of the UV-dependent levels of flavonoid metabolites in *Arabidopsis thaliana* (L.) Heynh. leaf rosettes: cumulative effects after a whole vegetative growth period. *Protoplasma* 243, 95–103.

Goulas, Y., Cerovic, Z.G., Cartelat, A. and Moya, I. (2004) Dualex: a new instrument for field measurements of epidermal ultraviolet absorbance by chlorophyll fluorescence. *Applied Optics* 43, 4488–4496.

Guidi, L., Degl'innocenti, E., Remorini, D., Biricolti, S., Fini, A. *et al.* (2011) The impact of UV-radiation on the physiology and biochemistry of *Ligustrum vulgare* exposed to different visible-light irradiance. *Environmental and Experimental Botany* 70, 88–95.

Hectors, K., Prinsen, E., Van Oevelen, S., Guisez, Y. and Jansen, M. (2009) Interactions between two UV-B protective responses: morphogenesis and UV screening. *Comparative Biochemistry and Physiology Part A: Molecular & Integrative Physiology* 153A, S202.

Hectors, K., Van Oevelen, S., Guisez, Y., Prinsen, E. and Jansen, M.A.K. (2012) The phytohormone auxin is a component of the regulatory system that controls UV-mediated accumulation of flavonoids and UV-induced morphogenesis. *Physiologia Plantarum* 145, 594–603.

Hectors, K., Van Oevelen, S., Geuns, J., Guisez, Y., Jansen, M.A.K. and Prinsen, E. (2014) Dynamic changes in plant secondary metabolites during UV acclimation in *Arabidopsis thaliana. Physiologia Plantarum* 152, 219–230.

Heisler, G.M., Grant, R.H. and Gao, W. (2003) Individual- and scattered-tree influences on ultraviolet irradiance. *Agricultural and Forest Meteorology* 120, 113–126.

Hofmann, R.W. and Jahufer, M.Z.Z. (2011) Tradeoff between biomass and flavonoid accumulation in white clover reflects contrasting plant strategies. *PloS ONE*, 6, e18949.

Izaguirre, M.M., Mazza, C.A., Svatos, A., Baldwin, I.T. and Ballaré, C.L. (2007) Solar ultraviolet-B radiation and insect herbivory trigger partially overlapping phenolic responses in *Nicotiana attenuata* and *Nicotiana longiflora. Annals of Botany* 99, 103–109.

Jacques, E., Hectors, K., Guisez, Y., Verbelen, J.P., Vissenberg, K., Prinsen, E. and Jansen, M. (2009) Leaf and cell development during UV-B acclimation in *Arabidopsis thaliana. Comparative Biochemistry and Physiology Part A: Molecular & Integrative Physiology* 153A, S203.

Jansen, M.A.K. and Bornman, J.F. (2012) UV-B radiation: from generic stressor to specific regulator. *Physiologia Plantarum* 145, 501–504.

Jenkins, G.I. (2014) The UV-B photoreceptor UVR8: from structure to physiology. *Plant Cell* 26, 21–37.

Kaiserli, E. and Jenkins, G.I. (2007) UV-B promotes rapid nuclear translocation of the *Arabidopsis* UV-B-specific signaling component UVR8 and activates its function in the nucleus. *Plant Cell* 19, 2662–2673.

Kim, S.-G., Yon, F., Gaquerel, E., Gulati, J. and Baldwin, I.T. (2011) Tissue specific diurnal rhythms of metabolites and their regulation during herbivore attack in a native tobacco, *Nicotiana attenuata. PLoS ONE* 6, e26214.

Kolb, C.A., Kaser, M.A., Kopecky, J., Zotz, G., Riederer, M. and Pfündel, E.E. (2001) Effects of natural intensities of visible and ultraviolet radiation on epidermal ultraviolet screening and photosynthesis in grape leaves. *Plant Physiology* 127, 863–875.

Kolb, C.A., Schreiber, U., Gademann, R. and Pfündel, E.E. (2005) UV-A screening in plants determined using a new portable fluorimeter. *Photosynthetica* 43, 371–377.

König, S., Feussner, K., Kaever, A., Landesfeind, M., Thurow, C., Karlovsky, P., Gatz, C., Polle, A. and Feussner, I. (2014) Soluble phenylpropanoids are involved in the defense response of *Arabidopsis* against *Verticillium longisporum. New Phytologist* 202, 823–837.

Kotilainen, T., Tegelberg, R., Julkunen-Tiitto, R., Lindfors, A., O'Hara, R.B. and Aphalo, P.J. (2010) Seasonal fluctuations in leaf phenolic composition under UV manipulations reflect contrasting strategies of alder and birch trees. *Physiologia Plantarum* 140, 297–309.

Krause, G.H., Galle, A., Gademann, R. and Winter, K. (2003a) Capacity of protection against ultraviolet radiation in sun and shade leaves of tropical forest plants. *Functional Plant Biology* 30, 533–542.

Krause, G.H., Grube, E., Virgo, A. and Winter, K. (2003b) Sudden exposure to solar UV-B radiation reduces net CO_2 uptake and photosystem I efficiency in shade-acclimated tropical tree seedlings. *Plant Physiology* 131, 745–752.

Krause, G.H., Grube, E., Koroleva, O.Y., Barth, C. and Winter, K. (2004) Do mature shade leaves of tropical tree seedlings acclimate to high sunlight and UV radiation? *Functional Plant Biology* 31, 743–756.

Krause, G.H., Jahns, P., Virgo, A., Garcia, M., Aranda, J., Wellmann, E. and Winter, K. (2007) Photoprotection, photosynthesis and growth of tropical tree seedlings under near-ambient and strongly reduced solar ultraviolet-B radiation. *Journal of Plant Physiology* 164, 1311–1322.

Kuhlmann, F. and Müller, C. (2010) UV-B impact on aphid performance mediated by plant quality and plant changes induced by aphids. *Plant Biology* 12, 676–684.

Kuhn, B.M., Geisler, M., Bigler, L. and Ringli, C. (2011) Flavonols accumulate asymmetrically and affect auxin transport in *Arabidopsis*. *Plant Physiology* 156, 585–595.

Lautenschlager-Fleury, D. (1955) Über die Ultraviolettdurchläessigkeit von Blattepidermen. *Berichte Schweiz Botanischen Gesellschaft* 65, 343–386.

Li, J., Yang, L., Jin, D., Nezames, C.D., Terzaghi, W. and Deng, X.W. (2013) UV-B-induced photomorphogenesis in Arabidopsis. *Protein & Cell* 4, 485–492.

Liakoura, V., Manetas, Y. and Karabourniotis, G. (2001) Seasonal fluctuations in the concentration of UV-absorbing compounds in the leaves of some Mediterranean plants under field conditions. *Physiologia Plantarum* 111, 491–500.

Lopez, M.L., Palancar, G.G. and Toselli, B.M. (2009) Effect of different types of clouds on surface UV-B and total solar irradiance at southern mid-latitudes: CMF determinations at Cordoba, Argentina. *Atmospheric Environment* 43, 3130–3136.

Louis, J., Meyer, S., Maunoury-Danger, F., Fresneau, C., Meudec, E. and Cerovic, Z.G. (2009) Seasonal changes in optically assessed epidermal phenolic compounds and chlorophyll contents in leaves of sessile oak (*Quercus petraea*): towards signatures of phenological stage. *Functional Plant Biology* 36, 732–741.

Madronich, S., Wagner, M. and Groth, P. (2011) Influence of tropospheric ozone control on exposure to ultraviolet radiation at the surface. *Environmental Science & Technology* 45, 6919–6923.

Mazza, C.A. and Ballaré, C.L. (2015) Photoreceptors UVR8 and phytochrome B cooperate to optimize plant growth and defense in patchy canopies. *New Phytologist* 207, 4–9.

Mazza, C.A., Boccalandro, H.E., Giordano, C.V., Battista, D., Scopel, A.L. and Ballaré, C.L. (2000) Functional significance and induction by solar radiation of ultraviolet-absorbing sunscreens in field-grown soybean crops. *Plant Physiology* 122, 117–125.

McClung, C.R. (2001) Circadian rhythms in plants. *Annual Review of Plant Physiology and Plant Molecular Biology* 52, 139–162.

Mewis, I., Schreiner, M., Chau Nhi, N., Krumbein, A., Ulrichs, C., Lohse, M. and Zrenner, R. (2012) UV-B irradiation changes specifically the secondary metabolite profile in broccoli sprouts: induced signaling overlaps with defense response to biotic stressors. *Plant and Cell Physiology* 53, 1546–1560.

Morales, L.O., Tegelberg, R., Brosché, M., Lindfors, A., Siipola, S. and Aphalo, P.J. (2011) Temporal variation in epidermal flavonoids due to altered solar UV radiation is moderated by the leaf position in *Betula pendula*. *Physiologia Plantarum* 143, 261–270.

Morales, L.O., Brosché, M., Vainonen, J., Jenkins, G.I., Wargent, J.J. *et al.* (2013) Multiple roles for UV RESISTANCE LOCUS8 in regulating gene expression and metabolite accumulation in *Arabidopsis* under solar ultraviolet radiation. *Plant Physiology* 161, 744–759.

Nenadis, N., Llorens, L., Koufogianni, A., Diaz, L., Font, J., Abel Gonzalez, J. and Verdaguer, D. (2015) Interactive effects of UV radiation and reduced precipitation on the seasonal leaf phenolic content/composition and the antioxidant activity of naturally growing *Arbutus unedo* plants. *Journal of Photochemistry and Photobiology B: Biology* 153, 435–444.

Neugart, S., Fiol, M., Schreiner, M., Rohn, S., Zrenner, R., Kroh, L.W. and Krumbein, A. (2013) Low and moderate photosynthetically active radiation affects the flavonol glycosides and hydroxycinnamic acid derivatives in kale (*Brassica oleracea* var. *sabellica*) dependent on two low temperatures. *Plant Physiology and Biochemistry* 72, 161–168.

Neugart, S., Zietz, M., Schreiner, M., Rohn, S., Kroh, L.W. and Krumbein, A. (2012) Structurally different flavonol glycosides and hydroxycinnamic acid derivatives respond differently to moderate UV-B radiation exposure. *Physiologia Plantarum* 145, 582–593.

Nybakken, L., Aubert, S. and Bilger, W. (2004a) Epidermal UV-screening of arctic and alpine plants along a latitudinal gradient in Europe. *Polar Biology* 27, 391–398.

Nybakken, L., Bilger, W., Johanson, U., Björn, L.O., Zielke, M. and Solheim, B. (2004b) Epidermal UV-screening in vascular plants from Svalbard (Norwegian Arctic). *Polar Biology* 27, 383–390.

Parisi, A.V. and Kimlin, M.G. (1999) Comparison of spectral biologically effective solar ultraviolet in adjacent tree and sun. *Physics in Medicine and Biology* 44, 2071–2080.

Peter, H.J., Krüger-Alef, C., Knogge, W., Brinkmann, K. and Weissenböck, G. (1991) Diurnal periodicity of chalcone-synthase activity during the development of oat primary leaves. *Planta* 183, 409–415.

Pollastrini, M., Di Stefano, V., Ferretti, M., Agati, G., Grifoni, D. *et al.* (2011) Influence of different light intensity regimes on leaf features of *Vitis vinifera* L. in ultraviolet radiation filtered condition. *Environmental and Experimental Botany* 73, 108–115.

Qi, Y., Heisler, G.M., Gao, W., Vogelmann, T.C. and Bai, S. (2010) Characteristics of UV-B radiation tolerance in broadleaf trees in southern USA. In: Gao, W., Schmoldt, D.L. and Slusser, J.R. (eds) *UV Radiation in Global Climate Change: Measurements, Modeling and Effects on Ecosystems.* Springer, Berlin, Germany.

Randriamanana, T.R., Nissinen, K., Moilanen, J., Nybakken, L. and Julkunen-Tiitio, R. (2015) Long-term UV-B and temperature enhancements suggest that females of *Salix myrsinifolia* plants are more tolerant to UV-B than males. *Environmental and Experimental Botany* 109, 296–305.

Ringli, C., Bigler, L., Kuhn, B.M., Leiber, R.M., Diet, A., Santelia, D., Frey, B., Pollmann, S. and Klein, M. (2008) The modified flavonol glycosylation profile in the *Arabidopsis rol*1 mutants results in alterations in plant growth and cell shape formation. *Plant Cell* 20, 1470–1481.

Rizzini, L., Favory, J.-J., Cloix, C., Faggionato, D., O'Hara, A. *et al.* (2011) Perception of UV-B by the *Arabidopsis* UVR8 protein. *Science* 332, 103–106.

Robson, T.M., Klem, K., Urban, O. and Jansen, M.A.K. (2015) Re-interpreting plant morphological responses to UV-B radiation. *Plant, Cell & Environment* 38, 856–866.

Schnitzler, J.P., Jungblut, T.P., Heller, W., Kofferlein, M., Hutzler, P. *et al.* (1996) Tissue localization of UV-B-screening pigments and of chalcone synthase mRNA in needles of Scots pine seedlings. *New Phytologist* 132, 247–258.

Schreiner, M., Mewis, I., Huyskens-Keil, S., Jansen, M.A.K., Zrenner, R. *et al.* (2012) UV-B-induced secondary plant metabolites – potential benefits for plant and human health. *Critical Reviews in Plant Sciences* 31, 229–240.

Searles, P.S., Flint, S.D. and Caldwell, M.M. (2001) A meta-analysis of plant field studies simulating stratospheric ozone depletion. *Oecologia* 127, 1–10.

Siipola, S.M., Kotilainen, T., Sipari, N., Morales, L.O., Lindfors, A.V., Robson, T.M. and Aphalo, P.J. (2015) Epidermal UV-A absorbance and whole-leaf flavonoid composition in pea respond more to solar blue light than to solar UV radiation. *Plant, Cell & Environment* 38, 941–952.

Snell, K.R.S., Kokubun, T., Griffiths, H., Convey, P., Hodgson, D.A. and Newsham, K.K. (2009) Quantifying the metabolic cost to an Antarctic liverwort of responding to an abrupt increase in UVB radiation exposure. *Global Change Biology* 15, 2563–2573.

Sullivan, J.H., Gitz, D.C., Liu-Gitz, L., Xu, C.P., Gao, W. and Slusser, J. (2007) Coupling short-term changes in ambient UV-B levels with induction of UV-screening compounds. *Photochemistry and Photobiology* 83, 863–870.

Thiel, S., Steiner, K. and Seidlitz, H.K. (1997) Modification of global erythemally effective irradiance by clouds. *Photochemistry and Photobiology* 65, 969–973.

Tilbrook, K., Arongaus, A.B., Binkert, M., Heijde, M., Yin, R. and Ulm, R. (2013) The UVR8 UV-B photoreceptor: perception, signaling and response. *The Arabidopsis Book* 11, e0164.

Turnbull, T.L., Barlow, A.M. and Adams, M.A. (2013) Photosynthetic benefits of ultraviolet-A to *Pimelea ligustrina*, a woody shrub of sub-alpine Australia. *Oecologia* 173, 375–385.

Veit, M., Bilger, T., Muhlbauer, T., Brummet, W. and Winter, K. (1996) Diurnal changes in flavonoids. *Journal of Plant Physiology* 148, 478–482.

Wargent, J.J., Elfadly, E.M., Moore, J.P. and Paul, N.D. (2011) Increased exposure to UV-B radiation during early development leads to enhanced photoprotection and improved long-term performance in *Lactuca sativa*. *Plant, Cell & Environment* 34, 1401–1413.

Wargent, J.J. and Jordan, B.R. (2013) From ozone depletion to agriculture: understanding the role of UV radiation in sustainable crop production. *New Phytologist* 197, 1058–1076.

Wargent, J.J., Nelson, B.C.W., McGhie, T.K. and Barnes, P.W. (2015) Acclimation to UV-B radiation and visible light in *Lactuca sativa* involves up-regulation of photosynthetic performance and orchestration of metabolome-wide responses. *Plant, Cell & Environment* 38, 929–940.

Williamson, C.E., Zepp, R.G., Lucas, R.M., Madronich, S., Austin, A.T. *et al.* (2014) Solar ultraviolet radiation in a changing climate. *Nature Climate Change* 4, 434–441.

Xie, L. and Hauser, M.-T. (2012) Induction of ARI12 upon broad band UV-B radiation is suppressed by UVR8 and cryptochromes. *Plant Signaling & Behavior* 7, 1411–1414.

Zavala, J.A., Mazza, C.A., Dillon, F.M., Chludil, H.D. and Ballaré, C.L. (2015) Soybean resistance to stink bugs (*Nezara viridula* and *Piezodorus guildinii*) increases with exposure to solar UV-B radiation and correlates with isoflavonoid content in pods under field conditions. *Plant, Cell & Environment* 38, 920–928.

7 The Effects of UV-B on the Biochemistry and Metabolism of Plants

Éva Hideg[1] and Åke Strid[2]*
[1]*Institute of Biology, University of Pécs, Ifjúság útja 6, H-7624 Pécs, Hungary;*
[2]*School of Science and Technology, Örebro University, SE-70182 Örebro, Sweden*

This chapter is dedicated to Prof. Jan M. Anderson (1932–2015) and to her lifetime achievements in photosynthesis and plant biology

List of Abbreviations

4CL: 4-coumarate-CoA ligase; 6,4PP: 6,4 pyrimidine-pyrimidone photoproduct; ANS: anthocyanidin synthase; BR: brassinosteroids; C3H: 4-coumarate-3-hydroxylase; C4H: cinnamate 4-hydroxylase; CHI: chalcone isomerase; CHS: chalcone synthase; COMT: caffeic acid *O*-methyltransferase; COP1: constitutively photomorphogenic 1; CPD: cyclobutane pyrimidine dimer; DFR: dihydroflavonol-4-reductase; ELIP: early light-inducible proteins; F3′,5′H: flavonoid 3′,5′-hydroxylase; F3H: flavanone 3-hydroxylase; F3′H: flavonoid 3′-hydroxylase; F5H: ferulate 5-hydroxylase; FLS: flavonol synthase; GA: giberellic acid; HR: homologous recombination; HY5: ELONGATED HYPOCOTYL5; HYH: HY5 HOMOLOG; MAPK: mitogen-activated protein kinase; NIC: nicotinamide; NO•: nitric oxide radical; PAL: phenylalanine ammonia-lyase; PAR: photosynthetically active radiation; PARP: poly(ADP-ribose)polymerase; PSII: photosystem II; RING: really interesting new gene; ROS: reactive oxygen species; SOD: superoxide dismutase; SSB: single strand break; TAL: tyrosine ammonia-lyase; TT: transparent testa; UFGT: UDP-flavonoid glucosyltransferase; UV-A: 315–400 nm ultraviolet radiation; UV-B: 280–315 nm ultraviolet radiation; UVI4: UV-B INSENSITIVE 4; UVR8: UV RESISTANCE LOCUS8.

UV-B in Plant Photobiology: Basic Concepts and Rules of Thumb

Plants perceive and integrate various stimuli from the environment and respond in a way that optimizes metabolism to the given conditions. Self-evidently, a large number of different biotic and abiotic factors influence this response, including light quality and light quantity. Also, since some of these factors may change rapidly, such as light intensity (shade/sun) or mechanical strain (wind speed and direction), there must be a temporal dampening of the responses to prevent wasting metabolic energy. In addition, additive, synergistic and antagonistic responses to combinations of environmental cues may

*Address for correspondence: ake.strid@oru.se

occur. Therefore it is obvious that the plants' responses to UV-B, UV-A and visible light are influenced by each other through the specific light receptors (i.e. UVR8, phototropins, cryptochromes, zeitlupes, phytochromes) that are responsible for absorption at the different wavelengths, and through subcellular signal integrating components, such as the Constitutively Photomorphogenic 1 (COP1) RING E3 ubiquitin ligase, which targets key transcription regulators for degradation by the ubiquitin-proteasome system (Lau and Deng, 2012) and ultimately influences gene expression, metabolism and morphology.

In the scenarios above, environmental stimuli vary within limits that are physiologically and evolutionarily sound for plants to be adapted to, and to which acclimation occurs. For such events, specific regulatory mechanisms have evolved (e.g. the UVR8 pathway for UV-B). More non-specific damaging alterations in metabolism may occur as an effect of rare catastrophic events, uncommon combination of stresses, or situations where plants are exposed to new anthropogenic environmental factors. For the case with UV-B as an environmental factor, non-specific damage may occur either under combination of UV-B with other stresses (Czégény *et al.*, 2014), or with individual plant species in extreme environments (such as in the Arctic or in monocultures of agricultural plants in the tropics), or even in inbred varieties of agricultural species where stress acclimation traits may have disappeared over breeding cycles. For such cases, severe oxidative stress ('distress'; Hideg *et al.*, 2013) and DNA damage (Teranishi *et al.*, 2004) are examples of primary outcomes.

Also, in cases where the natural environment has been exchanged for an artificial environment, acclimation processes may be disturbed. This includes plants grown in greenhouses for horticultural or ornamental purposes, post-harvest treatment of plants or plant tissue, or scientific experiments under light conditions that are skewed with respect to wavelength and intensities in comparison with natural conditions. Notwithstanding, such conditions may have high relevance from economical, biotechnological or mechanistic/scientific points of view, thus motivating research efforts that also may offer important clues to how plants perceive their environment and regulate their responses. However, this also means that the invoked responses may not apply under any other conditions than those used for the particular experiment, i.e. it is far from safe to draw any general conclusions about plant UV-B responses in an ecologically or agriculturally relevant fashion from experiments carried out in artificial environments.

From a plant UV-B photobiology perspective, it is particularly important to pay attention to the following: (i) the spectrum of the light source used to expose the plants, or the absorptive properties of the cladding material used in UV-B exclusion experiments (Aphalo *et al.*, 2012a); (ii) the duration and intensity of the UV-B used (maximally 4 hours UV per day centred around the solar noon should be used in experiments that are aimed at mimicking natural conditions); (iii) the intensity of the concomitant photosynthetically active radiation (PAR) used in the experiments; and (iv) thus, in particular, both the UV-B/PAR ratio and the PAR intensity may be of importance in relation to outdoor conditions; (v) other environmental stresses that might be inflicted on the plants during UV-B experiments; (vi) the spatial location of the experimental plants compared with the control plants so that the necessary repeats and relocations are performed to rule out positional effects. The UV-B/PAR ratio is not only of importance if conclusions from experiments are to be drawn about the situation outdoors. The UV-B/PAR ratio and the PAR intensity are also important factors for the interplay between different photoreceptors and their signal transduction pathways in eliciting an integrated response.

Thus, when reading, analysing, and judging plant UV-B photobiology literature, or performing studies on the effects of UV-B on plants and on the plant responses induced by UV-B, these aspects are important to bear in mind. For experimentalists new to the field we also recommend reading of a recent best practice handbook on UV photobiology edited by Aphalo *et al.* (2012b; see also Chapter 2, this volume).

Overview of UV-B Effects on Plants and Protective Responses

The high energy content of UV-B (280–315 nm), the highest of all wavelengths of the sun that penetrates the atmosphere of the earth, and the fact that many biomolecules that contain double bonds and delocalized π electronic systems can absorb radiation in the UV region, mean that UV-B radiation can be harmful to cells and organisms including plants. Therefore, high enough doses of UV-B have the potential to affect proteins, lipids, and nucleic acids by, for instance, knocking electrons out of their orbitals with chemical bond breakage or radical formation as the result. New chemical bonds can be formed leading to inter- or intra-crosslinking of molecules such as proteins and nucleic acids or exponential growth of the number of oxygen-centred radicals leading to lipid peroxidation. Eventually, non-functional proteins, leaky membranes, mutated DNA, and ribotoxicity (i.e. non-functional translation due to malfunctioning ribosomes) would be the result (Fig. 7.1).

In order to protect against catastrophic UV-B-induced events like this, plants have evolved a number of protective mechanisms to be able to cope with daily exposures to the sun's radiation. The first measures of protection include morphological alterations in leaf area and leaf thickness to optimize exposure, formation of UV-absorbing features at the leaf surfaces (wax layers or trichomes) that would reflect or diffract the radiation, and accumulation in the vacuoles of the epidermal cell layers of UV-absorbing pigments such as flavonoids, to minimize penetration of UV-B to deeper photosynthetically active cell layers. In addition, as a second measure of defence, enzymatic and small molecule antioxidant systems have evolved to neutralize the oxygen-centred free radicals and lipid peroxides that still can be formed by the low levels of UV-B that have not been removed by the first layer of protection. The second layer of protective strategies also

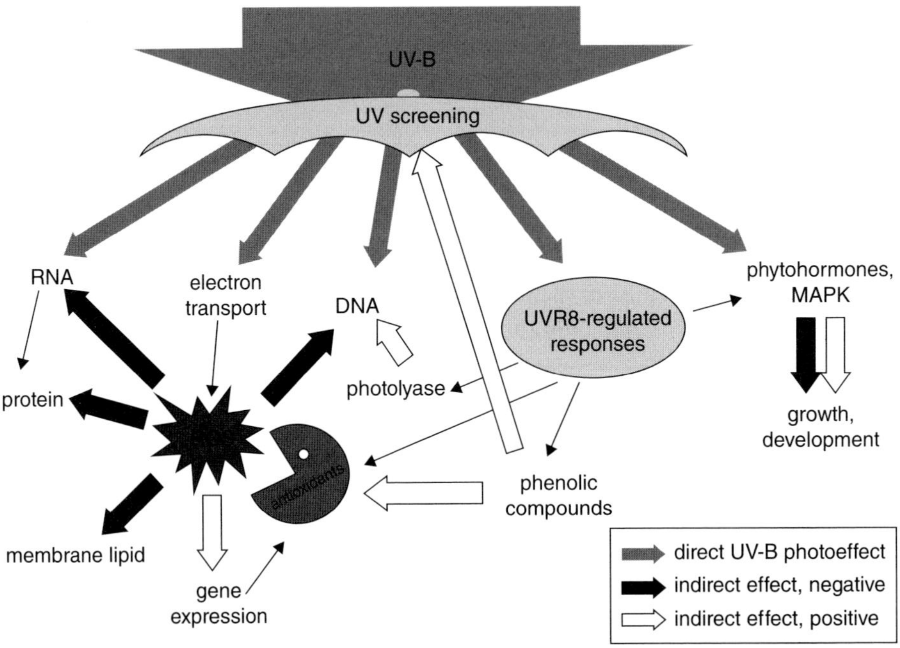

Fig. 7.1. A schematic representation of major UV-B induced pathways in plants after UV screening has reduced the penetration of UV into lower lying tissue. MAPK, mitogen-activated protein kinase; ROS, reactive oxygen species.

includes DNA repair systems, protein degradation systems to remove damaged polypeptides, and mechanisms for removal of faulty ribosomes or mRNAs.

This chapter deals with the research that has been carried out to understand some of the biochemical effects of UV-B and how the corresponding protective measures in plants function and are regulated.

UV Effects on Photosynthesis

Possible direct effects on electron transport/ proton motive force/photophosphorylation

There is a consensus in the scientific community about the fact that direct whole plant effects of UV-B on the components of the photosynthetic electron transport chain are rarely seen under natural conditions, be it in the agricultural context or in wild plants in ecologically relevant settings (Allen *et al.*, 1998). Notwithstanding, it has been shown in a large number of studies that different co-factors of photosystem II (PSII) are possible targets for UV-B. This includes the Mn cluster of the oxygen evolving complex, crucial tyrosine residues on the donor side and quinone acceptors, eventually giving rise to increased degradation of the core D1 and D2 proteins (see Jordan *et al.* (2016) for a recent review).

However, most of these deleterious events have been shown either in subfractionated leaf material (primarily in the case of the PSII studies) where both UV-B absorbing pigments and antioxidative components had been removed, or after acute exposure of non-acclimated plants, where the appropriate defence responses had not been fully developed beforehand. In general, the UV-B doses used in these studies, and that therefore were needed to obtain the effects, were well above those that would be expected in the natural environment, (with or without simulated ozone depletion). It is therefore doubtful if UV-B-induced lesions in the light and dark reactions of photosynthesis are significant from a physiological perspective.

Studies on membrane effects (partial uncoupling of thylakoid membranes measured as UV-B-dependent increases in the rates of relaxation of the single turnover flash-induced electrochromic shifts of thylakoid pigments) were also carried out in non-acclimated plants (Chow *et al.*, 1992; van Hasselt *et al.*, 1996; Strid *et al.*, 1996a, b). However, to obtain this effect the doses needed were lower than ambient and could be envisaged to occur under natural conditions (Strid *et al.*, 1994). Another UV-B effect that is the result of membrane permeabilisation is inhibition of the down-regulation of chloroplast diatoxanthin epoxidation that is normally controlled by the high ΔpH in the diatom *Phaeodactylum tricornutum*. Therefore, an increased rate of diatoxanthin epoxidation was seen (Mewes and Richter, 2002). However, what the physiological significance of this thylakoid membrane permeabilisation is under UV-B in higher plants (or under photoinhibitory conditions in white light for that matter; Tjus and Andersson, 1993), remains to be investigated. Interestingly, UV-B inhibits chilling-induced increases in relaxation of electrochromic shifts, possibly due to modification of membrane lipids, that leads to thylakoid leakiness at 22°C, but rather to increased fluidity from a crystalline state at 2°C (Strid *et al.*, 1996b). Also, the partial membrane uncoupling is independent of oxygen concentration and can therefore not easily be linked to a UV-B-dependent photooxidative process (van Hasselt *et al.*, 1996).

In addition to the activities and abundance of the ATP synthase being affected negatively by UV-B (Zhang *et al.*, 1994), a plasma membrane-associated K^+-ATPase isolated from cultured rose cells was shown to be sensitive to UV irradiation and the action spectrum of inhibition peaked in UV-B (Imbrie and Murphy, 1982). Moreover, a number of studies on phytoplankton species indicate major changes in the balance of the adenylate pool during UV-B exposure (summarized in Döhler *et al.*, 1997).

Direct effects on stomata opening and Calvin–Benson cycle enzymes

Photosynthesis is critically affected by stomatal movements controlling CO_2 influx. Stomatal opening and closing are regulated by a number of environmental factors (reviewed by Araújo *et al.*, 2011). Reports on UV-B effects on stomatal conductance are controversial as they include both negative and positive effects. For example, Martínez-Lüscher *et al.* (2013) found that supplemental UV-B radiation decreased stomatal conductance, and sub-stomatal CO_2 concentration in grapevine (*Vitis vinfera* cv. Tempranillo) leaves. Using UV-transmitting and UV-blocking films, Roro *et al.* (2016) found that solar UV radiation at high altitudes in the tropics increased stomata conductance in pea (*Pisum sativum*) leaves. The differences may be attributed to variations in UV-B fluence rates or diverse UV sensitivities of various species, but also signify the complexity of UV effects on stomata. Stomata opening largely depends on environmental factors other than UV, and Jansen and van den Noort (2000) showed that once *Vicia faba* stomata were exposed to UV-B, they were unable to re-adjust to changes in light or humidity.

In *Arabidopsis thaliana* low UV-B irradiance (0.18 μmol m^{-2} s^{-1} photon flux) was shown to induce stomata opening (Eisinger *et al.*, 2000), but higher (1.5–5.5 μmol m^{-2} s^{-1}) fluxes resulted in closure (He *et al.*, 2013; Tossi *et al.*, 2014). Two, possibly interconnected signalling pathways were identified in *Arabidopsis*. One involves GPA1, the Gα-subunit of an heterotrimeric G protein capable of eliciting H_2O_2 and subsequent nitric oxide radical ($NO^•$) production (He *et al.*, 2013). Another study demonstrated that the UV-B photoreceptor UVR8 and its signalling components COP1, HY5, and HYH are required to modulate UV-B-induced stomatal closure through H_2O_2 and $NO^•$ (Tossi *et al.*, 2014). The exact mechanism by which UV-B-dependent $NO^•$ induces stomatal closure is still unknown but an activation of ion channels via $NO^•$-induced cytoskeleton modulation has been hypothesised (He *et al.*, 2013).

UV-B was shown to affect not only the uptake of CO_2 through the stomata but also its processing in the Calvin–Benson cycle. Carboxylation is affected via UV-B-induced decline in Rubisco (ribulose-1,5-bisphosphate carboxylase) content or activity. Both mRNA levels and amounts of encoded protein subunits decreased in response to supplementary UV-B (Jordan *et al.*, 1992). Similarly to negative effects of low temperature or ozone, UV-B caused a decrease in the amount of Rubisco, as opposed to drought or high temperature that resulted in a down-regulation of the enzyme (Galmés *et al.*, 2013).

Nuclear-encoded genes were in general found to be more sensitive to UV-B than chloroplast-encoded ones (Jordan, 1996). Accordingly, a comparison of UV-B-induced changes in nuclear *rbcS* and chloroplast encoded *rbcL* genes (encoding the small and large subunits of Rubisco, respectively) in pea leaves showed a faster decrease in the former. However, the amounts of the corresponding Rubisco proteins did not follow the same order of degradation, and loss of chloroplast encoded large subunit was even observed before any decrease in rbcL mRNA levels (Mackerness *et al.*, 1997). Specifically, down-regulation of Rubisco is due to overall decreased carboxylation activity despite increased *in vivo* activation of the enzyme (Strid *et al.*, 1990), probably by a diminished pool of functional protein as a result of UV-B-induced crosslinking between small and large subunits: Wilson *et al.* (1995) showed that the 54 kDa large subunit could change into a 66 kDa protein by oxidative modification under UV-B radiation in a variety of C3 plants, and the effect was later attributed to tryptophan photo-oxidation (Gerhardt *et al.*, 1999), resulting in cross-linking between subunits (Ferreira *et al.*, 1996).

In addition to lowering carboxylation, UV-B also affects the regeneration of RuBP in the Calvin–Benson cycle. Allen *et al.* (1998) showed that UV-B irradiation induced a reduction in seduloheptulose 1,7-bisphosphatase and assumed that this was causing decreased RuBP regeneration in oilseed rape (*Brassica napus*) leaves. The authors observed this in the absence of any significant decrease in PSII photochemistry, and proposed the dark reactions of photosynthesis as the primary site of UV-B damage.

Indirect effects via reactive oxygen species

In leaves, the photosynthetic apparatus is a major source of various reactive oxygen species (ROS). Energy transduction from triplet chlorophylls to oxygen yields singlet oxygen, and electron transport may be diverted to oxygen to form the type-II photodynamic products superoxide radicals, hydrogen peroxide and hydroxyl radicals. These reactions can be brought about by a variety of stress conditions inhibiting or limiting photosynthetic electron transport (Asada, 2006). The threshold UV-B fluence rates that initiate photoreceptor-induced responses appear much lower than those that result in ROS production (Brosché and Strid, 2003; Jenkins, 2009).

A unique aspect of UV-B is that it has the capacity for photocleavage of hydrogen peroxide into hydroxyl radicals (Czégény *et al.*, 2014), which may contribute to the damaging effect of UV-B. Hydroxyl radicals are short-lived ROS that react at the site of their production, whereas H_2O_2 is capable of diffusing through membranes (Cheeseman, 2006) and thus may initiate oxidative damage via hydroxyl radicals upon absorption of a UV-B photon relatively far from its own production site. Hydroxyl radicals were identified in isolated chloroplasts exposed to high intensity UV-B (Hideg and Vass, 1996). Limited sensitivities of detection techniques and the presence of antioxidants did not allow direct detection of hydroxyl radicals in leaves exposed to more realistic UV-B fluxes. So far, the observed UV-B-induced increases in hydroxyl radical reactive antioxidants in *Arabidopsis thaliana* (Matxain *et al.*, 2009; Ristilä *et al.*, 2011;) and *Nicotiana tabacum* (Majer *et al.*, 2014a) leaves in response to supplementary UV-B radiation serve as indirect evidence.

Several studies demonstrated the potential vulnerability of both light- and dark-reactions of photosynthesis to ROS. Hydrogen peroxide and hydroxyl radicals were demonstrated to initiate the degradation of D1 protein in isolated PSII membranes in the absence of irradiation cleavage (Miyao *et al.*, 1995). The large subunit of Rubisco was shown to be prone to oxidation by ROS and two distinct pathways were suggested, both based on experiments including incubation with hydroxyl radical yielding Fenton reagents. One study showed that ROS first modified the protein, making it susceptible to proteolysis (Desimone *et al.*, 1998) and another demonstrated the possibility of direct fragmentation of the large subunit by hydroxyl radicals (Ishida *et al.*, 1999).

However, the above mentioned potential for UV-B-induced ROS does not appear to manifest as oxidative damage in intact plants exposed to solar UV-B (Allen *et al.*, 1998). Nevertheless, damaging effects of UV-B inducible ROS that are observed in laboratory experiments may be relevant to whole plant studies under natural conditions if the antioxidant network and/or repair systems are overwhelmed by the presence of additional stressors.

UV Effects on the Regulation of Photosynthesis

Antenna effects, non-photochemical quenching and ELIPs

Protection of the photosynthetic apparatus from oxidative stress includes efficient dissipation of excess excitation energy. Plants developing in sunlight are exposed to both PAR and UV, and UV-inducible metabolites may facilitate defence against photoinhibition by PAR. Experiments with barley plants acclimated to supplemental UV-B showed that UV-exposure contributes to the induction of tolerance to high PAR (Klem *et al.*, 2015). Repeated exposures of beech saplings over three growing seasons to 25% enhanced UV-B irradiation increased the ratio of zeaxanthin to the total xanthophyll pool and facilitated the non-radiative dissipation in the light harvesting complex (Šprtová *et al.*, 2003).

The early light-inducible proteins (ELIPs) are expressed in the thylakoid membrane during the greening processes of etiolated plants. ELIPs were shown to protect PSII against photoinhibition (Adamska and Kloppstech, 1991) and are also responsive to low UV-B fluxes (Sävenstrand *et al.*, 2004a). In fact, the ELIP gene is one of the few genes encoding a plastid-localised protein that is actually induced during UV-B exposure in pea. The large majority of genes encoding photosynthesis-related protein, independently of whether they are encoded in the plastid genome or in the nuclear genome, are down-regulated by UV-B. The induction of the ELIP gene thus most likely reflects the protective function of its corresponding protein.

UV-B induces the biosynthesis of flavonoids via the UVR8 photoreceptor regulation of gene expression, where the flavonoid biosynthesis genes are a few among a large number of up-regulated genes. Flavonoids, especially UV-B responsive flavonols, are efficient antioxidants *in vitro* (Rice-Evans *et al.*, 1995). Since flavonoids are also localized in chloroplasts (Saunders and McClure, 1976) it is assumed that they may contribute to keeping the plastid ROS concentrations low and thus to prevent photo-oxidative damage (Majer *et al.*, 2014b). In contrast, a high demand for UV-screening and antioxidative secondary metabolites may divert carbon from other biosynthetic pathways, as suggested by experiments using two Mediterranean species: comparing metabolites in sun-exposed and UV-filtered leaves of *Ligustrum vulgare* and *Phillyrea latifolia*, Guidi *et al.* (2016) found an increase in flavonoid concentrations and a decrease in the amounts of carotenoids, particularly of xanthophylls.

The 'PAR effect'

A very interesting regulatory aspect that has bearing on whether UV-B will act as a 'eustress' or a 'distress' in plants (Hideg *et al.*, 2013), is the concomitant or historical PAR level that the plants have been exposed to before and/or during UV-B exposure (Jordan *et al.*, 2016). Although firm overall conclusions on this are hard to draw due to very different UV sources, UV doses, and PAR levels used in the studies that have been reported in the literature, a high PAR level and/or a low UV-B/PAR ratio in general shifts the overall effect of UV-B on plants towards 'eustress' and a function of UV-B as a photomorphogenic regulator instead of a deleterious cue. Several aspects of this 'PAR effect', for instance of protection of both photosynthesis and growth against UV-B impact, have been reviewed by Krizek (2004). Therefore we will here only briefly highlight those particular 'PAR effects' that are directly related to the topic of this chapter:

1. One putative effect of UV-B on plants is an increased production of ROS (see above). Experiments with tobacco plants acclimated to supplementary UV-B in a growth chamber showed that the activation of peroxidase and SOD enzymes, and therefore consequently of leaf H_2O_2 concentrations, depended on the applied PAR and UV-B intensities rather than on the UV-B/PAR ratio (Czégény *et al.*, 2016a). The relevance of these data to field conditions, i.e. to higher PAR and closer to ambient UV-B fluxes are yet to be explored.
2. The PAR effect leading to protection of the mRNA levels for photosynthetic genes during UV-B exposure (Jordan *et al.*, 1991, 1994) has been shown to be dependent on photosynthesis itself, possibly photophosphorylation and/or the ATP content in particular, rather than on DNA photorepair or carbohydrate availability (Mackerness *et al.*, 1996, 1997; Jordan *et al.*, 2016). The exact molecular mechanism behind this has not conclusively been established.
3. In contrast to 'high dose effects' of UV-B, higher PAR does *not* protect from the 'low dose effect' of UV-B-induced changes in the relaxation of the electrochromic shifts in detached leaves (Strid *et al.*, 1996a), indicating a direct but hitherto unknown alteration by UV-B of the thylakoid lipids (see above).

The fact that the 'PAR effect' is seen for a multitude of plant processes (see also section below on homologous recombination) at all organisational levels, from molecule to whole plant, indicates that it is probably not conferred through one single mechanism but through a number of different pathways.

UV-inducible Metabolites and Protection against UV-B

Phenylpropanoids

One of the most fundamental and species-independent metabolic responses of UV-B-exposed plants is the increased biosynthesis and accumulation of various compounds of the phenylpropanoid and flavonoid biosynthetic pathway (Fig. 7.2). The products of this pathway include glycosylated and non-glycosylated flavonols, phenolics and anthocyanins. This class of multifunctional compounds (Agati and Tattini, 2010) mainly have two functions with regards to their capacity to protect plants against the deleterious effects of UV, namely as radiation-absorbing compounds and as antioxidants. Flavonoids and flavonoid glycosides are found in most plant tissues that are exposed to the sun's radiation and also in most compartments of plant cells (Agati *et al.*, 2012.). However, flavonoids residing in the cell walls or the vacuoles of epidermal tissue can be envisaged as the main pool of radiation-absorbing protecting substances (Kalbin *et al.*, 2001; Winkel-Shirley, 2002; Hectors *et al.*, 2014), whereas phenylpropanoids of the chloroplast envelope substantially contribute to ROS scavenging (Agati *et al.*, 2007, 2012). However, it is not clear how ROS other than possibly hydrogen peroxide would reach this pool of flavonoids. Nuclear localisation of quercetin derivatives has also been suggested (Hutzler *et al.*, 1998).

Although the exact speciation of the compounds that are produced as a result of UV exposure and that may contribute to UV

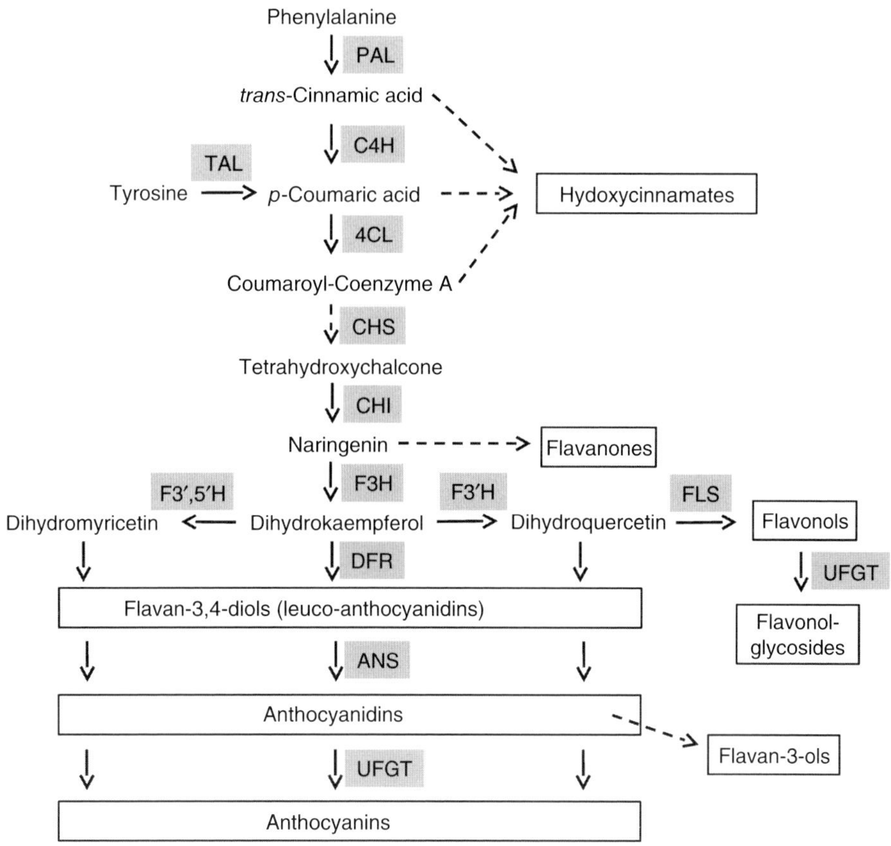

Fig. 7.2. Major components of the phenylpropanoid metabolic pathway. Enzyme acronymes (grey boxes) represent: 4CL, 4-coumarate-CoA ligase; ANS, anthocyanidin synthase; C4H, cinnamate 4-hydroxylase; CHI, chalcone isomerase; CHS, chalcone synthase; DFR, dihydroflavonol-4-reductase; F3′,5′H, flavonoid 3′,5′-hydroxylase; F3H, flavanone 3-hydroxylase; F3′H, flavonoid 3′-hydroxylase; PAL, phenylalanine ammonia-lyase; TAL, tyrosine ammonia-lyase; UFGT, UDP-flavonoid glucosyltransferase. Dashed arrows abbreviate several metabolic steps.

protection is dependent on taxon (see also Chapter 4, this volume), many studies have pointed out the flavonoid derivatives quercetin and/or kaempferol, and especially their glucosides (Fig. 7.3), as important UV-B-absorbing compounds (see e.g. Kolb *et al.*, 2001; Neugart *et al.*, 2012; Morales *et al.*, 2013; Hectors *et al.*, 2014). It is quite clear that properties of the flavonoid structures such as the number of hydroxyl substituents on the B ring, and the number, structure, and position (positions 7- on the A ring or 3- on the C ring) of the glycosyl moieties, contribute greatly to the antioxidant and UV-absorbing capacities of the compounds. For instance, compounds with a 4′,5′-dihydroxylated flavonol core (quercetin) are better antioxidants than 4′-monohydroxylated kaempferol (Rice-Evans *et al.*, 1995) but in comparison show a lower absorption of UV-B (Hectors *et al.*, 2014). Also, glycosylated flavonols absorb UV-B to a higher degree than the corresponding aglycones (Agati and Tattini, 2010; Hectors *et al.*, 2014). In addition, the presence of a hydroxycinnamic acid (such as 4-coumaric acid, caffeic acid, ferulic acid, hydroxyferulic acid or sinapic acid) acylated to a sugar moiety of flavonol glucosides presents further property variation (Neugart *et al.*, 2012; see also Chapter 4, this volume). For a more complete treatise on aspects of phenylpropanoid function and distribution, we refer the reader to Chapter 4 of this book.

The synthesis of the compounds that are of importance for UV protection in plants (flavonols, flavonol glucosides, and hydroxycinnamates) share the phenylpropanoid biosynthesis pathway since hydroxycinnamates are also the precursors of flavonoids. With the amino acid phenylalanine as the substrate, the enzyme phenylalanine ammonia lyase (PAL) catalyses the formation of cinnamic acid by non-oxidative deamination

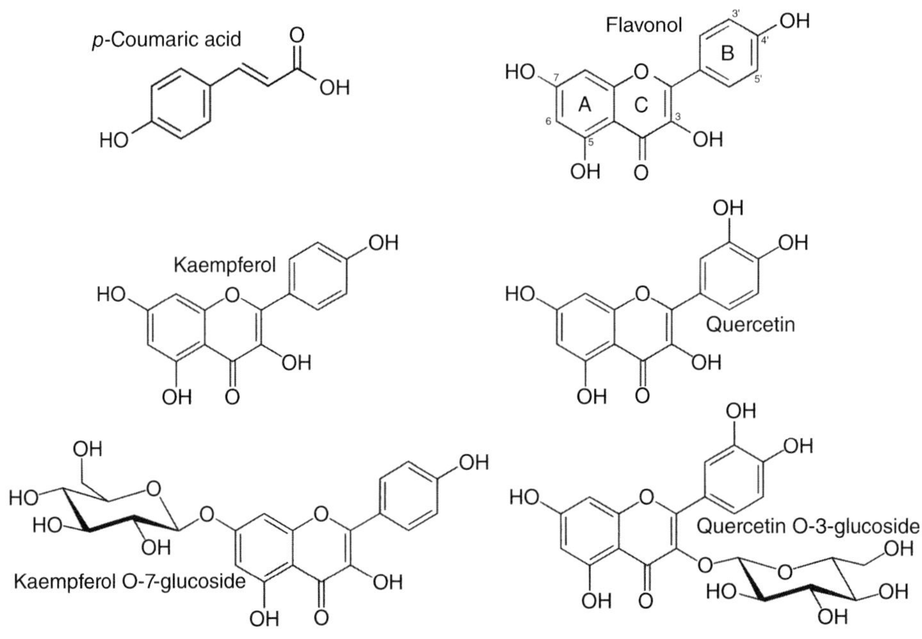

Fig. 7.3. The structure of p-coumaric acid as an example of a hydroxycinnamic acid. Also, the general chemical structure of flavonols, including naming of rings and indications of carbon atom numbering is shown. Furthermore, the structures of the kaempferol and quercetin aglycones and the O-7 and O-3 glucosides of quercetin and kaempferol, respectively, are depicted. Although the general structure for flavonols shown here is commonly agreed, the compound galangin, which normally is also classified as a flavonol, is lacking the hydroxyl group in the 4′ position of the B ring.

of the amino acid (Fig. 7.2). Hydroxylation of cinnamic acid is catalysed by cinnamate 4-hydroxylase (C4H) to *p*-coumaric acid (Figs. 7.2 and 7.3) which then is ligated with coenzyme A by 4-coumarate:CoA ligase (4CL) to provide the first precursor of the flavonoid biosynthetic pathway. Hydroxycinnamic acid can also be converted to caffeic acid by 4-coumarate-3-hydroxylase (C3H), which may be converted further to ferulic acid by caffeic acid *O*-methyltransferase (COMT), further again by ferulate 5-hydroxylase (F5H) to 5-hydroxyferulic acid, and finally to sinapic acid by COMT again. For entry into the flavonoid biosynthetic pathway, 4-coumarate-CoA is conjugated with 3 molecules of malonyl-CoA by chalcone synthase (CHS; TT4) to form 4,2′,4′,6′-tetra hydroxychalcone. This compound is converted to the flavanone naringenin by chalcone isomerase (CHI; TT5), and then flavanone 3-hydroxylase (F3H; TT6) turns naringenin into dihydrokaempferol. Whereas flavonol synthase (FLS) converts dihydrokaempferol into kaempferol, the enzyme flavonoid 3′-hydroxylase (F3′H; TT7) hydroxylates its B ring in the 3′ position to yield dihydroquercetin. Finally, FLS again converts dihydroquercetin to quercetin. Glycosylation is then carried out by different sugar-specific flavonol *O*-glycosyl transferases (Fig. 7.2) to form for instance kaempferol 3- or 7-*O*-rhamnosyls or quercetin 3-*O*-glucosyls (Hofer, 2016).

Chappell and Hahlbrock (1984) showed in their classic parsley cell suspension study a UV-induced transient expression of the *CHS* gene, followed by increased levels of the CHS protein, which in turn led to increased levels of flavonoids. Since then many studies in a large number of plant species have shown the induction of the different flavonoid biosynthesis genes as a response to UV-B. For instance, a number of DNA array studies in *Arabidopsis* have collectively shown increased expression of phenylpropanoid and flavonoid biosynthesis genes such as 4CL, CHS, CHI, F3H, and FLS (reviewed in Hideg *et al.*, 2013). The importance of a functional chain of flavonoid biosynthesis enzymes for UV tolerance was shown in an *Arabidopsis* study under environmentally controlled conditions using CHS and CHI mutants (*tt4* and *tt5*, respectively) where UV hypersensitivity was found in both mutants and which was most severe in *tt5* (Li *et al.*, 1993). The different *Arabidopsis* mutants accumulated flavonoid metabolites as expected: *tt5* accumulated 4,2′,4′,6′-tetrahydroxychalcone, *tt6* naringenin, and *tt7* kaempferol. The *tt4* mutant was devoid of flavonoids and the *tt3* mutant, which lacks the dihydroflavonol 4-reductase (DFR; TT3) converting dihydroquercetin and dihydrokaempferol to anthocyanin precursors (Fig. 7.2), excessively accumulated both quercetin and kaempferol (Peer *et al.*, 2001).

Antioxidants

In addition to their potential to initiate oxidative stress, ROS have important roles as signalling molecules (Apel and Hirt, 2004) and ROS were shown to be involved in the UV-B induction of several defence genes (Mackerness *et al.*, 2001). Cellular ROS concentrations high enough to initiate signal transduction, but not high enough to cause significant cellular damage, are expected to vary with species, developmental stage, and are expected to be controlled by antioxidants.

There is an array of chloroplast antioxidants to fulfil this role (Asada, 2006) and acclimation to solar UV-B (Martínez-Lüscher *et al.*, 2013; Guidi *et al.*, 2016) as well as to moderate UV-B photon fluxes applied in laboratory experiments (Majer *et al.*, 2014a, b) were found to enhance enzymatic defence and increase the amount of non-enzymatic antioxidants (Rao *et al.*, 1996; Kalbin *et al.*, 1997). Enzymatic defence in tobacco leaves under supplementary UV-B in growth chamber experiments was found to be centred on peroxidases (Majer *et al.*, 2014a), in accordance with the potential interaction of UV-B with hydrogen peroxide (Czégény *et al.*, 2014). The extent of UV-induced peroxidase and superoxide dismutase activation appear to affect the success of acclimation to UV-B in laboratory experiments performed on plants grown without UV radiation and then treated for a relatively short (4–14 day) period (Czégény *et al.*, 2016a). In these model experiments, a higher relative increase in peroxidase than in SOD activities seems to assure the control of the hydrogen peroxide load.

Plastid concentrations of ascorbate and the efficient regeneration of oxidised ascorbate are also essential in defence against oxidative stress by UV-B (Gao and Zhang, 2008; Czégény *et al.*, 2016b).

Hormones

The research carried out so far on the interaction between UV-B and hormonal status and action in plants shows a general dependence on methodological aspects such as species and developmental stage used, as well as on the quality and quantity of the UV that has been employed. This is an aspect that makes evaluation of the available literature cumbersome and general conclusions very hard to draw. To make substantial and definitive progress in the research on UV-B/hormonal effects, standardization of at least the irradiation conditions (both duration and levels of both UV-B and PAR) need to be carried out.

In a recent and thorough treatise on the interaction between UV-B exposure and responses of the different plant hormonal systems, Vanhaelewyn *et al.* (2016) examined the literature within this research field and came to the conclusion that different classes of hormones generally could be divided into two types: the ones that interact with UV-B where the radiation gives rise to stress (or 'distress' in the terms of Hideg *et al.*, 2013) and those that interact with UV-B in a way that gives rise to morphological changes in the plants for acclimation to the new light environment ('eustress'; Hideg *et al.*, 2013). Abscisic acid, jasmonates, salicylic acid and nitric oxide were considered to be stress hormones and their formation, signalling, and the physiological consequences of their action increased. Auxins, giberellic acids (GAs), and brassinosteroids (BRs) and their activities were down-regulated by UV-B with photomorphogenic and growth alterations as the result. For ethylene action the UV-B effects could be classified into both groups (Vanhaelewyn *et al.*, 2016). For cytokinins and strigolactones the number of studies are too few for any conclusions to be made. Of all plant hormones studies, the UV-B effects on auxin signalling appear to be the most reproducible. In this section we will briefly summarize the findings on the interactions of UV-B with auxins, GAs, BRs and ethylene.

A general UV-B photomorphogenic plant phenotype includes short stems and thick small leaves (Jansen, 2002). For instance, under a low dose UV-B regime, *Arabidopsis thaliana* displayed decreased leaf area and rosette size that reflected decreased levels of auxin (Hectors *et al.*, 2007, 2012). Also, UV-B down-regulated a large number of genes involved in auxin biosynthesis, auxin distribution, and auxin response (Hectors *et al.*, 2007). In addition to this, a good number of other studies on different species (reviewed in Vanhaelewyn *et al.*, 2016) substantiated the impact of UV-B on auxin action. UVR8 may be involved in this regulation (Hayes *et al.*, 2014) as was shown in an *Arabidopsis* study on UV-B-dependent inhibition of shade avoidance. Since the HY5/HYH transcription factors (Vanhaelewyn *et al.*, 2016) are involved both in UV-B and auxin signalling, regulation of these pathways may converge at this point.

Weller *et al.* (2009) have shown that GA biosynthesis in the pea plant is regulated through the photomorphogenic COP1/HY5 pathway. In *Arabidopsis*, UV-B-specific and gene-specific regulation of different GA oxidases was controlled by both UVR8 and HY5/HYH, and indicated that the decreased shade avoidance under UV-B was accomplished by alteration of both auxin and GA levels (Hayes *et al.*, 2014).

UV-B-dependent up-regulation of a number of *Arabidopsis* defence genes was dampened in BR function mutants (Sävenstrand *et al.*, 2004b) and the expression of genes related to BR signalling was altered in wild-type *Arabidopsis* exposed to UV-B (Hectors *et al.*, 2007). COP1 regulates the abundance of the BR-dependent transcription factor BZR1 in *Arabidopsis* in the dark through degradation of the inactive phosphorylated form of the protein and *cop1* mutants showed increased levels of this transcription factor

(Kim *et al.*, 2014). In addition, HY5 and active BZR1 interact with each other and thereby decrease BZR1-dependent gene regulating activity (Li and He, 2016). Therefore, a link between BR and light regulation of gene expression, and possibly also with UV-B regulation, has been established through the COP1/HY5 pair.

The bulk of studies showing UV-B-dependent ethylene production in plants were carried out at high UV intensities, indicating that this was a result of 'distress' (Vanhaelewyn *et al.*, 2016). This includes studies on ethylene-mediated stomatal closure in broad bean (He *et al.*, 2011). However, in *Vitis vinifera* and *Artemisia annua* UV-B at moderate levels gave rise to increased expression of genes associated with ethylene signalling (Pontin *et al.*, 2010; Pan *et al.*, 2014).

UV-B and the Central Dogma

DNA damage and consequences for replication

As outlined above, UV-B radiation that penetrates a living cell can give rise to DNA damage. The most common nucleic acid lesion is the cyclobutane pyrimidine dimer (CPD). UV absorption and concomitant inter-crosslinking between two adjacent thymine or cytosine bases (TT, TC, CT, or CC) in the nucleic acid are the causes for CPD formation. A less common event is the formation of another type of dimer, a so-called 6,4 pyrimidine-pyrimidone photoproduct (6,4PP). Upon continued exposure to UV-B, 6,4PP can be converted to Dewar isomer photoproducts (Takeuchi *et al.*, 1998). See the review by Taylor (2006) for structures of these DNA lesions. Obviously, accumulation of UV-induced DNA lesions would have serious cell biological consequences, particularly during replication. Depending on the severity of the accumulation of photoproducts, mutation, stalled replication forks, DNA strand breaks, cell cycle arrest, and programmed cell death may occur (Biever and Gardner, 2016).

Endo-reduplication (i.e. replication of the genome without mitosis, in turn leading to cellular polyploidy; de Veylder *et al.*, 2011) is another type of cell cycle-related event that can be induced by UV-B (Wargent *et al.*, 2009). Generally, endo-reduplication seems partly to be regulated by the *UV-B INSENSITIVE 4* (*UVI4*) gene in *Arabidopsis* (Hase *et al.*, 2006). The UVI4 protein is involved in promotion of mitosis and a *uvi4* mutant performs one extra round of hypocotyl endo-reduplication, particularly yielding increased amounts of hexadecaploid cells, in turn leading to increased plant size and to increased tolerance against UV-B. This is in agreement with the finding that tetraploid *Arabidopsis* plants are more tolerant to UV-B than diploid plants. Thus, together these findings suggest that increased polyploidy can be a protective response against UV-induced DNA damage (Hase *et al.*, 2006). Also, UV-B-induced endo-reduplication was shown to be at least partly regulated in a UVR8-dependent fashion (Wargent *et al.*, 2009).

The relevance of UV-B-induced damage as a major regulator of gene expression, metabolism, and morphology in plants has been debated for decades, especially before the discovery of the UV-B photoreceptor UVR8. In a study of CPDs and gene expression in pea, CPDs were formed to a greater extent in the epidermis than in mesophyll (Kalbin *et al.*, 2001). In addition, acclimation at low UV-B levels resulted in significantly higher basal levels of CPDs than in non-acclimated plants, both in the mesophyll and in the epidermis, and also increased damage in concomitant acute exposures. Importantly, there was also a lack of correlation between the number of CPDs and the levels of transcripts for several defence genes, which indicated that DNA damage does not control transcription of these genes. However, it seems that in particular examples, DNA damage inflicted by UV-B can be an important limiting factor for growth. For instance, in certain cultivars of rice, gene mutations that result in decreased activity of the CPD photolyase (Hidema *et al.*, 2000) have an impact on their growth and productivity (e.g. see Hidema and Kumagai, 2006 for a review). Likewise, UV-B-dependent inhibition of hypocotyl growth in etiolated *Arabidopsis* seedlings have been attributed to cell cycle arrest as a consequence of photodimer accumulation (Biever *et al.*, 2014). It was suggested that, during early plant development, before UVR8-dependent accumulation of protective flavonoids had reached sufficient levels, UV-B-induced DNA damage and the DNA repair mechanisms would have particular physiological and regulatory impacts on plants (Biever and Gardner, 2016). However, it is not clear how such a mechanism would interact with the well-known effect of UV-B on auxin levels (Hectors *et al.*, 2012) and this plant hormone's action on cell growth, cell division and cellular differentiation.

Under conditions primarily leading to single strand breakage (SSB) in DNA (UV-B, hydrogen peroxide, or any other stressor being the causal agent), the enzyme poly(ADP-ribose) polymerase (PARP) tags the SSB for repair by synthesizing and linking a poly(ADP) ribose chain to the location of the damage. During each round of PARP polymerase activity one molecule of nicotinamide (NIC) is released as a dissociated group. It was hypothesized that this NIC could function as a signalling molecule (Berglund, 1994) in plants linking DNA damage and gene expression. It was further shown that UV-B exposure of pea plants led to transient (48–60 hours) 2- to 3-fold increases in NIC levels (Kalbin *et al.*, 1997). The same UV-B exposures also led to induction of expression of the PAL and CHS genes, and to decreased levels of mRNA for photosynthetic genes. In addition, treatment of pea tissue cultures with NIC alone (i.e. without UV exposure) led to induction of CHS and glutathione reductase gene expression, primarily in root cultures (Berglund *et al.*, 1993). However, whether or not NIC is a true signalling intermediate between DNA damage and gene expression has not been ascertained.

Repair of UV-induced lesions in the genetic material

As a first-line repair pathway against UV-induced dimer photoproducts, plants possess two different blue light energy-dependent photolyases that restore the integrity of the DNA in an error-free fashion during a process called photoreactivation. One of these enzymes (PHR1/UVR2) is specific for removal of CPDs, the other (UVR3) is specific for 6,4PPs (Chen *et al.*, 1994; Nakajima *et al.*, 1998; Biever and Gardner, 2016). UV-B signalling components UVR8, HY5, and HYH all appear to be involved in PHR1 and UVR3 gene regulation (Brown and Jenkins, 2008; Castells *et al.*, 2010; Li *et al.*, 2015), although white light regulation and UV-B regulation of UVR3 has been questioned (Waterworth *et al.*, 2002). PHR1 is to a smaller extent also regulated by an unknown UVR8-independent pathway. This was found using the *Arabidopsis uvr8-6* mutant and studying the fraction of the regulation that could be correlated to the degree of CPD formation (Li *et al.*, 2015).

Repair mechanisms that are generally regarded as independent of light are alternatives for repair of UV-induced DNA lesions. These include base excision repair, nucleotide excision repair (Britt, 2002) and homologous recombination (HR) (Ries *et al.*, 2000; Molinier *et al.*, 2004), the latter possibly being employed when other repair mechanisms are less active. However, unexpectedly, Ries *et al.* (2000) found that UV-B-induced HR was dependent on the PAR levels, showing a 'protective PAR effect' (see above) of increased numbers of HR events. This 'PAR effect' was also UV-dependent and did not apply to genotoxic treatments using chemical mutagens (Ries *et al.*, 2000).

Moreover, there are indications of the presence of DNA replication polymerases in plants that can bypass non-repaired pyrimidine dimers and other DNA lesions to avoid consequences such as stalling of the replication fork, cell cycle arrest, and inhibition of mitosis (Sakamoto *et al.*, 2003). Our understanding of all the processes plants use to keep their genetic material intact and functional is incomplete and further development in this research field is expected.

Aspects of UV-B interaction with translation: ribotoxicity

One particular aspect of UV-induced crosslinking (see above) has interesting physiological implications due to the risk for ribotoxicity. Such UV-B-dependent ribotoxicity would involve crosslinking of either the rRNAs or mRNAs with each other, or with ribosomal proteins. The first account of UV-B-dependent alteration in ribosomal RNA composition and speciation was published by Wang and Strid (1998). In a controlled environment UV-B exposure study, the abundance of the pea chloroplast 23S rRNA decreased by 50% and a high molecular adduct of this rRNA species increased 14-fold at moderate or even low UV-B levels, while the cytoplasmic 18S rRNA and the chloroplastic 16S rRNA in essence remained unchanged. The authors hypothesized that the high molecular weight adduct could have been formed by crosslinking of the 23S rRNA with either smaller chloroplastic rRNAs or with ribosomal proteins which could ultimately lead to either premature degradation of chloroplastic mRNAs, or impairment of both *de novo* protein synthesis and replacement of damaged chloroplastic proteins, which in turn would affect photosynthetic function.

In a follow up UV-B study (Brosché *et al.*, 1999), three *P. sativum* mRNAs, in addition to the 23S rRNA, also showed increased abundance of high molecular weight adducts. Of the three protein-encoding genes, two would have their corresponding mRNA translated on cytoplasmic ribosomes (*PsLhcA4* and *PsUBC4*) and one on plastid ribosomes (*PsPsbA*). Two of these genes were photosynthetic (*PsLhcA4*, chlorophyll *a*/*b* binding protein of photosystem I; *PsPsbA*, D1 protein of photosystem II) and one was involved in protein degradation (*PsUBC4*; ubiquitin-conjugating enzyme). The best example of these is the *PsPsbA* gene were the presence of the high molecular weight mRNA adduct increased sharply from below the detection limit. The fact that *PsPsbA* is a single copy intronless gene excludes alternative explanations for the formation of the high molecular weight adduct such as alternative splicing or transcription of a second gene copy of different size. Isolated RNA that had been treated with the broad spectrum specificity protease Proteinase K gave identical results which excludes crosslinking with proteins as the reason for the appearance of large mRNA species for *PsPsbA* and 23S rRNA. For the two mRNAs encoding nuclear genes, crosslinking to the two larger ribosomal 18S and 26S rRNAs could also be excluded (Brosché *et al.*, 1999). Based on these studies, the most likely crosslinking partners appear to be small rRNAs or tRNAs.

However, in a more recent study carried out in maize, UV-B-induced ribosomal RNA crosslinking was shown to involve cytosolic ribosomal proteins S14, L23a, and L32, in addition to chloroplast ribosomal protein L29 (Casati and Walbot, 2004). Damaged ribosomes were eliminated and ribosomal function was restored within 16h after cessation of UV-B exposure. Thus, UV-B-dependent cross-linking of ribosomal RNA to both proteins and other RNA molecules seems possible. In addition, and with particular bearing on UV-damaged transcripts, the *Arabidopsis* ribosomal protein S27 was, in a UV-C study, proposed to be necessary for removing UV-damaged mRNAs (Revenkova *et al.*, 1999).

Outlook and Conclusion

The role for UV-B regulation of plant metabolism to a large extent is still enigmatic. Many studies show effects on one or the other of a number of metabolic pathways, whereas other studies do not. A general agreement has been reached that UV-B rarely poses a threat to plants under ambient radiation and most likely not under moderate elevation of UV-B levels either. Morphological effects on plant architecture, probably regulated by auxin, and increased levels of flavonoid pigments and antioxidants are plant responses towards the UV component of sunlight (see Chapter 5). The photoreceptor UVR8 regulates many of the UV-B responses but a growing number of studies also imply other factors/intercellular components, including those so far not discovered, as regulators of UV responses.

Results emerging from studies of UVR8 regulation of outdoor plant responses (Morales *et al.*, 2013; Findlay and Jenkins, 2016) are likely to both complicate and expand our knowledge on the impact of UV-B on plants. For instance, the studies on the fluctuation of the UVR8 monomer/dimer ratio in *Arabidopsis* plants grown under photoperiodic conditions indicates a more complex UV-B regulatory web than has previously been anticipated (Findlay and Jenkins, 2016) and this raises questions about how strong the impact of UVR8 function actually is under natural conditions and during what circumstances UVR8 regulation is most prominent.

Although developing tissues are known to react differently to environmental stimuli than mature ones, only a few UV studies have taken this into consideration. For instance, UV-B-regulation of photosynthetic genes is not functional in juvenile/etiolated tissue (Jordan *et al.*, 1994) but the question also arises about what other processes in developing tissue are in fact under UV control.

A number of intriguing findings may point out future directions for plant UV photobiology that may become very fruitful scientifically:

1. UVR8-dependent light escape in roots during development (Yokawa and Baluska 2015) and other root-associated morphological changes may indicate a prominent role for UV-B (Ge *et al.*, 2010; Krasylenko *et al.*, 2012).
2. UV-B/UVR8 involvement in development of shade avoidance through action on hormonal levels is another developmental issue that is highly interesting (Hayes *et al.*, 2014).
3. A physiological role of DNA damage as a growth regulator in juvenile plants has been proposed (Biever and Gardner, 2016) and this suggestion could add new fuel to the debate on whether DNA damage has any real role in UV regulatory processes.
4. There is a synergistic effect of simultaneous UV-B exposure and exposure to other environmental cues giving rise to more severe stress via enhanced ROS production (Czégény *et al.*, 2014). However, UV-B can also induce improved tolerance to drought stress in plants (Robson *et al.*, 2015). What are the determinants for such different outcomes of the combination of UV-B exposure with other environmental factors?

These are only a few examples of recent findings that could keep the plant UV photobiology community fully engaged with interesting challenges. The opportunities for new endeavours into plant UV research are plentiful for many years to come and we are certainly entering a new era of novel and exciting findings.

References

Adamska, I. and Kloppstech, K. (1991) Evidence for an association of the early light-inducible protein (ELIP) of pea with photosystem II. *Plant Molecular Biology* 16, 209–223.

Agati, G. and Tattini, M. (2010) Multiple functional roles of flavonoids in photoprotection. *New Phytologist* 186, 786–793.

Agati, G., Matteini, P., Goti, A. and Tattini, M. (2007) Chloroplast-located flavonoids can scavenge singlet oxygen. *New Phytologist* 174, 77–89.

Agati, G., Azzarello, E., Pollastri, S. and Tattini, M. (2012) Flavonoids as antioxidants in plants: location and functional significance. *Plant Science* 196, 67–76.

Allen, D.J., Nogués, S. and Baker, N.R. (1998) Ozone depletion and increased UV-B radiation: Is there a real threat to photosynthesis? *Journal of Experimental Botany* 49, 1775–1788.

Apel, K. and Hirt, H. (2004) Reactive oxygen species: metabolism, oxidative stress, and signal transduction. *Annual Review of Plant Biology* 55, 373–399.

Aphalo, P.J., Albert, A., McLeod, A.R., Heikkilä, A., Gómez, I. *et al.* (2012a) Manipulating UV radiation. In: Aphalo, A., Albert, A., Björn, L.O., McLeod, A.R., Robson, T.M. and Rosenqvist, E. (eds) *Beyond the Visible: A Handbook of Best Practice in Plant UV Photobiology*. University of Helsinki, Finland, pp. 35–70.

Aphalo, A., Albert, A., Björn, L.O., McLeod, A.R., Robson, T.M. and Rosenqvist, E. (eds) (2012b) *Beyond the Visible: A Handbook of Best Practice in Plant UV Photobiology*. University of Helsinki.

Araújo, W.L., Fernie, A.R. and Nunes-Nesi, A. (2011) Control of stomatal aperture: a renaissance of the old guard. *Plant Signaling and Behavior* 6, 1305–1311.

Asada, K. (2006) Production and scavenging of reactive oxygen species in chloroplasts and their functions. *Plant Physiology* 141, 391–396.

Berglund, T. (1994) Nicotinamide, a missing link in the early stress response in eukaryotic cells: a hypothesis with special reference to oxidative stress in plants. *FEBS Letters* 351, 145–149.

Berglund, T., Ohlsson, A.B., Rydström, J., Jordan, B.R. and Strid, Å. (1993) Effect of nicotinamide on gene expression and glutathione levels in tissue cultures of *Pisum sativum*. *Journal of Plant Physiology* 142, 676–684.

Biever, J.J. and Gardner, G. (2016) The relationship between multiple UV-B perception mechanisms and DNA repair pathways in plants. *Environmental and Experimental Botany* 124, 89–99.

Biever, J.J., Brinkman, D. and Gardner, G. (2014) UV-B inhibition of hypocotyl growth in etiolated *Arabidopsis thaliana* seedlings is a consequence of cell-cycle arrest initiated by photodimer accumulation. *Journal of Experimental Botany* 65, 2949–2961.

Britt, A. (2002) Repair of damaged bases. *The Arabidopsis Book* 1, e0005. DOI:http://dx.doi.org/10.1199/tab.0005

Brosché, M. and Strid, Å. (2003) Molecular events following perception of ultraviolet-B radiation by plants. *Physiologia Plantarum* 117, 1–10.

Brosché, M., Fant, C., Bergkvist, S.W., Strid, H., Svensk, A., Olsson, O. and Strid, Å. (1999) Molecular markers for UV-B stress in plants: alteration of the expression of four classes of genes in *Pisum sativum* and the formation of high molecular mass RNA adducts. *Biochimica et Biophysica Acta* 1447, 185–198.

Brown, B.A. and Jenkins, G.I. (2008) UV-B signaling pathways with different fluence-rate response profiles are distinguished in mature Arabidopsis leaf tissue by requirement for UVR8, HY5, and HYH. *Plant Physiology* 146, 576–588.

Casati, P. and Walbot, V. (2004) Crosslinking of ribosomal proteins to RNA in maize ribosomes by UV-B and its effects on translation. *Plant Physiology* 136, 3319–3332.

Castells, E., Molinier, J., Drevensek, S., Genschik, P., Barneche, F. and Bowler, C. (2010) *det1-1* induced UV-C hyposensitivity through *UVR3* and *PHR1* photolyase gene over-expression. *Plant Journal* 63, 392–404.

Chappell, J. and Hahlbrock, K. (1984) Transcription of plant defence genes in response to UV light or fungal elicitor. *Nature* 311, 76–78.

Cheeseman, J.M. (2006) Hydrogen peroxide concentrations in leaves under natural conditions. *Journal of Experimental Botany* 57, 2435–2444.

Chen, J.-J., Mitchell, D.L. and Britt, A.B. (1994) A light-dependent pathway for the elimination of UV-induced pyrimidine (6-4) pyrimidine photoproducts in Arabidopsis. *Plant Cell* 6, 1311–1317.

Chow, W.S., Strid, Å. and Anderson, J.M. (1992) Short-term treatment of pea plants with supplementary ultraviolet-B radiation: recovery time-courses of some photosynthetic functions and components. In: Murata, N. (ed.) *Research in Photosynthesis*, Vol. 4. Kluwer Academic Publishers, Dordrecht, The Netherlands, pp. 361–364.

Czégény, Gy., Wu, M., Dér, A., Eriksson, L.A., Strid, Å.and Hideg, É. (2014) Hydrogen peroxide contributes to the ultraviolet-B (280–315 nm) induced oxidative stress of plant leaves through multiple pathways. *FEBS Letters* 588, 2255–2261.

Czégény, Gy., Mátai, A. and Hideg, É. (2016a) UV-B effects on leaves – oxidative stress and acclimation in controlled environments. *Plant Science* 248, 57–63.

Czégény, Gy., Le Martret, B., Pávkovics, D., Dix, P.J. and Hideg, É. (2016b) Elevated ROS-scavenging enzymes contribute to acclimation to UV-B exposure in transplastomic tobacco plants, reducing the role of plastid peroxidases. *Journal of Plant Physiology* 201, 95–100.

Desimone, M., Wagner, E. and Johanningmeier, U. (1998) Degradation of active-oxygen-modified ribulose-1,5-bisphosphate carboxylase/oxygenase by chloroplastic proteases requires ATP-hydrolysis. *Planta* 205, 459–466.

de Veylder, L., Larkin, J.C. and Schnittger, A. (2011) Molecular control and function of endoreplication in development and physiology. *Trends in Plant Science* 16, 624–634.

Döhler, G., Drebes, G. and Lohmann, M. (1997) Effect of UV-A and UV-B radiation of pigments, free amino acids and adenylate content of *Dunaliella tertiolecta* Butcher (Chlorophyta). *Journal of Photochemistry and Photobiology B: Biology* 40, 126–131.

Eisinger, W., Swartz, T.E., Bogomolni, R.A. and Taiz, L. (2000) The ultraviolet action spectrum for stomatal opening in broad bean. *Plant Physiology* 122, 99–106.

Ferreira, R.M.B., Franco, E. and Teixeira, A.R.N. (1996) Covalent dimerization of ribulose bisphosphate carboxylase subunits by UV radiation. *Biochemical Journal* 318, 227–234.

Findlay, K.M.W. and Jenkins, G.I. (2016) Regulation of UVR8 photoreceptor dimer/monomer photoequilibrium in Arabidopsis plants grown under photoperiodic conditions. *Plant, Cell & Environment* 39, 1706–1714.

Galmés, J., Aranjuelo, I., Medrano, H. and Flexas, J. (2013) Variation in Rubisco content and activity under variable climatic factors. *Photosynthesis Research* 117, 73–90.

Gao, Q. and Zhang, L. (2008) Ultraviolet-B-induced oxidative stress and antioxidant defense system responses in ascorbate-deficient *vtc1* mutants of *Arabidopsis thaliana*. *Journal of Plant Physiology* 165, 138–148.

Ge, L., Peer, W., Robert, S., Swarup, R., Ye, S., Prigge, M. *et al.* (2010) *Arabidopsis ROOT UVB SENSITIVE2/WEAK AUXIN RESPONSE1* is required for polar auxin transport. *Plant Cell* 22, 1749–1761.

Gerhardt, K.E., Wilson, M.I. and Greenberg, B.M. (1999) Tryptophan photolysis leads to a UVB-induced 66 kDa photoproduct of ribulose-1,5 bisphosphate carboxylase/oxygenase (Rubisco) *in vitro* and *in vivo*. *Photochemistry and Photobiology* 70, 49–56.

Guidi, L., Brunetti, C., Fini, A., Agati, G., Ferrini, F., Gori, A. and Tattini M. (2016) UV radiation promotes flavonoid biosynthesis, while negatively affecting the biosynthesis and the de-epoxidation of xanthophylls: consequence for photoprotection? *Environmental and Experimental Botany* 127, 14–25.

Hase, Y., Trung, K.H., Matsunaga, T. and Tanaka, A. (2006) A mutation in the *uvi4* gene promotes progression of endo-reduplication and confers increased tolerance towards ultraviolet-B light. *Plant Journal* 46, 317–326.

Hayes, S., Velanis, C.N., Jenkins, G.I. and Franklin, K.A. (2014) UV-B detected by the UVR8 photoreceptor antagonizes auxin signaling and plant shade avoidance. *Proceedings of the National Academy of Sciences of the United States of America* 111, 11894–11899.

He, J.-M., Zhang, Z., Wang, R.-B., and Chen, Y.-P. (2011) UV-B-induced stomatal closure occurs via ethylene-dependent NO generation in *Vicia faba*. *Functional Plant Biology* 38, 293–302.

He, J.-M., Ma, X.-G., Zhang, Y., Sun, T.-F., Xu, F.-F. *et al.* (2013) Role and interrelationship of Gα protein, hydrogen peroxide, and nitric oxide in ultraviolet-B-induced stomatal closure in Arabidopsis leaves. *Plant Physiology* 161, 1570–1583.

Hectors, K., Prinsen, E., De Coen,W., Jansen, M.A.K. and Guisez, Y. (2007) *Arabidopsis thaliana* plants acclimated to low dose rates of ultraviolet-B radiation show specific changes in morphology and gene expression in the absence of stress symptoms. *New Phytologist* 175, 255–270.

Hectors, K., van Oevelen, S., Guisez, Y., Prinsen, E. and Jansen, M.A.K. (2012) The phytohormone auxin is a component of the regulatory system that controls UV-mediated accumulation of flavonoids and UV-induced morphogenesis. *Physiologia Plantarum* 145, 594–603.

Hectors, K., van Oevelen, S., Geuns, J., Guisez, Y., Jansen, M.A.K. and Prinsen, E. (2014) Dynamic changes in plant secondary metabolites during UV acclimation in *Arabidopsis thaliana*. *Physiologia Plantarum* 152, 219–230.

Hideg, É. and Vass, I. (1996) UV-B induced free radical production in plant leaves and isolated thylakoid membranes. *Plant Science* 115, 251–260.

Hideg, É., Jansen, M.A.K. and Strid, Å. (2013) UV-B exposure, ROS, and stress: inseparable companions or loosely linked associates? *Trends in Plant Science* 18, 107–115.

Hidema, J. and Kumagai, T. (2006) Sensitivity of rice to ultraviolet-B radiation. *Annals of Botany* 97, 933–942.

Hidema, J., Kumagai, T. and Sutherland, B.M. (2000) UV radiation-sensitive Norin 1 rice contains defective cyclobutane pyrimidine dimer photolyase. *Plant Cell* 12, 1569–1578.

Hofer, B. (2016) Recent developments in the enzymatic *O*-glycosylation of flavonoids. *Applied Microbiology and Biotechnology* 100, 4269–4281.

Hutzler, P., Fischbach, R., Heller, W., Jungblut, T.P., Reuber, S. *et al.* (1998) Tissue localization of phenolic compounds in plants by confocal laser scanning microscopy. *Journal of Experimental Botany* 49, 953–965.

Imbrie, C.W. and Murphy, T.M. (1982) UV action spectrum (254–405 nm) for inhibition of a K^+-stimulated adenosine triphosphatase from the plasma membrane of *Rosa damascene*. *Photochemistry and Photobiology* 36, 537–542.

Ishida, H., Makino, A. and Mae, T. (1999) Fragmentation of the large subunit of ribulose-1,5-bisphosphate carboxylase by reactive oxygen species occurs near Gly-329. *Journal of Biological Chemistry* 274, 5222–5226.

Jansen, M.A.K. (2002) Ultraviolet-B radiation effects on plants: induction of morphogenic responses. *Physiologia Plantarum* 116, 423–429.

Jansen, M.A.K. and van den Noort, R.E. (2000) Ultraviolet-B radiation induces complex alterations in stomatal behaviour. *Physiologia Plantarum* 110, 189–194.

Jenkins, G.I. (2009) Signal transduction in responses to UV-B radiation. *Annual Review of Plant Biology* 60, 407–431.

Jordan, B.R. (1996) The effects of UV-B radiation on plants: a molecular prospective. In: Callow, J.A. (ed.) *Advances in Botanical Research*. Academic Press, Oxford, UK, pp. 97–162.

Jordan, B.R., Chow, W.S., Strid, Å. and Anderson, J.M. (1991) Reduction in *cab* and *psb A* RNA transcripts in response to supplementary ultraviolet-B radiation. *FEBS Letters* 284, 5–8.

Jordan, B.R., He, J., Chow, W.S. and Anderson, J.M. (1992) Changes in mRNA levels and polypeptide subunits of ribulose 1,5-bisphosphate carboxylase in response to supplementary ultraviolet-B radiation. *Plant Cell and Environment* 15, 91–98.

Jordan, B.R., James, P., Strid, Å. and Anthony, R. (1994) The effect of supplementary UV-B radiation on gene expression and pigment composition in etiolated and green pea leaf tissue: UV-B induced changes are gene-specific and dependent upon the developmental stage. *Plant Cell and Environment* 17, 45–54.

Jordan, B.R., Strid, Å. and Wargent, J.J. (2016) What role does UV-B play in determining photosynthesis? In: Pessarakli, M. (ed.) *Handbook of Photosynthesis*, 3rd edition. CRC Press, Boca Raton, Florida, pp. 275–286.

Kalbin, G., Ohlsson, A.B., Berglund, T., Rydström, J. and Strid, Å. (1997) Ultraviolet-B-radiation-induced changes in nicotinamide and glutathione metabolism and gene expression in plants. *European Journal of Biochemistry* 249, 465–472.

Kalbin, G., Hidema, J., Brosché, M., Kumagai, T., Bornman J.F. and Strid, Å. (2001) UV-B-induced DNA damage and expression of defence genes under UV-B stress: tissue-specific molecular marker analysis in leaves. *Plant Cell and Environment* 24, 983–990.

Kim, B., Jeong, Y.J., Corvalan, C., Fujioka, S., Cho, S., Park, T. and Choe, S. (2014) Darkness and *gulliver2/phyB* mutation decrease the abundance of phosphorylated BZR1 to activate brassinosteroid signaling in Arabidopsis. *The Plant Journal* 77, 737–747.

Klem, K., Holub, P., Štroch, M., Nezval, J., Špunda, V., Tříska, J. *et al.* (2015) Ultraviolet and photosynthetically active radiation can both induce photoprotective capacity allowing barley to overcome high radiation stress. *Plant Physiology and Biochemistry* 93, 74–83.

Kolb, C.A., Köser, M.A., Kopecký, J., Zotz, G., Riederer, M. and Pfündel, E.E. (2001) Effects of natural intensities of visible and ultraviolet radiation on epidermal ultraviolet screening and photosynthesis in grape leaves. *Plant Physiology* 127, 863–875.

Krasylenko, Y.A., Yemets, A.I., Sheremet, Y.A. and Blume, Y.B. (2012) Nitric oxide as a critical factor for perception of UV-B irradiation by microtubules in *Arabidopsis*. *Physiologia Plantarum* 145, 505–515.

Krizek, D.T. (2004) Influence of PAR and UV-A in determining plant sensitivity and photomorphogenic responses to UV-B radiation. *Photochemistry and Photobiology* 79, 307–315.

Lau, O.S. and Deng, W.X. (2012) The photomorphogenic repressors COP1 and DET1: 20 years later. *Trends in Plant Science* 17, 584–593.

Li, J., Ou-Lee, T.-M., Raba, R., Amundson, R.G. and Last, R.L. (1993) Arabidopsis flavonoid mutants are hypersensitive to UV-B irradiation. *Plant Cell* 5, 171–179.

Li, N., Teranishi, M., Yamaguchi, H., Matsushita, T., Watahiki, M.K. *et al.* (2015) UV-B-induced CPD photolyase gene expression is regulated by UVR8-dependent and -independent pathways in Arabidopsis. *Plant Cell Physiology* 56, 2014–2023.

Li, Q.-F. and He, J.-X., (2016) BZR1 interacts with HY5 to mediate brassinosteroid and light-regulated cotyledon opening in *Arabidopsis* in darkness. *Molecular Plant* 9, 113–125.

Mackerness, S.A.H., Butt, J.P. and Jordan, B.R. (1996) Amelioration of ultraviolet-B-induced down-regulation of mRNA transcripts for chloroplast proteins, by high irradiance, is mediated by photosynthesis. *Journal of Plant Physiology* 148, 100–106.

Mackerness, S.A.H., Surplus, S.L., Jordan, B.R. and Thomas, B. (1997) Ultraviolet-B effects on transcript levels for photosynthetic genes are not mediated through carbohydrate metabolism. *Plant, Cell and Environment* 20, 1431–1437.

Mackerness, S.A.H., John, C.F., Jordan, B. and Thomas, B. (2001) Early signaling components in UV-B responses: distinct roles for different reactive oxygen species and nitric oxide. *FEBS Letters* 489, 237–242.

Majer, P., Czégény, Gy., Sándor, Gy., Dix, P.J. and Hideg, É. (2014a) Antioxidant defence in UV-irradiated tobacco leaves is centred on hydrogen-peroxide neutralization. *Plant Physiology and Biochemistry* 82, 239–243.

Majer, P., Neugart, S., Krumbein, A., Schreiner, M. and Hideg, É. (2014b) Singlet oxygen scavenging by leaf flavonoids contributes to sunlight acclimation in *Tilia platyphyllos*. *Environmental and Experimental Botany* 100, 1–9.

Martínez-Lüscher, J., Morales, F., Delrot, S., Sánchez-Díaz, M., Gomés, E., Aguirreolea, J. and Pascual, I. (2013) Short- and long-term physiological responses of grapevine leaves to UV-B radiation. *Plant Science* 213, 114–122.

Matxain, J.M., Padro, D., Ristilä, M., Strid, Å. and Eriksson, L.A. (2009) Evidence of high ·OH radical quenching efficiency by Vitamin B6. *Journal of Physical Chemistry B* 113, 9629–9632.

Mewes, H. and Richter, M. (2002) Supplementary ultraviolet-B radiation induces a rapid reversal of the diadinoxanthin cycle in the strong light-exposed diatom *Phaeodactylum tricornutum*. *Plant Physiology* 130, 1527–1535.

Miyao, M., Ikeuchi, M., Yamamoto, N. and Ono, T. (1995) Specific degradation of the D1 protein of Photosystem II by treatment with hydrogen peroxide in darkness: implications for the mechanism of degradation of the D1 protein under illumination. *Biochemistry* 34, 10019–10026.

Molinier, J., Ramos, C., Fritsch, O. and Hohn, B. (2004) CENTRIN2 modulates homologous recombination and nucleotide excision repair in Arabidopsis. *Plant Cell* 16, 1633–1643.

Morales, L.O., Brosché, M., Vainonen, J., Jenkins, G.I., Wargent, J.J. *et al.* (2013) Multiple roles for UV RESISTANCE LOCUS 8 in regulating gene expression and metabolite accumulation in *Arabidopsis* under solar UV radiation. *Plant Physiology* 161, 744–759.

Nakajima, S., Sugiyama, M., Iwai, S., Hitomi, K., Otoshi, E. *et al.* (1998) Cloning and characterization of a gene (*UVR3*) required for photorepair of 6-4 photoproducts in *Arabidopsis thaliana*. *Nucleic Acids Research* 26, 638–644.

Neugart, S., Zietz, M., Schreiner, M., Rohn, S., Kroh, L.W. and Krumbein, A. (2012) Structurally different flavonol glycosides and hydroxycinnamic acid derivatives respond differently to moderate UV-B radiation exposure. *Physiologia Plantarum* 145, 582–593.

Pan, W.S., Zheng, L.P., Tian, H., Li, W.Y. and Wang, J.W. (2014) Transcriptome responses involved in artemisinin production in *Artemisia annua* L. under UV-B radiation. *Journal of Photochemistry and Photobiology B: Biology* 140, 292–300.

Peer, W.A., Brown, D.E., Tague, B.W., Muday, G.K., Taiz, L. and Murphy, A.S. (2001) Flavonoid accumulation patterns of transparent testa mutants of Arabidopsis. *Plant Physiology* 126, 536–548.

Pontin, M.A., Piccoli, P.N., Francisco, R., Bottini, R., Martinez-Zapater, J.M. and Lijavetzky, D. (2010) Transcriptome changes in grapevine (*Vitis vinifera* L.) cv. Malbec leaves induced by ultraviolet-B radiation. *BMC Plant Biology* 10, 224.

Rao, M.V., Paliyath, C. and Ormrod, D.P. (1996) Ultraviolet-B- and ozone-induced biochemical changes in antioxidant enzymes of *Arabidopsis thaliana*. *Plant Physiology* 110, 125–136.

Revenkova, E., Masson, J., Koncz, C., Afsar, K., Jakovleva, L. and Paszkowski, J. (1999) Involvement of *Arabidopsis thaliana* ribosomal protein S27 in mRNA degradation triggered by genotoxic stress. *EMBO Journal* 18, 490–499.

Rice-Evans, C.A., Miller, N.J., Bolwell, P.G., Bramley, P.M. and Pridham, J.B. (1995) The relative antioxidant activities of plant-derived polyphenolic flavonoids. *Free Radical Research* 22, 375–383.

Ries, G., Buchholz, G., Frohnmeyer, H. and Hohn, B. (2000) UV-damage-mediated induction of homologous recombination in *Arabidopsis* is dependent on photosynthetically active radiation. *Proceedings of the National Academy of Sciences of the United States of America* 97, 13425–13429.

Ristilä, M., Strid, H., Eriksson, L.A., Strid, Å. and Sävenstrand, H. (2011) The role of the pyridoxine (vitamin B6) biosynthesis enzyme PDX1 in ultraviolet-B radiation responses in plants. *Plant Physiology and Biochemistry* 49, 284–292.

Robson, T.M., Hartikainen, S.M. and Aphalo, P.J. (2015) How does solar ultraviolet-B radiation improve drought tolerance of silver birch (*Betula pendula* Roth.) seedlings? *Plant Cell and Environment* 38, 953–967.

Roro, A.G., Terfa, M.T., Solhaug, K.A., Tsegaye, A., Olsen, J.E. and Torre, S. (2016) The impact of UV radiation at high altitudes close to the equator on morphology and productivity of pea (*Pisum sativum*) in different seasons. *South African Journal of Botany* 106, 119–128.

Sakamoto, A., Lan, V., Hase, Y., Shikazono, N., Matsunaga, T. and Tanaka, A. (2003) Disruption of the AtRev3 gene causes hypersensitivity to UV-B light and γ-rays in Arabidopsis: implication of the presence of a translesion synthesis mechanism in plants. *Plant Cell* 15, 2042–2057.

Saunders, J.A. and McClure, J.W. (1976) The distribution of flavonoids in chloroplasts of twenty-five species of vascular plants. *Phytochemistry* 15, 809–810.

Sävenstrand, H., Olofsson, M., Samuelsson, M. and Strid, Å. (2004a) Induction of early light-inducible protein gene expression in *Pisum sativum* after exposure to low levels of UV-B irradiation and other environmental stresses. *Plant Cell Reports* 22, 532–536.

Sävenstrand, H., Brosché, M. and Strid, Å. (2004b) Ultraviolet-B signalling: *Arabidopsis* brassinosteroid mutants are defective in UV-B regulated defence gene expression. *Plant Physiology and Biochemistry* 42, 687–694.

Šprtová, M., Špunda, V., Kalina, J. and Marek, M.V. (2003) Photosynthetic UV-B response of beech (*Fagus sylvatica* L.) saplings. *Photosynthetica* 41, 533–543.

Strid, Å., Chow, W.S. and Anderson, J.M. (1990) Effects of supplementary UV-B radiation on photosynthesis in *Pisum sativum. Biochimica et Biophysica Acta* 1020, 260–268.

Strid, Å., Chow, W.S., and Anderson, J.M. (1994) UV-B damage and protection at the molecular level in plants. *Photosynthesis Research* 39, 475–489.

Strid, Å., Chow, W.S. and Anderson, J.M. (1996a) Changes in the relaxation of electrochromic shifts of photosynthetic pigments and in the levels of mRNA transcripts in leaves of *Pisum sativum* as a result of exposure to supplementary UV-B radiation: the dependency on the intensity of the photosynthetically active radiation. *Plant Cell Physiology* 37, 61–67.

Strid, Å., Chow, W.S. and Anderson J.M. (1996b) Temperature dependency of changes in the relaxation of electrochromic shifts, of chlorophyll fluorescence, and in the levels of mRNA transcripts in detached leaves from *Pisum sativum* exposed to supplementary UV-B radiation. *Plant Science* 115, 199–206.

Takeuchi, Y., Murakami, M., Nakajima, N., Kondo, N. and Nikaido, O. (1998) The photorepair and photoisomerization of DNA lesions in etiolated cucumber cotyledons after irradiation by UV-B depends on wavelength. *Plant Cell Physiology* 39, 745–750.

Taylor, J.-S. (2006) Structure and properties of DNA photoproducts. In: Siede, W., Kow, Y.W., Doetsch, P.W. (eds) *DNA Damage Recognition*. Taylor & Francis, New York, pp. 67–94.

Teranishi, M., Iwamatsu, Y., Hidema, J. and Kumagai, T. (2004) Ultraviolet-B sensitivities in Japanese lowland rice cultivars: cyclubutane pyrimidine dimer photolyase activity and gene mutation. *Plant Cell Physiology* 45, 1848–1856.

Tjus, S.E. and Andersson, B. (1993) Loss of the trans-thylakoid proton gradient is an early event during photoinhibitory illumination of chloroplast preparations. *Biochimica et Biophysica Acta* 1183, 315–322.

Tossi, V., Lamattina, L., Jenkins, G.I. and Cassia, R.O. (2014) Ultraviolet-B-induced stomatal closure in Arabidopsis is regulated by the UV RESISTANCE LOCUS8 photoreceptor in a nitric oxide-dependent mechanism. *Plant Physiology* 164, 2220–2230.

van Hasselt, P.R., Chow, W.S. and Anderson, J.M. (1996) Short-term treatment of pea leaves with supplementary UV-B at different oxygen concentrations: impacts on chloroplast and plasma membrane bound processes. *Plant Science* 120, 1–9.

Vanhaelewyn, L., Prinsen, E., van der Straeten, D. and Vandenbussche, F. (2016) Hormone-controlled UV-B responses in plants. *Journal of Experimental Botany* 67, 4469–4482.

Wang, X.M. and Strid, Å. (1998) Low doses of UV-B radiation cause formation of a high molecular weight adduct of the chloroplastic 23S rRNA in *Pisum sativum*. Comparison with mRNA levels for plastidic and extraplastidic proteins. In: Garab, G. (ed.) *Photosynthesis: Mechanisms and Effects*, Vol. 3. The Kluwer Academic, Dordrecht, The Netherlands, pp. 2353–2356.

Wargent, J.J., Gegas,, V.C., Jenkins, G.I., Doonan, J.H. and Paul, N.D. (2009) UVR8 in *Arabidopsis thaliana* regulates multiple aspects of cellular differentiation during leaf development in response to ultraviolet-B radiation. *New Phytologist* 183, 315–326.

Waterworth, W.M., Jiang, Q., West, C.E., Nikaido, M. and Bray C.M. (2002) Characterization of *Arabidopsis* photolyase enzymes and analysis of their role in protection from ultraviolet-B radiation. *Journal of Experimental Botany* 53, 1005–1015.

Weller, J.L., Hecht, V., Schoor, J.K.V., Davidson, S.E. and Ross, J.J. (2009) Light regulation of gibberellin biosynthesis in pea is mediated through the COP1/HY5 pathway. *Plant Cell* 21, 800–813.

Wilson, M.I., Ghosh, S., Gerhardt, K.E., Holland, N., Babu, T.S. *et al.* (1995) *In vivo* photomodification of ribulose 1,5-bisphosphate carboxylase oxygenase holoenzyme by ultraviolet-B radiation: formation of a 66-kilodalton variant of the large subunit. *Plant Physiology* 109, 221–229.

Winkel-Shirley, B. (2002) Biosynthesis of flavonoids and effects of stress. *Current Opinion in Plant Biology* 5, 218–223.

Yokawa, K. and Baluska, F. (2015) Pectins, ROS homeostasis and UV-B responses in plant roots. *Phytochemistry* 112, 80–83.

Zhang, J., Henkow, L., Hu, X., Jordan, B.R and Strid, Å. (1994) The effects of ultraviolet-B radiation on the CF_oF_1-ATPase. *Biochimica et Biophysica Acta* 1185, 295–302.

8 Discovery and Characterization of the UV-B Photoreceptor UVR8

Gareth I. Jenkins*

Institute of Molecular, Cell and Systems Biology, College of Medical, Veterinary and Life Sciences, University of Glasgow, UK

Introduction

Light is a key regulator of plant growth and development. This phenomenon, termed photomorphogenesis, is vital for the productivity, survival and reproductive capability of plants as it enables them to modulate their development to optimize light capture for photosynthesis, to compete with their neighbours and to control the timing of physiological processes. In addition, light modifies metabolic activity to produce various compounds that provide sunscreen protection and deter pests and pathogens. Pivotal to the whole of photomorphogenesis is the ability of plants to sense different aspects of their light environment, its spectral quality, intensity, incident direction and duration of the photoperiod. For this purpose they have evolved several different photoreceptors, which, in turn, are coupled to signal transduction networks to initiate the relevant physiological responses (Kami *et al.*, 2010). The most extensively characterized photoreceptors are the phytochromes, which detect principally red and far-red light. In addition, plants possess several UV-A/blue light photoreceptors, notably the cryptochromes and phototropins. Extensive research has provided a detailed understanding of the molecular structures, signalling mechanisms and physiological roles of these photoreceptors (Kami *et al.*, 2010).

As explained below, UV-B wavelengths also initiate photomorphogenic responses in plants, but discovering the underlying molecular mechanisms proved to be an onerous task. Several decades elapsed between the realization that plants possess a UV-B photoreceptor and its actual discovery. One of the barriers to progress was the lack of a characteristic spectral signature for the putative photoreceptor. All proteins, and many other compounds in plants, absorb UV-B wavelengths, making biochemical identification of a UV-B photoreceptor very difficult. In contrast, the red/far-red photoreversibility of phytochrome facilitated its purification from plant material (Briggs and Rice, 1972), and the spectral properties of UV-A/blue light-absorbing flavin chromophores helped in the characterization of cryptochromes (Lin *et al.*, 1995) and phototropins (Christie *et al.*, 1998). These latter photoreceptors were discovered through mutant screens in Arabidopsis (Ahmad and Cashmore, 1993; Liscum and Briggs, 1995), taking advantage of specific UV-A/blue light responses that

*Corresponding author: Gareth.Jenkins@glasgow.ac.uk

they initiated, but this approach was more difficult for the putative UV-B photoreceptor because of the lack of a simple response, specific to the photoreceptor, which could facilitate mutant isolation.

This chapter describes how the elusive UV-B photoreceptor, UV RESISTANCE LOCUS 8 (UVR8) was eventually discovered, and focuses on the structure and mechanism of action of the protein. The events in signal transduction leading to physiological responses are presented in Chapter 9. Further discussion of UVR8 structure and function can be found in several recent reviews, including Jenkins (2009, 2014a, b), Heijde and Ulm (2012), Jiang *et al.* (2012), Li *et al.* (2013), Huang and Deng (2013), Tilbrook *et al.* (2013), Ulm and Jenkins (2015) and Yang *et al.* (2015).

Discovery of the UV-B Photoreceptor UVR8

Photomorphogenic responses to UV-B

A number of studies, principally during the last century, showed that ambient levels of UV-B regulate various aspects of plant development and biochemical composition (Klein, 1978; Tevini and Teramura, 1989; Jordan, 1996). Importantly, research in the 1970s identified responses to low doses of UV-B that were not mediated by phytochrome and were not caused by DNA damage (Wellmann, 1976, 1983). The action spectra for these 'photomorphogenic' UV-B responses had maxima at 295–300 nm, whereas action spectra for DNA damage peak at 260 nm (Wellmann, 1976; Ensminger, 1993; Jenkins, 2009; Jiang *et al.*, 2012). Numerous photomorphogenic UV-B responses have now been characterized, including the promotion of cotyledon opening, the suppression of hypocotyl extension and the stimulation of flavonoid biosynthesis (Wellmann, 1976; Ballaré *et al.*, 1995; Boccalandro *et al.*, 2001; Ryan *et al.*, 2001; Suesslin and Frohnmeyer, 2003). Furthermore, it is now clear that the regulation of transcription of hundreds of genes by low doses of UV-B underpins the observed photomorphogenic responses (Ulm *et al.*, 2004; Brown *et al.*, 2005; Brown and Jenkins, 2008; Favory *et al.*, 2009).

An elusive UV-B photoreceptor

The above research spawned the hypothesis that plants possess a UV-B-specific photoreceptor. However, for 30 years no real progress was made in discovering its molecular identity. Nevertheless, experiments with both mutant plants and plant cell cultures showed unequivocally that photomorphogenic UV-B responses are distinct to those mediated by the phytochrome and cryptochrome photoreceptors (Ballaré *et al.*, 1995; Christie and Jenkins, 1996; Frohnmeyer *et al.*, 1998; Boccalandro *et al.*, 2001; Suesslin and Frohnmeyer, 2002). Furthermore, it was demonstrated that photomorphogenic UV-B responses are not mediated by activation of pathogen defence or wound-response signalling pathways (Boccalandro *et al.*, 2001; Jenkins *et al.*, 2001; Ulm *et al.*, 2004; Jenkins, 2009; Gonzalez Besteiro *et al.*, 2011), whereas these pathways, along with DNA damage signalling and general stress signalling pathways, are involved in some responses to relatively high doses of UV-B (Jenkins *et al.*, 2001, 2009; Kilian *et al.*, 2007; Gonzalez Besteiro *et al.*, 2011; Hideg *et al.*, 2013). However, although the above experiments identified processes that are not involved in photomorphogenic UV-B perception and signalling, they gave little positive indication of what components might actually mediate the responses. The pharmacological experiments with cell cultures provided evidence for the involvement of cellular calcium and redox activity (Christie and Jenkins, 1996; Frohnmeyer *et al.*, 1997, 1999; Long and Jenkins, 1998), but no specific signalling proteins were identified. Other experiments raised the possibility that the putative UV-B photoreceptor might possess pterin or flavin chromophores (Ensminger and Schäfer 1992; Ballaré *et al.*, 1995) and some authors speculated that UV-B could be absorbed by aromatic amino acids in a protein (Ensminger, 1993; Ballaré *et al.*, 1995; Gerhardt *et al.*, 2005), but no specific insights into the molecular basis of UV-B photoreception were obtained.

Thus, it was evident that a different strategy was needed to discover the mechanisms of photomorphogenic UV-B photoreception and signal transduction, and hence the research turned to employing a genetic approach in Arabidopsis, which had proved very successful in studies of other aspects of photomorphogenesis. The initial challenge was deciding which type of genetic screen would have the best prospect of isolating UV-B photoreception or signalling mutants. Screens for mutants altered in sensitivity to UV-B had succeeded in identifying proteins involved in DNA repair (Harlow *et al.*, 1994; Jiang *et al.*, 1997; Landry *et al.*, 1997) or protective sunscreen biosynthesis (Lois and Buchanan, 1994), but were not targeted to photoreception and signalling (Jenkins and Brown, 2007). A screen for altered suppression of hypocotyl extension by UV-B had isolated the *uv-b light insensitive 3* (*uli3*) mutant (Suesslin and Frohnmeyer, 2003), but it was not clear if the candidate ULI3 protein had a specific role in UV-B perception. In theory, the best prospect for isolating mutants specific to photomorphogenic UV-B photoreception and signalling was likely to be a transgene expression screen, in which the promoter of a gene induced by photomorphogenic UV-B perception drives expression of an easily screenable reporter. Initial attempts using the promoter of the *CHALCONE SYNTHASE* (*CHS*) gene, an extensively studied marker of photomorphogenic UV-B signalling (Jenkins *et al.*, 2001) fused to the β-glucuronidase reporter showed promise (Jackson *et al.*, 1995), but it was evident that a different reporter would be needed to enable a large scale screen to be undertaken.

Discovery and initial characterization of the *uvr8* mutant

One of the mutants found in a screen for plants hyper-sensitive to UV-B had a particularly interesting phenotype. In contrast to the DNA repair and sunscreen mutants previously isolated in the screen, this mutant (*uv resistance locus 8-1*) was altered in gene regulation following UV-B exposure (Kliebenstein *et al.*, 2002). In particular, the mutant was impaired in UV-B induction of *CHS* gene expression (Fig. 8.1) and had reduced levels of protective flavonoid compounds. In addition, the mutant displayed elevated expression of the PATHOGENESIS RELATED 1 (PR1) and PR5 proteins, which are induced by activation of a pathogen defence signalling pathway. Thus, the reduced viability of the *uvr8-1* mutant under UV-B was thought to be due to impaired UV-B perception and a consequent lack of protective responses, and the induction of PR1 and PR5 was likely to be the result of stress.

Sequencing of the *UVR8* gene (Kliebenstein *et al.*, 2002) showed that it encodes a 7-bladed β-propeller protein with moderate sequence similarity to human Regulator of Chromosome Condensation 1 (RCC1), which is involved in nucleocytoplasmic transport and the regulation of cell cycle progression and mitosis (Renault *et al.*, 1998). However, it became clear that UVR8 and RCC1 are not functional homologues (Brown *et al.*, 2005). Thus, although UVR8 was proposed to have a role in UV-B signalling (Kliebenstein *et al.*, 2002), how it might function was not evident from the initial molecular characterization.

Additional *uvr8* alleles were isolated through an extensive transgene expression

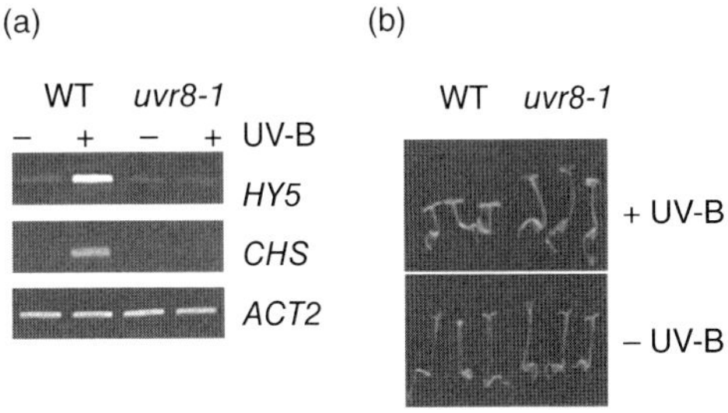

Fig. 8.1. Arabidopsis *uvr8* mutant phenotype. (a) *HY5* and *CHS* transcript levels relative to control *ACTIN2* transcripts in wild-type and *uvr8-1* mutant plants exposed (+) or not (–) to UV-B. Transcripts were assayed by RT-PCR as described by Brown and Jenkins (2008). (b) Hypocotyl length in wild-type and *uvr8-1* mutant seedlings grown in the presence (+) or absence (–) of UV-B as described by Heilmann *et al.* (2016).

screen using the *CHS* promoter fused to the firefly luciferase reporter (Brown *et al.*, 2005). Four mutants defective in the photomorphogenic response to UV-B were obtained by screening 50,000 plants and, remarkably, all were found to be *uvr8* mutants in allelism tests with *uvr8-1*. Subsequently, more *uvr8* alleles were isolated in a different transgene expression screen (Favory *et al.*, 2009).

Further experiments were undertaken to characterize the *uvr8* mutant phenotype. Importantly, Brown *et al.* (2005) showed that the *uvr8* mutant has a UV-B specific phenotype; whereas the induction of *CHS* expression by UV-B was absent, *CHS* induction by several other stimuli was retained. Furthermore, by using microarray analysis, Brown *et al.* (2005) found that UVR8 regulates expression of over 70 genes in response to UV-B. The set of UVR8-regulated genes includes those involved in flavonoid biosynthesis, other metabolic pathways, chloroplast function, DNA repair and protection against oxidative damage. Therefore, the inability of *uvr8* mutant plants to stimulate UV-protective gene expression accounts for their UV-hypersensitive phenotype. A later microarray study involving narrowband UV-B exposure of seedlings (Favory *et al.*, 2009) found that UVR8 regulates potentially hundreds of genes, emphasizing the importance of UVR8 in a range of plant processes.

UVR8 was found to regulate gene expression over a broad range of UV-B fluence rates, from the very low fluence rates characteristic of photomorphogenic responses to above-ambient levels (Brown and Jenkins, 2008). In contrast, UV-B signalling pathways independent of UVR8 operate over the higher fluence range (Brown and Jenkins, 2008; Jenkins, 2009). Thus, gene expression responses mediated by UVR8 are crucial in enabling plants to acclimate to UV-B so they can tolerate exposure to relatively high levels of ambient UV-B (Favory *et al.*, 2009; Morales *et al.*, 2013).

UV-B stimulates the expression of various transcription factors likely to mediate the large-scale transcriptomic responses to UV-B (Ulm *et al.*, 2004; Kilian *et al.*, 2007). Among these are two transcription factors prominently involved in photomorphogenesis, ELONGATED HYPOCOTYL 5 (HY5) and HY5 HOMOLOG (HYH). Ulm *et al.* (2004) found that *hy5* mutant plants are defective in the UV-B-induced expression of several genes. Significantly, Brown *et al.* (2005) discovered that UVR8 regulates UV-B stimulated *HY5* and *HYH* gene expression (Fig. 8.1) and, moreover, that HY5 is required for the expression of many UVR8-regulated genes. Thus, they proposed that HY5 is a key transcriptional effector of UVR8 regulated gene expression. Indeed, *hy5* mutant plants showed hypersensitivity to UV-B, similar to *uvr8* mutants. Brown and Jenkins (2008) further showed that HY5 and HYH act redundantly to regulate the expression of a number of UVR8 target genes, although not all UVR8-regulated gene expression responses are controlled by HY5/HYH (Feher *et al.*, 2011; Hayes *et al.*, 2014).

Research in Roman Ulm's laboratory discovered the role of the CONSTITUTIVELY PHOTOMORPHOGENIC 1 (COP1) protein in photomorphogenic UV-B responses. COP1, bound to a SPA protein, acts as a substrate receptor for an E3 ubiquitin ligase complex to mediate the targeted proteolytic degradation of effectors of photomorphogenesis, such as HY5, in darkness (Osterlund *et al.*, 2000; Lau and Deng, 2012). However, COP1 was required for the stimulation of *HY5* gene expression and accumulation of the HY5 protein in response to UV-B exposure (Oravecz *et al.*, 2006; Favory *et al.*, 2009). Furthermore, transcriptome analysis revealed that COP1 regulated largely the same set of genes as UVR8, indicating that UVR8 and COP1 function together in the same pathway (Favory *et al.*, 2009). Consistent with these findings, UV-B exposure was found to stimulate a physical interaction between UVR8 and COP1 (Favory *et al.*, 2009). It was later shown that COP1-SPA switches its association from the E3 ubiquitin ligase complex to UVR8 following UV-B exposure (Huang *et al.*, 2013).

Thus, a model for photomorphogenic UV-B responses emerged from the above studies of UVR8, HY5 and COP1: UV-B exposure induced an interaction between UVR8 and COP1, causing the rapid stimulation of expression and accumulation of the HY5/HYH transcription factors, which in turn regulated the transcription of a large number of genes in the UVR8 pathway. The outstanding question was how UV-B initiated the above sequence of events.

Evidence that UVR8 is a UV-B photoreceptor

Several lines of evidence suggested that UVR8 might be a UV-B photoreceptor (Favory *et al.*, 2009; Jenkins, 2009). Firstly, it is required for photomorphogenic responses initiated by low fluence rates of UV-B, including gene expression, flavonoid biosynthesis and hypocotyl growth suppression (Kliebenstein *et al.*, 2002; Brown *et al.*, 2005; Brown and Jenkins, 2008; Favory *et al.*, 2009; Fig. 8.1). Second, it acts in a UV-B specific manner (Brown *et al.*, 2005). Third, experiments using transgenic plants expressing a GFP-UVR8 fusion showed that UV-B exposure stimulated the rapid accumulation of UVR8 in the nucleus (Kaiserli and Jenkins, 2007), analogous to the nuclear accumulation of phytochrome in response to inductive light treatments (Kircher *et al.*, 1999; Yamaguchi *et al.*, 1999). Fourth, UVR8 is involved in rapid (within 5 minutes) responses to UV-B, both nuclear accumulation and interaction with COP1 (Kaiserli and Jenkins, 2007; Favory *et al.*, 2009). Fifth, extensive genetic screens had failed to isolate any other component that could be a candidate UV-B photoreceptor (Brown *et al.*, 2005; Favory *et al.*, 2009).

Research was therefore directed towards testing whether UVR8 could function as a UV-B photoreceptor. Ultimately to demonstrate whether a protein acts as a photoreceptor it is necessary to show that direct absorption of light of the appropriate wavelengths causes a change in the protein and that this change initiates a relevant *in vivo* response; furthermore, a mutation of the protein that modifies light absorption should have a corresponding effect on its activity.

Rizzini *et al.* (2011) made the important discovery that UVR8 exists as a homo-dimer in plants not exposed to UV-B and that UV-B illumination triggers rapid conversion of the dimer into monomers. Moreover, a conformational change occurs during monomerisation that exposes the C-terminus of the protein, making it available for detection by a specific antibody. This change probably facilitates the interaction of UVR8 with COP1, which involves a region of UVR8 located near to the C-terminus (Cloix *et al.*, 2012; Yin *et al.*, 2015). Rizzini *et al.* (2011) further showed that the UV-B-induced interaction between COP1 and UVR8 previously seen in plants (Favory *et al.*, 2009) also occurs when plant extracts are exposed to UV-B, provided that UVR8 is present.

The above experiments are consistent with the hypothesis that UVR8 acts as a UV-B photoreceptor to initiate dimer dissociation and subsequent interaction with COP1, but since the experiments were undertaken with plant material it is difficult to completely rule out the possibility that some other plant protein, dependent on the presence of UVR8, actually functions as the photoreceptor. Therefore, the demonstration that UVR8 initiates a response to UV-B in non-plant cells was key to concluding that it is a UV-B photoreceptor. Arabidopsis UVR8 expressed in yeast converts from the dimer to the monomer following UV-B exposure and, furthermore, when COP1 is also expressed in the cells, UV-B induces an interaction between UVR8 and COP1 (Rizzini *et al.*, 2011). Similar results are obtained with mammalian cells (Rizzini *et al.*, 2011; Crefcoeur *et al.*, 2013). There is no likelihood that an endogenous UV-B perception pathway could initiate the UV-B response; neither yeast nor mammalian cells possess a UVR8 protein and the response requires the expression of functional UVR8. Moreover, activation of the DNA damage signalling pathway in yeast does not induce the interaction between UVR8 and COP1 (Cloix *et al.*, 2012). The response to UV-B is impaired by mutations of UVR8 (see below), consistent with it acting as the photoreceptor, but does not require an intact COP1, since UVR8 will interact in a UV-B-dependent manner with truncated COP1 containing only the WD40-repeat domain (Rizzini *et al.*, 2011; Cloix *et al.*, 2012).

Thus the experiments in heterologous systems, taken together with those in plants, provide powerful evidence that UVR8 acts as a UV-B photoreceptor. UV-B causes a physical change to the protein, namely monomerisation, which initiates the interaction with COP1, leading to physiological responses in plants (see Fig. 8.2). It was subsequently demonstrated that purified, dimeric UVR8 protein undergoes UV-B-induced monomerisation (Christie *et al.*, 2012; Wu *et al.*, 2012; Fig. 8.3a) and, as discussed below, mutations that affect photoreception have an equivalent effect *in vivo*. Therefore UVR8 satisfies all the criteria required to demonstrate its photoreceptor function.

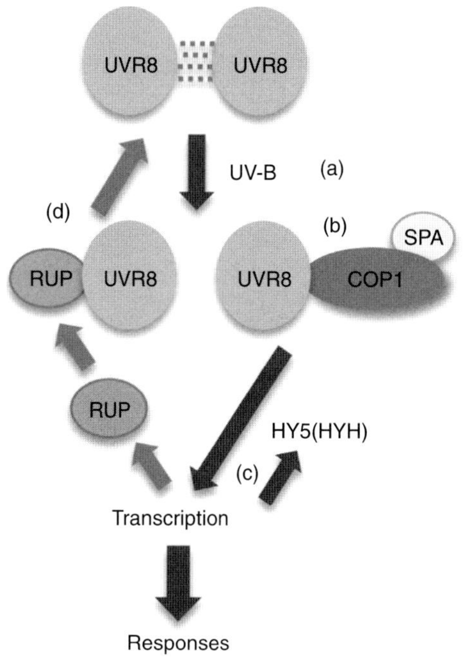

Fig. 8.2. A model of UVR8 action. Dimeric UVR8 monomerises following UV-B exposure (a). Monomeric UVR8 interacts with COP1 bound to a SPA protein (b) to initiate transcription of target genes (c) and physiological responses. UVR8/COP1 stimulate expression and accumulation of the HY5 and HYH transcription factors, which mediate transcription of many UVR8 regulated genes. Among the genes induced by UVR8/COP1/HY5 are those encoding RUP1 and RUP2 proteins, which interact with UVR8 monomers to promote re-dimerisation (d). The figure is modified from Jenkins (2014b).

UVR8 Structure and Mechanism of Action

UVR8 crystal structure

Procedures were devised to express UVR8 in heterologous systems, notably *E. coli*, and to purify sufficient amounts of the recombinant protein for *in vitro* studies. This enabled both the structure of UVR8 to be elucidated by x-ray crystallography and the mechanism of UVR8 action to be explored through mutational studies (Christie *et al.*, 2012; Wu *et al.*, 2012). Unfortunately, attempts to crystallize intact UVR8 were unsuccessful, because it appears that the N- and C-terminal regions of the protein are flexible and their presence impairs crystallization. Hence, a truncated form of UVR8 was produced and high resolution crystal structures were reported for the protein lacking 11 amino acids at the N-terminus and 59 amino acids at the C-terminus (Christie *et al.*, 2012; Wu *et al.*, 2012). Nevertheless, this truncated protein is still active in UV-B photoreception.

The crystal structures were obtained only for the dimeric form of UVR8, since monomeric UVR8 re-associates to form dimers under the conditions used for crystallization. Moreover, attempts to induce monomerisation in crystals of the dimer simply led to fracturing of the crystals (Wu *et al.*, 2012). Thus, no structural information is presently available for the monomer, nor for the C-terminal region of UVR8. This is a significant gap in knowledge, because a detailed understanding of the monomer structure and of the UV-B-induced conformational changes associated with its formation (Rizzini *et al.*, 2011; Heilmann *et al.*, 2014; Miyamori *et al.*, 2015) will indicate how the signalling-active state is generated.

As expected from the predicted amino acid sequence (Kliebenstein *et al.*, 2002) and homology models based on the structure of proteins with sequence similarity to UVR8 (Rizzini *et al.*, 2011; Wu *et al.*, 2011), the UVR8 structure is that of a 7-bladed β-propeller protein (Christie *et al.*, 2012; Wu *et al.*, 2012; Fig. 8.3b). The protein has a doughnut shape with a water-filled depression at the rear face

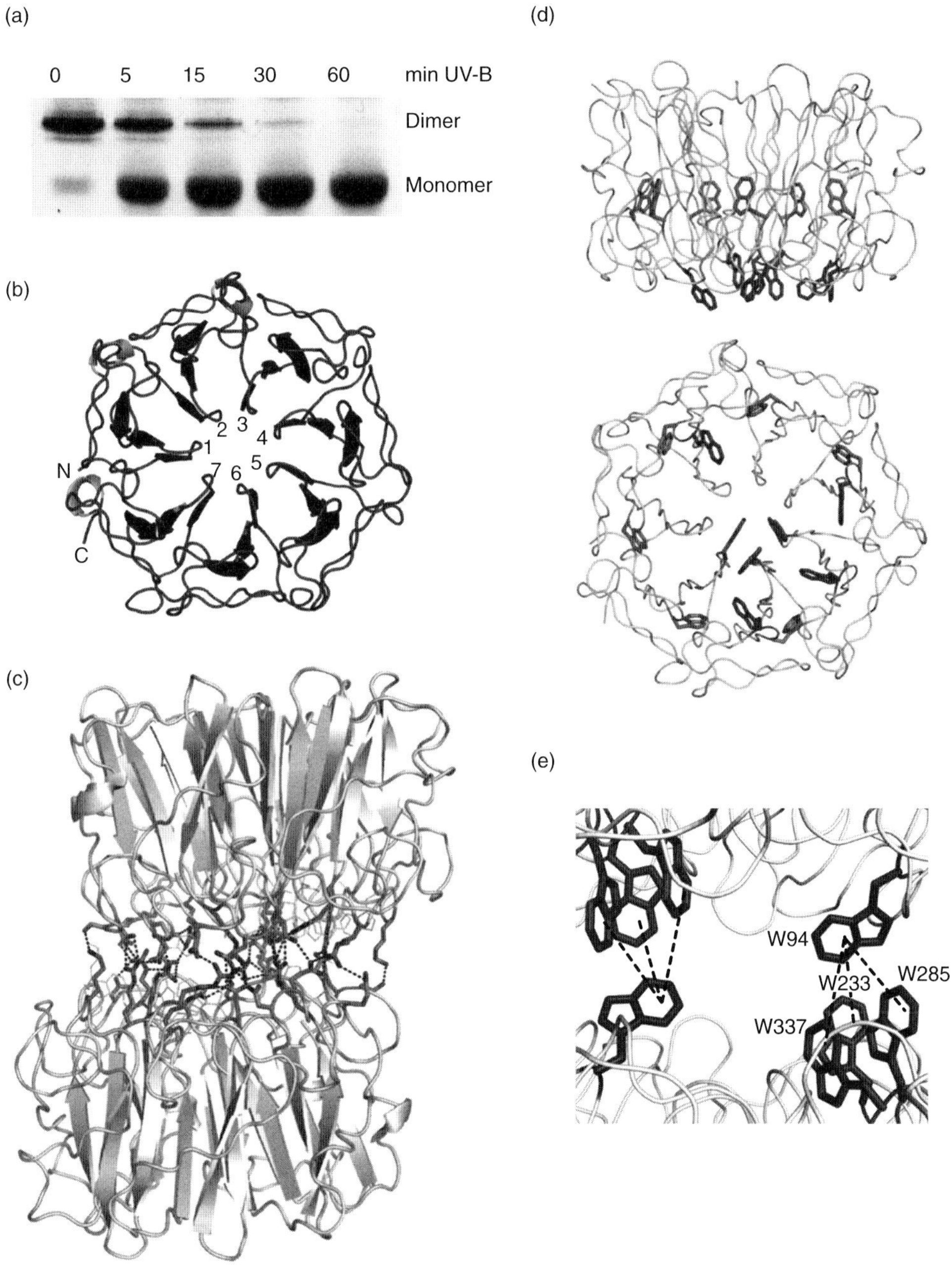

Fig. 8.3. UVR8 protein structure. (a) UV-B exposure of purified UVR8 converts the dimer to the monomer, as shown by SDS-PAGE with non-boiled protein samples. (b) 7-bladed β-propeller structure of the UVR8 monomer. (c) Salt bridge interactions between charged amino acids across the dimer interface maintain the dimeric structure. (d) Images of the UVR8 monomer from the side (upper image) and from the dimer interface (lower image) showing the 6 tryptophans in the β-propeller core and 7 tryptophans in the dimer interface. (e) Pyramidal arrangement of tryptophans across the dimer interface. (a) is from Christie *et al.* (2012), (b)–(e) are modified from Jenkins (2014b).

and has several features that make it distinct from related proteins (Jenkins, 2014b). UVR8 molecules in the dimer come into contact through the surface at the front face, termed the dimer interface (Fig. 8.3c).

Charged amino acids maintain the dimer

The dimer interface is key to UVR8 structure and function. This interface contains several charged amino acids that are crucial for maintaining the dimer structure in close proximity to aromatic amino acids that are concerned with UV-B photoreception. The charged amino acids, which are mainly basic arginine and acidic aspartate and glutamate residues, are arranged in patches of electrostatic potential that facilitate the formation of salt bridge interactions between the monomers (Fig. 8.3c). The network of salt bridges across the dimer interface is sufficiently strong to hold the monomers together even in the presence of high concentrations of sodium dodecyl sulfate (SDS), provided the sample is not boiled (Rizzini *et al.*, 2011; Christie *et al.*, 2012; Wu *et al.*, 2012). However, conditions that neutralize the salt bridges such as increased ionic strength (Wu *et al.*, 2012) or decreased pH (Christie *et al.*, 2012) result in dissociation of the monomers.

The crystal structure indicates that some salt bridge interactions across the dimer interface are likely to be stronger than others. Hence, to test the relative importance of particular charged amino acids in maintaining the dimer structure, mutant proteins were produced by site-directed mutagenesis. The dimer/monomer status of purified mutant proteins was then examined (Christie *et al.*, 2012; Wu *et al.*, 2012; Heilmann *et al.*, 2016). A key amino acid is arginine (R) 286 (R286), which forms single and double hydrogen-bonded salt bridges with aspartic acid (D) residues D96 and D107, respectively, on the opposing monomer. Mutation either of R286 to alanine (UVR8^{R286A}), or of D96 and D107 to structurally similar asparagine (N) (UVR8^{D96N,D107N}) results in UVR8 becoming constitutively monomeric, demonstrating the importance of these amino acids in dimer formation. However, if R286 is conservatively mutated to lysine, UVR8 appears dimeric and monomerises in response to UV-B exposure, highlighting the importance of the positive charge of this residue. A further important residue is R338, which interacts with both D44 and glutamic acid (E) E43 on the opposing monomer. Wu *et al.* (2012) reported that UVR8^{R338A} is constitutively monomeric, indicating that R338 is also important in maintaining the dimer. However, the dimer/monomer status of this mutant is dependent on salt concentration, appearing monomeric at high salt concentrations and in a dimer/monomer equilibrium at lower concentrations (Heilmann *et al.*, 2016). Computational studies support the conclusion that both R286 and R338 are important residues for dimer formation (Wu *et al.*, 2013).

An important point is that specific mutations may weaken the dimer without leading to complete dissociation. Hence the method of determining dimer/monomer status is critical. SDS-PAGE with non-boiled samples is commonly used to examine UVR8 dimer/monomer status, and with this method wild-type UVR8 appears dimeric prior to UV-B exposure and monomeric following UV-B exposure (Fig. 8.3a). However, this method detects even slight weakening of the dimer and many mutants appear monomeric in the above assay. In contrast, size exclusion chromatography (SEC) enables dimer/monomer status to be assessed more rigorously. For instance, the alanine mutant of R146, which forms a double hydrogen bonded salt bridge with E182, appears constitutively monomeric in the non-boiled SDS-PAGE assay, but under SEC is a dimer that monomerises in response to UV-B (Christie *et al.*, 2012; Wu *et al.*, 2012; Heilmann *et al.*, 2016).

Specific tryptophans act in UV-B photoreception

Whereas photoreceptor proteins normally bind non-proteinaceous cofactor molecules ('chromophores') to enable them to detect particular wavelengths of light, purified UVR8 does not have a bound chromophore to facilitate UV-B absorption. It is therefore likely that specific amino acids in the primary sequence act in UV-B photoreception, the obvious candidates being tryptophan (W) amino acids, which strongly absorb UV-B wavelengths. Arabidopsis UVR8 has 14 tryptophans, one in the C-terminal region, 6 in the β-propeller core of the protein, and 7 in the dimer interface (Fig. 8.3d). It is likely that the core tryptophans help to maintain the β-propeller structure, since they form hydrogen bonds and hydrophobic interactions between adjacent propeller blades. Mutation of some of these core tryptophans to alanine causes UVR8 to be unstable or non-functional *in vivo*, whereas mutation to amino acids that are still able to form hydrophobic interactions (phenylalanine or tyrosine) produces stable proteins (O'Hara and Jenkins, 2012).

Several lines of evidence show that particular tryptophans in the dimer interface are intimately concerned with UV-B photoreception. Three of these tryptophans, the 'triad' of W233, W285 and W337, are sufficiently close that their electronic orbitals overlap. Moreover, the triad of tryptophans in each monomer of the dimer is in close proximity to W94 of the opposing monomer. Thus, each dimer has two pyramidal arrangements of tryptophans across the dimer interface (Christie *et al.*, 2012; Fig. 8.3e). The excitonic coupling of tryptophans gives rise to a characteristic signal in the far-UV circular dichroism (CD) spectrum, showing a peak at approximately 234 nm and a trough at approximately 221 nm (Grishina and Woody, 1994). Such a signal was observed for UVR8 (Christie et al., 2012) and, moreover, mutation of the pyramid tryptophans to either alanine or phenylalanine (F) reduced the CD signal, supporting the hypothesis that the pyramid tryptophans are electronically coupled. Mutation of the different tryptophans decreased the CD signal to varying extents, with the greatest reduction observed for $UVR8^{W233F}$.

Importantly, the CD signal was much reduced following exposure of wild-type UVR8 protein to UV-B, indicating that it provides a spectral signature for photoreception (Christie *et al.*, 2012). Several factors may contribute to this spectral change, including a potential loss of excited electrons from the tryptophan pyramid (Mathes *et al.*, 2015), changes in the relative orientation of electronically coupled tryptophan indole side chains (Zeng *et al.*, 2015), and disruption of the cross-dimer pyramid following monomerisation. Consistent with the latter possibility, constitutively monomeric UVR8 mutants were found to have a substantially reduced CD signal (Christie *et al.*, 2012). Furthermore, mutant analysis provided insights into the importance of particular tryptophans in UV-B photoreception. The reduction in the CD signal following UV-B exposure was not observed when either W233 or W285 was mutated to alanine or phenylalanine, whereas mutation of W337 or W94 had little and no effect, respectively. Thus, W233 and W285 are required for the photoactivation of UVR8 by UV-B. Moreover, mutation of W285 to phenylalanine alters the spectral sensitivity of UVR8 (Christie *et al.*, 2012). Phenylalanine absorbs UV-C wavelengths, and thus UV-C exposure of $UVR8^{W285F}$ initiates monomerisation and reduces the CD signal, albeit weakly. This observation highlights the key role of W285 in UVR8 photoreception.

Further evidence that W285 and W233 are crucial amino acids in UV-B photoreception by UVR8 was provided by fluorescence spectroscopy. Tryptophans fluoresce following UV-B absorption and changes in fluorescence emission can provide insights into the identity and behaviour of chromophore tryptophans. Wu *et al.* (2012) observed a rise in fluorescence at 335 nm following UV-B excitation of UVR8, which was followed by a gradual decline. However, phenylalanine mutants of W285 and W233 did not show this fluorescence increase, indicating that they are impaired in UV-B photoreception. In contrast, several other tryptophan mutants, including $UVR8^{W337F}$ and $UVR8^{W94F}$ had similar fluorescence emission to wild-type UVR8. These observations concur with the CD spectroscopy data (Christie *et al.*, 2012) and support the hypothesis that W285 and W233 are the principal UV-B chromophores for UVR8. Further support for this hypothesis is provided by assays of the dimer/monomer status of the tryptophan mutants. SEC shows that phenylalanine and alanine mutants of W285 and W233 are dimers that are non-responsive to UV-B, whereas the equivalent mutants of W94 and W337 are dimers that monomerise in response to UV-B (Christie *et al.*, 2012; Wu *et al.*, 2012).

Examination of the fluorescence emission spectrum of wild-type UVR8 following excitation by UV-B shows that the peak of emission shifts to longer wavelengths (327 to 335 nm, Heilmann *et al.*, 2014; 332 to 337 nm, Liu *et al.*, 2014). Such a red-shift in fluorescence is observed when tryptophans become exposed to a more polar environment, so a likely explanation is that alteration of charge networks at the interface during monomerisation modifies the local protein environment of tryptophans. Observations with salt bridge mutants support this interpretation. In the constitutively monomeric R286A mutant the emission is already red-shifted prior to UV-B exposure and shows little change thereafter (Heilmann *et al.*, 2014).

The mechanism of UVR8 photoreception

Since UVR8 is the only known photoreceptor that does not employ one or more bound chromophores for light perception, there is considerable interest in elucidating its unique mechanism of tryptophan-mediated photoreception. The researchers who obtained the first crystal structures of UVR8 suggested possible mechanisms. Wu *et al.* (2012) highlighted the network of cation-π interactions between the aromatic rings of tryptophans at the dimer interface and the side chains of adjacent arginines, including the key salt bridging residue R286. They suggested that disruption of these interactions following photoreception could lead to dimer dissociation. Christie *et al.* (2012) proposed that excitation of electrons in the tryptophan triad following UV-B absorption could lead to the transfer of an electron to a salt bridging arginine, such as R286, thus neutralising the salt bridges and causing monomerisation.

Subsequent computational studies support the hypothesis that electron transfer from chromophore tryptophans to salt bridging amino acids initiates dimer dissociation, but they differ in details of the proposed mechanism (Jenkins, 2014b; Li *et al.*, 2014; Voityuk *et al.*, 2014; Wu *et al.*, 2014; Yang *et al.*, 2015). From calculations of the excited state properties of selected tryptophans, Voityuk *et al.* (2014) proposed that W233 and W285 undergo charge separation to produce the radical ion pair [$W233^{-}$ $W285^{+}$] and suggested that the relatively large dipole moment of this charge transfer state could alter the electrostatic potential between salt bridge amino acids, facilitating proton transfer from R286 to D96 and thus neutralizing the salt bridge. Moreover, the large dipole moment could cause structural changes leading to the physical separation of residues involved in salt bridge formation or, as suggested by Zeng *et al.* (2015) the physical re-orientation of chromophore tryptophans.

Importantly, there is now experimental evidence that proton coupled electron transfer is involved in the photoreception mechanism of UVR8. Time-resolved fluorescence measurements following UV-B excitation of UVR8 (Liu *et al.*, 2014) revealed rapid (~150 ps) fluorescence quenching, most likely associated with photochemical events within the tryptophan triad because it was much reduced in the $UVR8^{W285F}$ mutant. Mathes *et al.* (2015) used time-resolved fluorescence and absorption spectroscopy to monitor processes occurring over the femtosecond to microsecond range following UV-B exposure. They observed three components in the decay of fluorescence, each with distinct temporal and spectral characteristics, the most rapid of which (150–400 ps) probably corresponds to the 150 ps component observed by Liu *et al.* (2014). Absorption spectroscopy revealed that the initial excited state decays over the several hundred picosecond to nanosecond timescale to form a tryptophan neutral triplet species, with the subsequent appearance of a tryptophan neutral radical that is stable for hundreds of microseconds. One of the complications of this type of analysis is that multiple tryptophans, and not only those in the pyramid, can potentially contribute to the spectral changes observed, and it is therefore difficult to assign spectral changes to specific tryptophans. Nevertheless, based on the spectral signatures and lifetimes of the intermediates identified, the authors proposed a model for the molecular mechanism of UVR8 photoreception in which electron transfer from W285 to R286 is accompanied by proton transfer to D96, via a water molecule, giving rise to the observed neutral radical of W285 and neutralisation of the R286-D96 salt bridge (Fig. 8.4).

Zeng *et al.* (2015) obtained further novel insights into the mechanism of photoreception by using low temperature dynamic X-ray crystallography to identify structural changes associated with UV-B photoreception. They detected a change in the relative orientation of the indole rings of W233 and W285 following UV-B absorption, which they suggested is triggered by the rapid initial charge separation of [W233⁻ W285⁺] proposed by Voityuk *et al.* (2014). Furthermore, rotation of the W285 indole ring causes the water molecule involved in hydrogen bonding between W285, R286 and D96 to be ejected, hence weakening the critical salt bridge interactions involving R286. Zeng *et al.* (2015) further observed large-scale motions in the UVR8 β-propeller, particularly in the blades housing W285 and W233, which might also promote dimer dissociation.

Other researchers have also noted that structural changes accompany photoactivation of UVR8. Heilmann *et al.* (2014) found differences in elements of secondary structure between the UVR8 dimer and monomer by FTIR spectroscopy, although there was no evidence for major changes in secondary structure in the protein. Using transient grating spectroscopy, Miyamori *et al.* (2015) observed a UV-B-induced change in conformation of UVR8 within approximately 50 ms of UV-B excitation followed by monomerisation within 200 ms. The changes observed by Miyamori *et al.* (2015) and Heilmann *et al.* (2014) did not involve the C-terminal region of the protein as they were still present in C-terminally truncated mutant proteins. Nevertheless, experiments involving antibody detection (Rizzini *et al.*, 2011) and limited proteolysis (Heilmann *et al.*,

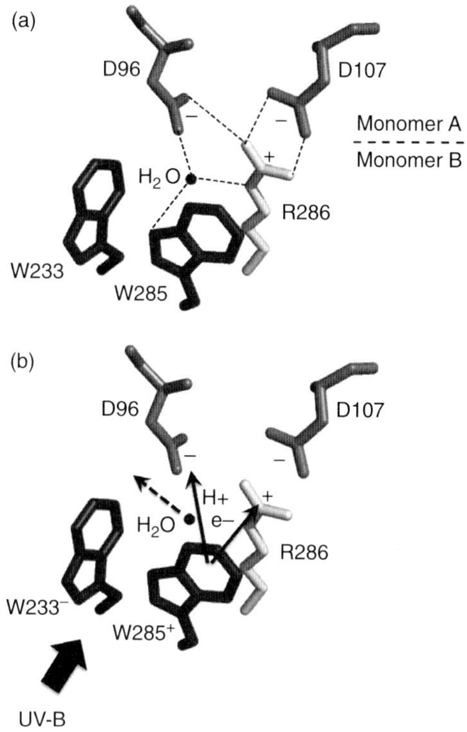

Fig. 8.4. Proton coupled electron transfer in the UVR8 photoreception mechanism. (a) The chromophore tryptophans W233 and W285 are adjacent to R286, which forms salt bridges across the dimer interface with D96 and D107 of the associated monomer. A water molecule mediates hydrogen bonding between W285, R286 and D96. Hydrogen bonds are shown as dotted lines. (b) Exposure to UV-B is proposed to cause: charge separation between W233 and W285 generating [W233- W285+] (Voityuk *et al.*, 2014); electron transfer from W285 to R286 coupled with proton transfer from W285 to D96 via the water molecule resulting in formation of the W285 neutral radical (Mathes *et al.*, 2015); rotation of the indole rings of W233 and W285 causing ejection of the water molecule (Zeng *et al.*, 2015). These changes lead to disruption of the salt bridge/hydrogen bond network that is key to maintaining the dimer, resulting in dimer dissociation.

2014) show that the C-terminal region of UVR8 becomes more accessible in the monomer. Thus, several different approaches reveal that UVR8 undergoes modest structural changes to the core of the protein as a result of photoreception, and that although these changes do not require the C-terminus of the protein the formation of the monomer leads to an increase in availability of the C-terminal region required for interaction with COP1. A caveat to these studies with purified UVR8 is that no other protein is present, whereas there is evidence that both the dimer and monomer can interact with two closely related proteins, REPRESSOR OF UV-B PHOTOMORPHOGENESIS 1 (RUP1) and RUP2 that bind to the C-terminal region (Gruber *et al.*, 2010; Cloix *et al.*, 2012). Hence the presence of the RUPs could modify the conformation of UVR8 and influence some of its structural and dynamic properties.

A number of questions regarding the photoactivation mechanism of UVR8 remain to be answered. One of the most important is the role of the non-pyramid tryptophans, whose number and positions are highly conserved in UVR8 sequences. As mentioned above, the tryptophans in the β-propeller core are likely to have a structural role, but the function of the non-pyramid tryptophans and other aromatic amino acids in the dimer interface is unclear. W198, W250 and W302 are positioned at the periphery of the dimer interface together with F305 and tyrosine (Y) Y201 and Y253. Several studies have suggested that non-pyramid tryptophans may act as a UV-B 'antenna', channelling excitation energy to the chromophore tryptophans, thereby increasing the efficiency and potentially the wavelength range of UV-B detection (Liu *et al.*, 2014; Voityuk *et al.*, 2014; Zeng *et al.*, 2015; Wu *et al.*, 2015). Voityuk *et al.* (2014) suggested that W233 is likely to be the terminal acceptor for energy transfer because its excited state is strongly stabilised by the local protein environment. Wu *et al.* (2015) came to the same conclusion from calculations of the absorption spectra of individual tryptophans in UVR8, which suggested that W233 has the longest wavelength of absorption. These authors additionally proposed that the overlapping absorption spectra of interface tryptophans, and in particular W94, may extend the wavelength range over which UVR8 can function.

Experimental evidence in support of excitation energy transfer between tryptophans in UVR8 was reported by Liu *et al.* (2014). The authors observed that fluorescence emission following UV-B exposure of UVR8 is much greater in $UVR8^{W285F}$ than in the wild-type protein, and suggested that W285 strongly quenches fluorescence following resonant energy transfer from multiple tryptophans. Measurements of time-resolved fluorescence emission identified components decaying over 1–2 ns, which were proposed to be associated with resonant energy transfer from tryptophans in the β-propeller and dimer interface to the triad tryptophans. However, Mathes *et al.* (2015) did not find any evidence for resonant energy transfer between tryptophans in measurements of the anisotropy of fluorescence emission. Hence, further studies are needed to test the hypothesis of excitation energy transfer between UVR8 tryptophans, and in this respect *in vivo* experimentation will be particularly valuable. It is important to note that plants growing in sunlight only perceive wavelengths greater than approximately 295 nm because of absorption by the ozone layer, so absorption of shorter wavelengths by UVR8 tryptophans, whether observed *in vitro* or predicted by calculations, is irrelevant *in vivo*. Measurements of UVR8-mediated gene expression show that the photoreceptor is active up to at least 310 nm (Brown *et al.*, 2009).

Regeneration of the UVR8 dimer

It is important to consider how the dimeric photoreceptor is regenerated following monomerisation. In principle the monomers could be degraded and new dimers synthesized, but experiments using inhibitors of protein synthesis and degradation indicate that rapid turnover of UVR8 does not occur (Heijde and Ulm, 2013; Heilmann and Jenkins, 2013). With purified UVR8, monomers reassociate to form the dimer following UV-B exposure, but the process takes over 24 hours for completion (Christie *et al.*, 2012; Wu *et al.*, 2012). However, *in vivo* re-dimerisation occurs within an hour (Heijde and Ulm, 2013; Heilmann and Jenkins, 2013), suggesting that it is facilitated by other proteins. Indeed, RUP1 and RUP2 stimulate re-dimerisation

(Heijde and Ulm, 2013; Fig. 8.2). The RUPs interact with the same region of UVR8 as COP1 (Cloix *et al.*, 2012), which dissociates from UVR8 during re-dimerisation (Heijde and Ulm, 2013). Thus the RUP proteins act as negative regulators of UVR8 signalling (Gruber *et al.*, 2010; see Chapter 9).

In vivo structure–function studies of UVR8

Several studies have investigated the mechanism of action of UVR8 *in vivo*. Inevitably the molecular environment of UVR8 in cells will differ from that used in experiments with the purified protein. Apart from differences in, for example, ionic composition and the presence of various small organic molecules, the presence of interacting proteins, notably COP1 and the RUP proteins will be a key factor. UVR8 may constitutively bind RUP proteins *in vivo* (Gruber *et al.*, 2010; Cloix *et al.*, 2012), but this will not occur with the purified protein. It is therefore essential to extend hypotheses developed in studies of the purified protein to the *in vivo* situation.

Most studies to investigate the mechanism of UVR8 action *in vivo* have focused on the role of tryptophans in photoreception. Interestingly, these studies were initiated prior to the availability of a crystal structure as it was anticipated that tryptophans would have a key role in UVR8 function. The first mutational studies of UVR8 involved expression in yeast cells. When $UVR8^{W285F}$ is expressed in yeast it forms a dimer that does not respond to UV-B (Rizzini *et al.*, 2011; Huang *et al.*, 2014), consistent with experiments with the purified protein (Christie *et al.*, 2012; Wu *et al.*, 2012). In contrast, $UVR8^{W337F}$ is dimeric and monomerises in response to UV-B, highlighting the functional difference between W285 and W337 (Rizzini *et al.*, 2011). $UVR8^{W233F}$, $UVR8^{W233A}$, $UVR8^{W285A}$ and $UVR8^{W337A}$ are all reported to be constitutively monomeric in yeast when examined by SDS-PAGE with non-boiled samples (Rizzini *et al.*, 2011; Huang *et al.*, 2014). However, as mentioned previously, the more rigorous SEC method shows that the purified proteins form dimers that either respond ($UVR8^{W337A}$) or have an impaired response ($UVR8^{W233F}$, $UVR8^{W233A}$, $UVR8^{W285A}$) to UV-B (Christie *et al.*, 2012; Wu *et al.*, 2012). Nevertheless, it is clear that the above mutations weaken the dimer structures. Further experiments with proteins expressed in yeast examined the UV-B-induced interaction of UVR8 with COP1. Rizzini *et al.* (2011) and O'Hara and Jenkins (2012) reported that phenylalanine and tyrosine mutants of the triad tryptophans do not interact with COP1, indicating an inability of the mutants to form the signalling-active monomeric state, although Huang *et al.* (2014) reported interaction between $UVR8^{W233F}$ and COP1. In contrast, alanine mutants of the triad tryptophans interact with COP1 in both the presence and absence of UV-B (Rizzini *et al.*, 2011; O'Hara and Jenkins, 2012; Huang *et al.*, 2014), most likely because changes in the conformation of the protein caused by the introduction of alanine, which have been demonstrated for $UVR8^{W285A}$ by crystallography (Wu *et al.*, 2012), result in constitutive exposure of the region that binds COP1.

Structure–function studies were undertaken in Arabidopsis by transformation of UVR8 tryptophan mutants into *uvr8* null mutant plants (Christie *et al.*, 2012; O'Hara and Jenkins, 2012; Heijde *et al.*, 2013; Huang *et al.*, 2013, 2014). In addition to studies of dimer/monomer status and interaction with COP1, it was possible to test functional complementation of the impaired UV-B-induced hypocotyl growth suppression and gene expression phenotypes of the *uvr8* mutant. $UVR8^{W285F}$ forms a dimeric protein that does not respond to UV-B, does not interact with COP1 and does not complement the impaired responses to UV-B of *uvr8-1* (O'Hara and Jenkins, 2012; Heijde *et al.*, 2013; Huang *et al.*, 2013, 2014). In contrast, $UVR8^{W285A}$ appears constitutively monomeric and interacts constitutively with COP1, consistent with the findings in yeast (O'Hara and Jenkins, 2012; Heijde *et al.*, 2013; Huang *et al.*, 2013, 2014). Huang *et al.* (2013, 2014) and Heijde *et al.* (2013) reported that $UVR8^{W285A}$ plants exhibit a *cop* mutant phenotype, with short hypocotyls and expanded cotyledons in darkness. This phenotype was most obvious in plants strongly expressing the transgene (30–40 fold over-expression), consistent with sequestration of COP1 by binding to UVR8^{W285A} (Heijde *et al.*, 2013). These over-expressers accumulated relatively high levels of the HY5 transcription factor, a target of COP1 E3 ubiquitin ligase activity, and hence displayed photomorphogenic UV-B responses in the absence of UV-B.

Both phenylalanine and alanine mutants of W233 are impaired in responses to UV-B in Arabidopsis, although activity is detectable (O'Hara and Jenkins, 2012; Huang *et al.*, 2014). Huang *et al.* (2014) found that UVR8^{W233F} and UVR8^{W233A} are monomeric and show little interaction with COP1, whereas O'Hara and Jenkins (2012) reported that UVR8^{W233A} shows constitutive binding to COP1 and forms a weak dimer that monomerises in response to UV-B. Since the dimer is evidently weakened by the mutation, the extent to which it is present may depend on the developmental stage of the plants and conditions of growth and illumination. Similar factors may influence the extent of binding of UVR8^{W233A} to COP1, which is also observed in yeast (Rizzini *et al.*, 2011; O'Hara and Jenkins, 2012; Huang *et al.*, 2014).

The alanine mutant of the remaining triad tryptophan, UVR8^{W337A}, forms weak dimers in Arabidopsis and binds COP1 constitutively. Nevertheless, the protein responds to UV-B to produce monomers and shows little loss of function compared to wild-type UVR8 (O'Hara and Jenkins, 2012). UVR8^{W94A}, along with mutants in several tryptophans located either in the β-propeller core of the protein or at the periphery of the dimer interface, shows no apparent loss of function (O'Hara and Jenkins, 2012). It should be noted, however, that these assays were undertaken with saturating UV-B exposures in growth cabinets and further analysis should be undertaken using a range of UV-B fluence rates and growth conditions.

Mutations of the charged amino acids responsible for dimer formation also impact on UVR8 function *in vivo*. Mutation of the key salt bridging residue R286 to alanine results in constitutive monomerisation in Arabidopsis, no interaction with COP1 and lack of response to UV-B (Huang *et al.*, 2014; Heilmann *et al.*, 2016). However, conservative mutation of R286 to lysine results in formation of a protein with a weakened dimer that mediates UV-B responses (Heilmann *et al.*, 2016). Interestingly, when the two aspartate amino acids that form salt bridges with R286 are conservatively mutated to asparagine, a UVR8 protein (UVR8^{D96N,D107N}) is produced that is strongly impaired in dimer formation and constitutively interacts with COP1, but shows a response to UV-B that is very similar to that of wild-type UVR8 over a range of UV-B doses (Heilmann *et al.*, 2016). Since the photomorphogenic activity of this essentially monomeric mutant does not show any correlation with UV-B-induced dimer dissociation, the most likely explanation is that photoreception by monomeric UVR8^{D96N,D107N} mediates the response to UV-B. The concept that monomeric UVR8 can act in photoreception is supported by biophysical studies. Monomeric mutants of UVR8 exhibit spectroscopic signals associated with UV-B photoreception, similar to the wild-type (Heilmann *et al.*, 2014; Mathes *et al.*, 2015; Miyamori *et al.*, 2015). The above findings therefore raise the question of whether wild-type UVR8 monomers can mediate responses to UV-B *in vivo*.

Mutation of another important charged amino acid, R338 to alanine also impacts on UVR8 function. UVR8^{R338A} is impaired in response but retains some activity, is monomeric and binds COP1 constitutively (Huang *et al.*, 2014; Heilmann *et al.*, 2016). Huang *et al.* (2014) reported that plants expressing UVR8^{R338A} have a weak *cop1* phenotype, although it is less evident than in plants expressing UVR8^{W285A}.

Further *in vivo* studies of UVR8 structure-function have focused on the interaction with COP1, which involves a 27 amino acid region in the C-terminus of the protein (Cloix *et al.*, 2012; Yin *et al.*, 2015). These studies will be discussed in Chapter 9. Taken together, the *in vivo* studies of UVR8 structure–function relationships are generally consistent with those undertaken with the purified protein. In particular, they support the main conclusions for the roles of specific tryptophans in UV-B photoreception and the importance of key salt bridge amino acids in maintaining the dimer structure.

Evolutionary Conservation of UVR8 Structure and Function

The UVR8 amino acid sequence is highly conserved among diverse plant taxa, ranging from green algae, through Bryophytes and lycopods to the Angiosperms (Rizzini *et al.*, 2011; Wu *et al.*, 2011). The number and position of key residues, including the tryptophans and salt bridging arginines, are conserved, suggesting that all UVR8 sequences have a similar molecular mechanism. A recent study of *Chlamydomonas reinhardtii* UVR8 supports this contention. The protein monomerizes on exposure to UV-B, and interacts with Chlamydomonas COP1 (Tilbrook *et al.*, 2016). This interaction induces transcriptome changes associated with acclimation to UV-B in the algal cells. The presence of a functional UVR8 in Chlamydomonas shows that the protein appeared early in plant evolution, most likely to provide protection against the relatively high levels of UV-B that prevailed at that time (Rozema *et al.*, 1997) and which could damage the photosynthetic machinery. Remarkably, Chlamydomonas UVR8 complements the Arabidopsis *uvr8* mutant, demonstrating a conservation of function despite massive evolutionary separation.

Interestingly, examination of *UVR8* sequences from various species deposited in public databases shows that some species have more than one *UVR8* gene, in contrast to Arabidopsis which has just one. This raises the possibility that *UVR8* genes may be subject to differential expression and the encoded proteins could conceivably differ in physiological roles. Studies of UVR8 in species other than Arabidopsis have recently started to appear, and provide evidence that some *UVR8* genes are subject to regulation by various factors (Wu *et al.*, 2016; Zhang *et al.*, 2016). This interesting aspect is likely to be explored extensively in future research.

Summary and Perspective

The finding that UVR8 is a UV-B photoreceptor with a novel mechanism of action represents a major advance in photobiology. Furthermore, the discovery of UVR8 provides a new perspective on plant responses to UV-B light and suggests new approaches to investigate these responses. The challenge now is to discover how UVR8 functions in cells through interaction with other proteins and signalling pathways, how its activity is regulated and what responses it mediates in diverse species of plants growing in a range of environments.

Substantial progress has been made in understanding the unique tryptophan-based mechanism of UVR8 photoreception as a result of the discovery of the crystal structure, biophysical and computational approaches, and structure–function analyses both with the purified protein and *in vivo*. Further experimentation will reveal additional details of the mechanism and how it leads to dimer dissociation. A major gap in knowledge concerns the conformational changes to the protein that occur following photoreception and how these changes promote interaction with the signalling partner COP1. In addition, it is important to understand how the RUP proteins return UVR8 to the dimeric state.

Research with Arabidopsis has highlighted an increasing number of physiological responses regulated by UVR8 (Jenkins, 2014a; see also Chapter 9, this volume). Nevertheless, it is important to extend these studies to other species, including commercially important species, and non-vascular plants, where knowledge of photomorphogenic UV-B responses is less well advanced. In addition, most studies with Arabidopsis have employed rather artificial plant growth conditions and it is essential to explore how UVR8 functions, and is regulated, in more realistic environments. Initial studies show that UVR8 regulates gene expression and biochemical composition in Arabidopsis exposed to natural sunlight (Morales *et al.*, 2013). Moreover, research has shown that under photoperiodic conditions UVR8 does not behave as a simple, UV-B-activated on/off switch, but rather a photoequilibrium is established between the dimer and monomer forms, regulated by the RUP proteins (Findlay and Jenkins, 2016). Hence it is important to understand how the UVR8 photoequilibrium is regulated in natural growth environments and how it is coupled to physiological responses.

References

Ahmad, M. and Cashmore, A.R. (1993) HY4 gene of *A. thaliana* encodes a protein with characteristics of a blue-light photoreceptor. *Nature* 366, 162–166.

Ballaré, C.L., Barnes, P.W. and Flint, S.D. (1995) Inhibition of hypocotyl elongation by ultraviolet-B radiation in de-etiolating tomato seedlings. 1. The photoreceptor. *Physiologia Plantarum* 93, 584–592.

Boccalandro, H.E., Mazza, C.A., Mazzella, M.A., Casal, J.J. and Ballaré, C.L. (2001) Ultraviolet-B radiation enhances a phytochrome-B-mediated photomorphogenic response in Arabidopsis. *Plant Physiology* 126, 780–788.

Briggs, W.R. and Rice, H.V. (1972) Phytochrome: chemical and physical properties and mechanism of action. *Annual Review of Plant Physiology* 23, 293–334.

Brown, B.A. and Jenkins, G.I. (2008) UV-B signaling pathways with different fluence-rate response profiles are distinguished in mature Arabidopsis leaf tissue by requirement for UVR8, HY5, and HYH. *Plant Physiology* 146, 576–588.

Brown, B.A., Cloix, C., Jiang, G.H., Kaiserli, E., Herzyk, P. *et al.* (2005) A UV-B-specific signaling component orchestrates plant UV protection. *Proceedings of the National Academy of Sciences of the United States of America* 102, 18225–18230.

Brown, B.A., Headland, L.R. and Jenkins, G.I. (2009) UV-B action spectrum for UVR8-mediated HY5 transcript accumulation in Arabidopsis. *Photochemistry and Photobiology* 85, 1147–1155.

Christie, J.M. and Jenkins, G.I. (1996) Distinct UV-B and UV-A/blue light signal transduction pathways induce chalcone synthase gene expression in Arabidopsis cells. *Plant Cell* 8, 1555–1567.

Christie, J.M., Reymond, P., Powell, G.K., Bernasconi, P., Raibekas, A.A. *et al.* (1998) Arabidopsis NPH1: a flavoprotein with the properties of a photoreceptor for phototropism. *Science* 282, 1698–1701.

Christie, J.M., Arvai, A.S., Baxter, K.J., Heilmann, M., Pratt, A.J. *et al.* (2012) Plant UVR8 photoreceptor senses UV-B by tryptophan-mediated disruption of cross-dimer salt bridges. *Science* 335, 1492–1496.

Cloix, C., Kaiserli, K., Heilmann, M., Baxter, K.J., Brown, B.A. *et al.* (2012) The C-terminal region of the UV-B photoreceptor UVR8 initiates signaling through interaction with COP1. *Proceedings of the National Academy of Sciences of the United States of America* 109, 16366–16370.

Crefcoeur, R.P., Yin, R., Ulm, R. and Halazonetis, T.D. (2013) UV-B-mediated induction of protein–protein interactions in mammalian cells. *Nature Communications* 4, 1779.

Ensminger, P.A. (1993) Control of development in plants and fungi by far-UV radiation. *Physiologia Plantarum* 88, 501–508.

Ensminger, P.A. and Schäfer, E. (1992) Blue and ultraviolet-B light photoreceptors in parsley cells. *Photochemistry and Photobiology* 55, 437–447.

Favory, J.J., Stec, A., Gruber, H., Rizzini, L., Oravecz, A. *et al.* (2009) Interaction of COP1 and UVR8 regulates UV-B-induced photomorphogenesis and stress acclimation in *Arabidopsis*. *EMBO Journal* 28, 591–601.

Feher, B., Kozma-Bognar, L., Kevei, E., Hajdu, A., Binkert, M. *et al.* (2011) Functional interaction of the circadian clock and UV RESISTANCE LOCUS 8-controlled UV-B signaling pathways in *Arabidopsis thaliana*. *Plant Journal* 67, 37–48.

Findlay, K.M.W. and Jenkins, G.I. (2016) Regulation of UVR8 photoreceptor dimer/monomer photo-equilibrium in Arabidopsis plants grown under photoperiodic conditions. *Plant, Cell & Environment* 39, 1706–1714.

Frohnmeyer, H., Bowler, C. and Schäfer, E. (1997) Evidence for some signal transduction elements involved in UV-light-dependent responses in parsley protoplasts. *Journal of Experimental Botany* 48, 739–750.

Frohnmeyer, H., Bowler, C., Xhu, G., Yamagata, H., Schäfer, E. and Chua, N.-H. (1998) Different roles for calcium and calmodulin in phytochrome- and UV-regulated expression of chalcone synthase. *Plant Journal* 13, 763–772.

Frohnmeyer, H., Loyall, L., Blatt, M.R. and Grabov, A. (1999) Millisecond UV-B irradiation evokes prolonged elevation of cytosolic-free Ca^{2+} and stimulates gene expression in transgenic parsley cell cultures. *Plant Journal* 20, 109–117.

Gerhardt, K.E., Wilson, M.I. and Greenberg, B.M. (2005) Ultraviolet wavelength dependence of photomorphological and photosynthetic responses in *Brassica napus* and *Arabidopsis thaliana*. *Photochemistry and Photobiology* 81, 1061–1068.

Gonzalez Besteiro, M.A., Bartels, S., Albert, A. and Ulm, R. (2011) Arabidopsis MAP kinase phosphatase 1 and its target MAP kinases 3 and 6 antagonistically determine UV-B stress tolerance, independent of the UVR8 photoreceptor pathway. *Plant Journal* 68, 727–737.

Grishina, I.B. and Woody, R.W. (1994) Contributions of tryptophan side chains to the circular dichroism of globular proteins: exciton couplets and coupled oscillators. *Faraday Discussions* 99, 245–262.

Gruber, H., Heijde, M., Heller, W., Albert, A., Seidlitz, H.K. and Ulm, R. (2010) Negative feedback regulation of UV-B-induced photomorphogenesis and stress acclimation in Arabidopsis. *Proceedings of the National Academy of Sciences of the United States of America* 107, 20132–20137.

Harlow, G.R., Jenkins, M.E., Pittalwala, T.S. and Mount, D.W. (1994) Isolation of uvh1, an Arabidopsis mutant hypersensitive to ultraviolet-light and ionizing-radiation. *Plant Cell* 6, 227–235.

Hayes, S., Velanis, C.N., Jenkins, G.I. and Franklin, K.A. (2014) UV-B detected by the UVR8 photoreceptor antagonizes auxin signaling and plant shade avoidance. *Proceedings of the National Academy of* Sciences *of the United States of America* 111, 11894–11899.

Heijde, M. and Ulm, R. (2012) UV-B photoreceptor-mediated signaling in plants. *Trends in Plant Science* 17, 230–237.

Heijde, M. and Ulm, R. (2013) Reversion of the Arabidopsis UV-B photoreceptor UVR8 to the homodimeric ground state. *Proceedings of the National Academy of* Sciences *of the United States of America* 110, 1113–1118.

Heijde, M., Binkert, M., Yin, R., Ares-Orpel, F., Rizzini, L. *et al.* (2013) Constitutively active UVR8 photoreceptor variant in Arabidopsis. *Proceedings of the National Academy of Sciences of the United States of America* 110, 20326–20331.

Heilmann, M. and Jenkins, G.I. (2013) Rapid reversion from monomer to dimer regenerates the ultraviolet-B photoreceptor UV RESISTANCE LOCUS8 in intact Arabidopsis plants. *Plant Physiology* 161, 547–555.

Heilmann, M., Christie, J.M, Kennis, J.T., Jenkins, G.I. and Mathes, T. (2014) Photoinduced transformation of UVR8 monitored by vibrational and fluorescence spectroscopy. *Photochemical and Photobiological Sciences* 14, 252–257.

Heilmann, M., Velanis, C.N., Cloix, C., Smith, B.O., Christie, J.M. and Jenkins, G.I. (2016) Dimer/monomer status and in vivo function of salt-bridge mutants of the plant UV-B photoreceptor UVR8. *Plant Journal* 88, 71–81. DOI:10.1111/tpj.13260.

Hideg, É., Jansen, M.A.K. and Strid, A. (2013) UV-B exposure, ROS, and stress: inseparable companions or loosely linked associates? *Trends in Plant Science* 18, 107–115.

Huang, X. and Deng, X.W. (2013) Organization of protein complexes under photomorphogenic UV-B in *Arabidopsis*. *Plant Signaling & Behaviour* 8, e27206.

Huang, X., Ouyang, X., Yang, P., Lau, O.S., Chen, L. *et al.* (2013) Conversion from CUL4-based COP1–SPA E3 apparatus to UVR8–COP1–SPA complexes underlies a distinct biochemical function of COP1 under UV-B. *Proceedings of the National Academy of Sciences of the United States of America* 110, 16669–16674.

Huang, X., Yang, P., Ouyang, X., Chen, L. and Deng, X.W. (2014) Photoactivated UVR8-COP1 module determines photomorphogenic UV-B signaling output in Arabidopsis. *PLOS Genetics* 10, e1004218.

Jackson, J.A., Fuglevand, G., Brown, B.A., Shaw, M.J. and Jenkins, G.I. (1995) Isolation of *Arabidopsis* mutants altered in the light-regulation of chalcone synthase gene expression using a transgenic screening approach. *Plant Journal* 8, 369–380.

Jenkins, G.I. (2009) Signal transduction in responses to UV-B radiation. *Annual Reviews of Plant Biology* 60, 407–31.

Jenkins, G.I. (2014a) The UV-B photoreceptor UVR8: from structure to physiology. *Plant Cell* 26, 21–37.

Jenkins, G.I. (2014b) Structure and function of the UV-B photoreceptor UVR8. *Current Opinion in Structural Biology* 29, 52–57.

Jenkins, G.I. and Brown, B.A. (2007) UV-B perception and signal transduction. In: Whitelam, G.C. and Halliday, K.J. (eds) *Light and Plant Development*. Blackwell Publishing, Oxford, UK, pp.155–182.

Jenkins, G.I., Long, J.C., Wade, H.K., Shenton, M.R. and Bibikova, T.N. (2001) UV and blue light signaling: pathways regulating chalcone synthase gene expression in *Arabidopsis*. *New Phytologist* 151, 121–131.

Jiang, C.Z., Yee, J., Mitchell, D.L. and Britt, A.B. (1997) Photorepair mutants of Arabidopsis. *Proceedings of the National Academy of Sciences of the United States of America* 94, 7441–7445.

Jiang, L., Wang, Y., Björn, L.O., He, J.-X. and Li, S.-S. (2012) Sensing of UV-B radiation by plants. *Plant Signaling & Behavior* 7, 1–5.

Jordan, B.R. (1996) The effects of ultraviolet-B radiation on plants: a molecular perspective. *Advances in Botanical Research* 22, 97–162.

Kaiserli, E. and Jenkins, G.I. (2007) UV-B promotes rapid nuclear translocation of the UV-B-specific signaling component UVR8 and activates its function in the nucleus. *Plant Cell* 19, 2662–2673.

Kami, C., Lorrain, S., Hornitschek, P. and Fankhauser, C. (2010) Light-regulated plant growth and development. *Current Topics in Developmental Biology* 91, 29–66.

Kilian, J., Whitehead, D., Horak, J., Wanke, D., Weinl, S. *et al.* (2007) The AtGenExpress global stress expression data set: protocols, evaluation and model data analysis of UV-B light, drought and cold stress responses. *Plant Journal* 50, 347–363.

Kircher, S., Kozma-Bognar, L., Kim, L., Adam, E., Karter, K. *et al.* (1999) Light quality-dependent nuclear import of the plant photoreceptors phytochrome A and B. *Plant Cell* 11, 1445–1456.

Klein, R.M. (1978) Plants and near-ultraviolet radiation. *Botanical Reviews* 44, 1–127.

Kliebenstein, D.J., Lim, J.E., Landry, L.G. and Last, R.L. (2002) Arabidopsis UVR8 regulates ultraviolet-B signal transduction and tolerance and contains sequence similarity to human Regulator of Chromatin Condensation 1. *Plant Physiology* 130, 234–243.

Landry, L.G., Stapleton, A.E., Lim, J., Hoffman, P., Hays, J.B. *et al.* (1997) An Arabidopsis photolyase mutant is hypersensitive to ultraviolet-B radiation. *Proceedings of the National Academy of Sciences of the United States of America* 94, 328–332.

Lau, O.S. and Deng, X.W. (2012) The photomorphogenic repressors COP1 and DET1: 20 years later. *Trends in Plant Science* 17, 584–593.

Li, J., Yang, L., Jin, D., Nezames, C.D., Terzaghi, W. and Deng, X.-W. (2013) UV-B-induced photomorphogenesis in Arabidopsis. *Protein Cell* 4, 485–492.

Li, X., Chung, L.W., Morokuma, K. and Li, G. (2014) Theoretical study on the UVR8 photoreceptor: sensing ultraviolet-B by tryptophan and dissociation of homodimer. *Journal of Chemical Theory and Computation* 10, 3319–3330.

Lin, C., Robertson, D.E., Ahmad, M., Raibekas, A.A., Jorns, M.S. *et al.* (1995) Association of flavin adenine dinucleotide with the Arabidopsis blue light receptor CRY1. *Science* 269, 968–970.

Liscum, E. and Briggs, W.R. (1995) Mutations in the NPH1 locus of Arabidopsis disrupt the perception of phototropic stimuli. *Plant Cell* 7, 473–485.

Liu, Z., Li, X., Zhong, F.W., Li, J., Wang, L., Shi, Y. and Zhong, D. (2014) Quenching dynamics of ultraviolet-light perception by UVR8 photoreceptor. *Journal of Physical Chemistry Letters* 5, 69–72.

Lois, R. and Buchanan, B.B. (1994) Severe sensitivity to ultraviolet-radiation in an Arabidopsis mutant deficient in flavonoid accumulation. 2. Mechanisms of UV-resistance in Arabidopsis. *Planta* 194, 504–509.

Long, J.C. and Jenkins, G.I. (1998) Involvement of plasma membrane redox activity and calcium homeostasis in the UV-B and UV-A/blue light induction of gene expression in Arabidopsis. *Plant Cell* 10, 2077–2086.

Mathes, T., Heilmann, M., Pandit, A., Zhu, J., Ravensbergen, J. *et al.* (2015) Proton-coupled electron transfer constitutes the photoactivation mechanism of the plant photoreceptor UVR8. *Journal of the American Chemical Society* 137, 8113–8120.

Miyamori, T., Nakasone, Y., Hitomi, K., Christie, J.M., Getzoff, E.D. and Terazima, M. (2015) Reaction dynamics of the UV-B photosensor UVR8. *Photochemical and Photobiological Sciences* 14, 995–1004.

Morales, L.O., Brosche, M., Vainonen, J., Jenkins, G.I., Wargent, J.J. *et al.* (2013) Multiple roles for UV RESISTANCE LOCUS8 in regulating gene expression and metabolite accumulation in Arabidopsis under solar ultraviolet radiation. *Plant Physiology* 161, 744–759.

O'Hara, A. and Jenkins, G.I. (2012) *In vivo* function of tryptophans in the Arabidopsis UV-B photoreceptor UVR8. *Plant Cell* 24, 3755–3766.

Oravecz, A., Baumann, A., Máté, Z., Brzezinska, A., Molinier, J. *et al.* (2006) CONSTITUTIVELY PHOTOMORPHOGENIC1 is required for the UV-B response in Arabidopsis. *Plant Cell* 18, 1975–1990.

Osterlund, M.T., Hardtke, C.S., Wei, N. and Deng, X.W. (2000) Targeted destabilization of HY5 during light-regulated development of Arabidopsis. *Nature* 405, 462–466.

Renault, L., Nassar, N., Vetter, I., Becker, J., Klebe, C. *et al.* (1998) The 1.7 A crystal structure of the regulator of chromatin condensation (RCC1) reveals a seven-bladed propeller. *Nature* 392, 97–101.

Rizzini, L., Favory, J.-J., Cloix, C., Faggionato, D., O'Hara, A. *et al.* (2011) Perception of UV-B by the *Arabidopsis* UVR8 protein. *Science* 332, 103–106.

Rozema, J., van de Staaij, J., Björn, L.O. and Caldwell, M. (1997) UV-B as an environmental factor in plant life: stress and regulation. *Trends in Ecology and Evolution* 12, 22–28.

Ryan, K.G., Swinny, E.E., Winefield, C. and Markham, K.R. (2001) Flavonoids and UV photoprotection in Arabidopsis mutants. *Zeitschrift für Naturforschung C* 56, 745–754.

Suesslin, C. and Frohnmeyer, H. (2003) An *Arabidopsis* mutant defective in UV-B light-mediated responses. *Plant Journal* 33, 591–601.

Tevini, M. and Teramura, A.H. (1989) UV-B effects on terrestrial plants. *Photochemistry and Photobiology* 50, 479–487.

Tilbrook, K., Arongaus, A.B., Binkert, M., Heijde, M., Yin, R. and Ulm, R. (2013) The UVR8 UV-B photoreceptor: perception, signaling and response. *The Arabidopsis Book* 11, e0164. DOI:10.1199/tab.0164

Tilbrook, K., Dubois, M., Crocco, C.D., Yin, R., Chappuis, R. *et al.* (2016) UV-B perception and acclimation in *Chlamydomonas reinhardtii*. *Plant Cell* 28, 966–983.

Ulm, R. and Jenkins, G.I. (2015) Q&A: How do plants sense and respond to UV-B radiation? *BMC Biology* 13, 45.

Ulm, R., Baumann, A., Oravecz, A., Mate, Z., Adam, E. *et al.* (2004) Genome-wide analysis of gene expression reveals function of the bZIP transcription factor HY5 in the UV-B response of *Arabidopsis*. *Proceedings of the National Academy of Sciences of the United States of America* 101, 1397–1402.

Voityuk, A.A., Marcus, R.A. and Michel-Beyerle, M. (2014) On the mechanism of photoinduced dimer dissociation in the plant UVR8 photoreceptor. *Proceedings of the National Academy of Sciences of the United States of America* 111, 5219–5224.

Wellmann, E. (1976) Specific ultraviolet effects in plant morphogenesis. *Photochemistry and Photobiology* 24, 659–660.

Wellmann, E. (1983) UV radiation in photomorphogenesis. In: Shropshire, W. Jr and Mohr, H. (eds) *Encyclopedia of Plant Physiology. New Series, Vol. 16B.* Springer, Berlin, pp. 745–756.

Wu, D., Hu, Q., Yan, Z., Chen, W., Yan, C. *et al.* (2012) Structural basis of ultraviolet-B perception by UVR8. *Nature* 484, 214–219.

Wu, M., Grahn, E., Eriksson, L.A. and Strid, A. (2011) Computational evidence for the role of *Arabidopsis thaliana* UVR8 as a UV-B photoreceptor and identification of its chromophore amino acids. *Journal of Chemical Information and Modeling* 51, 1287–1295.

Wu, M., Strid, A. and Eriksson, L.A. (2013) Interactions and stabilities of the UV RESISTANCE LOCUS8 (UVR8) protein dimer and its key mutants. *Journal of Chemical Information and Modeling* 53, 1736–1746.

Wu, M., Strid, A. and Eriksson, L.A. (2014) Photochemical reaction mechanism of UV-B-induced monomerization of UVR8 dimers as the first signaling event in UV-B-regulated gene expression in plants. *Journal of Physical Chemistry B* 118, 951–965.

Wu, Q., Huang, B., Niehaus, T.A., Yang, X., Fan, J. and Zhang, R.-Q. (2015) The role of tryptophans in the UV-B absorption of a UVR8 photoreceptor – a computational study. *Physical Chemistry Chemical Physics* 17, 10786.

Wu, Q., Su, N., Zhang, X., Liku, Y., Cui, J. and Liang, Y. (2016) Hydrogen peroxide, nitric oxide and UV RESISTANCE LOCUS8 interact to mediate UV-B-induced anthocyanin biosynthesis in radish sprouts. *Science Reports* 6, 29164.

Yamaguchi, R., Nakamura, M., Mochizuki, N., Kay, S.A. and Nagatani, A. (1999) Light-dependent translocation of a phytochrome B-GFP fusion protein to the nucleus in transgenic *Arabidopsis*. *Journal of Cell Biology* 145, 437–445.

Yang, X., Montano, S. and Ren, Z. (2015) How does photoreceptor UVR8 perceive a UV-B signal? *Photochemistry and Photobiology* 91, 993–1003.

Yin, R., Arongaus, A.B., Binkert, M. and Ulm, R. (2015) Two distinct domains of the UVR8 photoreceptor interact with COP1 to initiate UV-B signaling in Arabidopsis. *Plant Cell* 27, 202–213.

Zeng, X., Ren, Z., Wu, Q., Fan J., Peng, P.-P. *et al.* (2015) Dynamic crystallography reveals early signalling events in ultraviolet photoreceptor UVR8. *Nature Plants* 1, 14006.

Zhang, H.-N., Li, W.-C., Wang, H.-C., Shi, S.-Y., Shu, B. *et al.* (2016) Transcriptome profiling of light-regulated anthocyanin biosynthesis in the pericarp of Litchi. *Frontiers in Plant Science* 7, 963.

9 UV-B Signal Transduction from Photoperception to Response

Melanie Binkert and Roman Ulm*
Department of Botany and Plant Biology, University of Geneva, Switzerland

Introduction

Sunlight fuels photosynthesis in plants and is an important environmental trigger, but it is also a potential environmental stress factor (e.g. high light, UV-B radiation). Light captured by specific photoreceptors affects plant development throughout the life cycle, in many cases optimizes photosynthesis, and also protects the organism from potential light stress. Photoreceptors perceive photons of specific wavelength and convert signals into cellular signalling cascades. Various photoreceptors have evolved in plants that detect and respond to changes in the light spectrum in terms of light quality, quantity, direction and duration. These include the red/far-red light-perceiving phytochromes, the blue/UV-A light-perceiving cryptochromes, phototropins, Zeitlupe family members, and the UV-B photoreceptor UV RESISTANCE LOCUS 8 (UVR8) (Heijde and Ulm, 2012; Galvao and Fankhauser, 2015). Visible light photoreceptors absorb light with a bound chromophore. Red/far-red light perception is based on a linear tetrapyrrole (bilin) prosthetic group, and blue/UV-A photons are absorbed by flavin-derived chromophores (Galvao and Fankhauser, 2015). The perception mechanism of UVR8 is unique. Instead of a bound chromophore, specific intrinsic tryptophan residues absorb UV-B photons and trigger a signalling cascade (Rizzini *et al.*, 2011; Christie *et al.*, 2012; Wu *et al.*, 2012; Zeng *et al.*, 2015; see also Chapter 8, this volume).

Photoreceptor-based UV-B Responses

The profound effects of UV-B on plant growth and development were studied long before the UVR8 photoreceptor was identified (Wellmann, 1983; Ulm, 2006; see also Chapter 5, this volume). A specific UV-B photomorphogenic effect was postulated about 45 years ago for the UV-B-induced biosynthesis of particular flavonoids, an important class of plant secondary metabolites (Wellmann, 1971). In many cases, it is still not known whether the UV-B response studied is UVR8 photoreceptor-dependent or not. Given the availability of *Arabidopsis uvr8* mutants (Kliebenstein *et al.*, 2002; Brown *et al.*, 2005; Favory *et al.*, 2009) and the potential to generate UVR8 knockouts in other plant species (Belhaj *et al.*, 2015), the relationship between UV-B phenotypes and UVR8

*Corresponding author: roman.ulm@unige.ch

signalling should be rather easy to establish. Where a particular UV-B effect is found to be independent of UVR8, identification of the initial signal and the signalling cascade remains challenging. The possibilities to be investigated include (i) an (as yet) unknown photoreceptor; (ii) the indirect perception of a specific cellular UV-B effect, or (iii) a more general UV-B stress response (e.g. Ulm *et al.*, 2004; Brown and Jenkins, 2008; González Besteiro *et al.*, 2011; González Besteiro and Ulm, 2013; Hideg *et al.*, 2013; Biever *et al.*, 2014; Horak and Farre, 2015; Li *et al.*, 2015; Robson *et al.*, 2015; Xie *et al.*, 2015).

Arabidopsis uvr8 knockout mutants are apparently indistinguishable from wild-type seedlings under visible light in the absence of UV-B (e.g. Kliebenstein *et al.*, 2002; Brown *et al.*, 2005; Favory *et al.*, 2009). The UV-B phenotypes most used in the study of UVR8 signalling are hypocotyl growth inhibition, flavonol and anthocyanin accumulation, and changes in gene expression or protein accumulation (e.g. Kliebenstein *et al.*, 2002; Brown *et al.*, 2005; Favory *et al.*, 2009; Morales *et al.*, 2013; Huang *et al.*, 2014b) (Fig. 9.1). In addition, *uvr8* mutants do not develop UV tolerance under UV-B-containing growth conditions, whereas UVR8 overexpression lines show enhanced UV-B acclimation (Favory *et al.*, 2009; González Besteiro *et al.*, 2011). In agreement with the phenotypic data, UVR8 orchestrates UV-B-induced expression of genes for flavonoid biosynthesis, DNA repair, and protection against oxidative stress and photoinhibition (Brown *et al.*, 2005; Favory *et al.*, 2009; Stracke *et al.*, 2010; Davey *et al.*, 2012; Li *et al.*, 2015; Tilbrook *et al.*, 2016). Similar to *Arabidopsis*, UV-B acclimation and tolerance is also apparent in the green alga *Chlamydomonas reinhardtii*, where it is also linked to the UVR8 signalling pathway (Tilbrook *et al.*, 2016). Several additional UV-B effects have been associated with UVR8 photoreceptor activity, including entrainment of the circadian clock (Feher *et al.*, 2011), phototropism (Vandenbussche *et al.*, 2014), downward leaf curling (Fierro *et al.*, 2015), and stomata opening (Tossi *et al.*, 2014). Moreover, UVR8 has been implicated in UV-B effects on defence responses (Demkura and Ballare, 2012), leaf development (Wargent *et al.*, 2009), growth inhibition under salt stress (Fasano *et al.*, 2014), inhibition of shade avoidance (Hayes *et al.*, 2014; Mazza and Ballare, 2015), and auxin signalling (Hayes *et al.*, 2014; Vandenbussche *et al.*, 2014). Finally, given its specificity and sensitivity, UVR8 has been added to the expanding optogenetic toolkit for engineering specific UV-B responses in non-plant systems (Chen *et al.*, 2013; Crefcoeur *et al.*, 2013; Muller *et al.*, 2013).

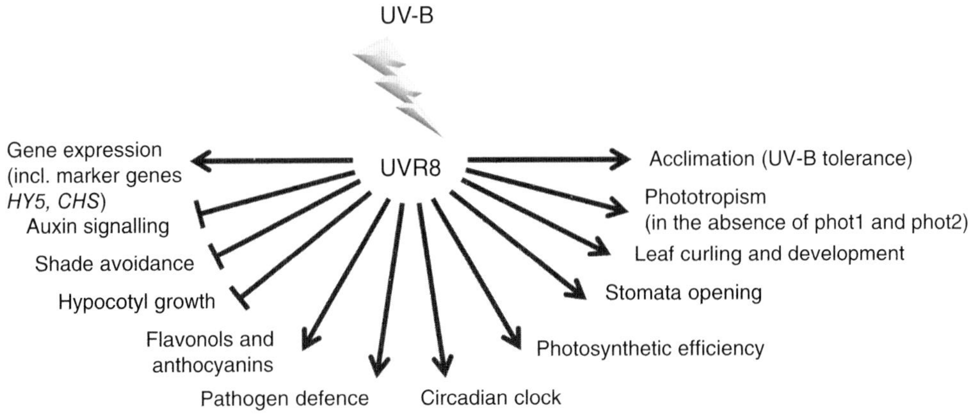

Fig. 9.1. Physiological and molecular UV-B responses that have been linked to UVR8-mediated signalling. Further information and references are provided in the text.

UVR8-mediated UV-B Signalling

In the last decade, several components of the UVR8 signalling pathway have been identified and characterized and insights into the UV-B perception mechanism obtained (Li *et al.*, 2013; Tilbrook *et al.*, 2013; Jenkins, 2014; Ulm and Jenkins, 2015; Yang *et al.*, 2015). Downstream of the UVR8 photoreceptor and central to UV-B-regulated responses is a signalling cascade connecting perception of the UV-B signal to gene expression. Our present understanding of how UV-B photoreception by UVR8 leads to transcriptional changes in the nucleus is far from satisfactory, but important steps towards this goal have been made recently and these will be discussed here.

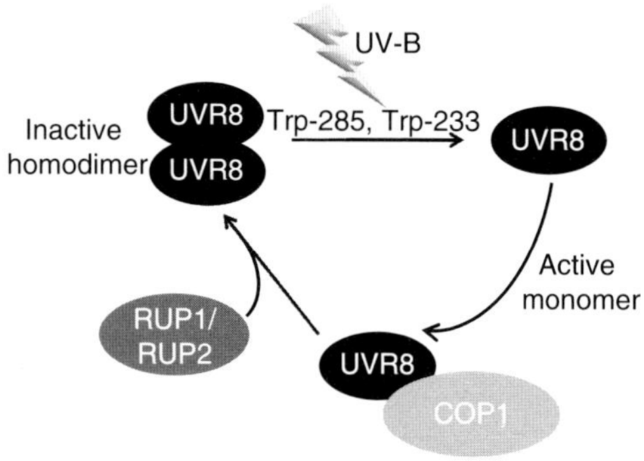

Fig. 9.2. The UVR8 photocycle. UVR8 forms a homodimer in its inactive ground state. Tryptophan-based UV-B absorption results in UVR8 monomerization. Active UVR8 monomers interact with COP1 and initiate UV-B signalling. Redimerization of UVR8 to its homodimeric ground state is facilitated by the action of RUP1 and RUP2. This process disrupts the UVR8-COP1 interaction and inactivates the signalling pathway. Regenerated UVR8 homodimers are again UV-B perception competent.

The UVR8-COP1-RUP1/RUP2 core of the UV-B receptor photocycle

The core of the UVR8 photocycle includes UVR8 monomerization upon UV-B absorption by specific intrinsic tryptophan residues, interaction with CONSTITUTIVELY PHOTOMORPHOGENIC 1 (COP1), and redimerization of UVR8 facilitated by REPRESSOR OF UV-B PHOTOMORPHOGENESIS 1 (RUP1) and RUP2 (Favory *et al.*, 2009; Rizzini *et al.*, 2011; Heijde and Ulm, 2013) (Fig. 9.2). This signalling pathway, as well as its importance for UV-B acclimation, seems to be largely conserved evolutionarily from green algae to higher plants (Tilbrook *et al.*, 2013, 2016).

UVR8 forms a homodimer that monomerizes in response to UV-B

UVR8 is a protein of 440 amino acids (Kliebenstein *et al.*, 2002) with a core domain in the form of a seven-bladed β-propeller (Christie *et al.*, 2012; Wu *et al.*, 2012; Zeng *et al.*, 2015). In its ground state, UVR8 is present as a symmetric homodimer held together by a network of hydrogen bonds and electrostatic interactions between charged amino acids across the dimer interaction surface (Rizzini *et al.*, 2011; Christie *et al.*, 2012; Wu *et al.*, 2012; Zeng *et al.*, 2015; Heilmann *et al.*, 2016). Specific tryptophans at the dimer interface (Trp-233 and Trp-285) were shown by mutational studies to be crucial for photoreception (Rizzini *et al.*, 2011; Christie *et al.*, 2012; O'Hara and Jenkins, 2012; Wu *et al.*, 2012; Heijde *et al.*, 2013; Huang *et al.*, 2014b). UV-B absorption by Trp-233 and Trp-285 leads to dissociation of the UVR8 homodimer (Rizzini *et al.*, 2011; Christie *et al.*, 2012; Wu *et al.*, 2012), but the underlying mechanism is not yet fully understood. UVR8 monomerization may involve (i) proton-coupled electron transfer from UV-B-excited tryptophans to adjacent charged key amino acids that maintain the homodimer and (ii) ejection of an adjacent water molecule due to reorientation of the epicentre Trp-233 and Trp-285. Together this may weaken hydrogen bonds between the monomers (Mathes *et al.*, 2015; Zeng *et al.*, 2015). Although conditions can be established in the laboratory under which UVR8 can switch between fully homodimeric and fully monomeric forms, under natural conditions UVR8 establishes a dimer/monomer photo-equilibrium that is regulated by UV-B (Findlay and Jenkins,

2016). Further details of UV-B photoreception and early structural events at the UVR8 photoreceptor level may be found in Chapter 8 of this book and recent specific reviews (Jenkins, 2014; Yang *et al.*, 2015).

UV-B-activated UVR8 interacts with COP1

Following UV-B photon absorption and monomerization, UVR8 interacts with COP1 (Favory *et al.*, 2009; Rizzini *et al.*, 2011). COP1 forms an E3 ubiquitin ligase complex with the SUPPRESSOR OF *phyA-105* (SPA) 1 to SPA4 protein quartet (Menon *et al.*, 2016). The COP1-SPA complex is a repressor of photomorphogenesis in darkness by ubiquitination of several positive regulators of photomorphogenesis, which tags them for proteasomal degradation (Lau and Deng, 2012; Menon *et al.*, 2016). The COP1 protein comprises a RING domain, a coiled-coil, and a WD40-repeat domain (Lau and Deng, 2012). COP1 is also a crucial signalling component of the UV-B response in *Arabidopsis* (Oravecz *et al.*, 2006; Favory *et al.*, 2009). Intriguingly, expression of early UV-B-regulated genes depends largely on COP1 activity (Oravecz *et al.*, 2006; Favory *et al.*, 2009).

UV-B activated UVR8 interacts with the WD40-repeat domain of COP1 (Favory *et al.*, 2009), which like UVR8 also forms a seven-bladed β-propeller structure (Uljon *et al.*, 2016). Two separate domains of UVR8 are involved in the interaction with COP1: a domain of maximum 27 amino acids (C27) in the UVR8 C-terminus and the UVR8 β-propeller core (Cloix *et al.*, 2012; Yin *et al.*, 2015) (Fig. 9.3). The UVR8 core lacking the C27 domain can interact with COP1 in a UV-B-dependent fashion, although not as strongly as wild-type UVR8 (Yin *et al.*, 2015). In the same way, expression of a UVR8 core fused to a DNA-binding domain together with a COP1 WD40-domain fused to a transcriptional activation domain were also sufficient to establish a UV-B-responsive transcriptional system in mammalian cells (Muller *et al.*, 2013). An intriguing hypothesis is that the freed β-propeller surface

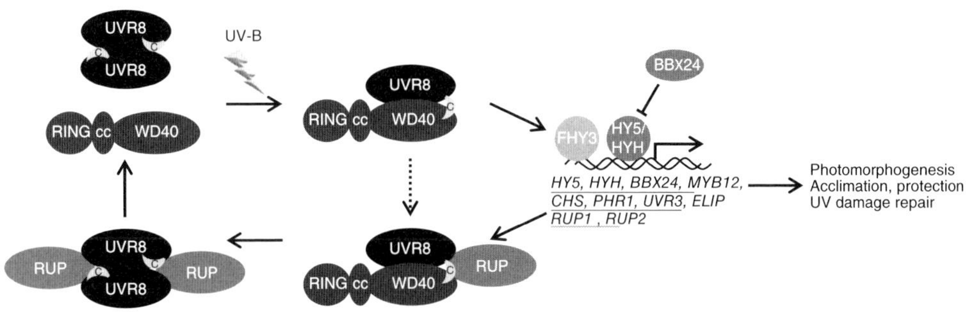

Fig. 9.3. Working model of UVR8 signalling. UVR8 homodimers dissociate upon UV-B perception, which allows interaction between the 7-bladed β-propeller domains of UVR8 and COP1 (indicated as RING-cc-WD40, with the WD40 domain as the interaction domain). This interaction is further stabilized by structural changes in UVR8 that allow binding of the UVR8 C27 domain ($UVR8^{397-423}$, indicated by the crescent labelled 'C') to the COP1 WD40 domain, which also initiates UV-B signalling. The activated UVR8-COP1 signalling pathway involves HY5/HYH and FHY3 transcription factor-mediated gene expression, including genes associated with UV-B-induced photomorphogenesis, acclimation and UV-damage repair. The UV-B-induced genes also include *BBX24* as well as *RUP1* and *RUP2* that form negative feedback loops. The B-box protein BBX24 functions as a negative regulator of UV-B responses by interacting with HY5. The WD40-repeat proteins RUP1 and RUP2 are phylogenetically and structurally related to COP1 and also interact with UVR8, but only with the C27 domain. Interaction of RUP1 and RUP2 with UVR8 facilitates UVR8 redimerization and disruption of the UVR8-COP1 interaction. RUP1 and RUP2 are still able to interact with the C27 domain of homodimeric UVR8, in contrast to COP1. The COP1-interacting SPA proteins are omitted from this model. cc, coiled coil; RING, Really Interesting New Gene; WD40, WD40 repeat domain.

of UVR8 after monomerization is replaced by the structurally related COP1 β-propeller; however, this has not been tested experimentally. Although it interacts UV-B-dependently with COP1, the UVR8 core without a C-terminus containing a C27 domain is apparently not active in *Arabidopsis* (Cloix *et al.*, 2012; Yin *et al.*, 2015). This is also true for the *uvr8-2* mutant allele that contains a premature stop codon at Trp-400 (Brown *et al.*, 2005; Cloix *et al.*, 2012; Yin *et al.*, 2015).

The UVR8 C27 contributes to COP1 interaction with UVR8 and contains a so-called VP domain (Cloix *et al.*, 2012; Wu *et al.*, 2013; Yin *et al.*, 2016). The VP domain has the core sequence V-P-E/D-φ-G (where φ = a hydrophobic residue) that was found previously to mediate interaction of several COP1 substrate proteins with the COP1 WD40-repeat domain (Holm *et al.*, 2001; Holm *et al.*, 2002; Datta *et al.*, 2006; Uljon *et al.*, 2016). The isolated UVR8 C-terminus including the C27 domain was found to interact constitutively with COP1 in transgenic plants and in yeast (Cloix *et al.*, 2012; Yin *et al.*, 2015). This suggests that the C27 domain in the UV-B light-activated wild-type UVR8 is released from structural constraints by the UVR8 core, which then allows its interaction with COP1. As in the case of other COP1-interacting proteins, mutation in the UVR8 VP core of Val-410 and Pro-411 to Ala-Ala (AA) abrogates interaction of the isolated C-terminal domain with COP1 (Yin *et al.*, 2015). The same mutations in the UVR8 full-length protein UVR8$^{VP\text{-}AA}$ still permit interaction with COP1 through the UVR8 β-propeller core domain (Yin *et al.*, 2015). *Arabidopsis* plants expressing UVR8$^{VP\text{-}AA}$ still show UV-B-specific UVR8$^{VP\text{-}AA}$-COP1 interaction but are impaired in downstream UV-B responses such as UV-B-induced gene expression and hypocotyl growth inhibition (Yin *et al.*, 2015). Although the C-terminal domain was not included in the published UVR8 crystal structures (Christie *et al.*, 2012; Wu *et al.*, 2012; Zeng *et al.*, 2015), the experimental data suggest that two COP1 interaction surfaces are exposed in monomeric UVR8 upon UV-B perception that are not accessible in the UVR8 homodimer. Interaction of COP1 with the β-propeller domain of UVR8 allows UV-B-dependent UVR8-COP1 interaction, whereas the C-terminal UVR8 domain is thought to further stabilize the interaction and to affect COP1 activity (Yin *et al.*, 2015).

The COP1 substrate basic leucine zipper (bZIP) transcription factor ELONGATED HYPOCOTYL 5 (HY5) is stabilized upon UVR8 interaction with COP1, indicating that COP1 is inactivated by UV-B (Favory *et al.*, 2009; Huang *et al.*, 2013). COP1-SPA forms a CULLIN4 - DAMAGED DNA BINDING PROTEIN 1 (CUL4-DDB1)-based E3 ubiquitin ligase complex, in which COP1-SPAs act as substrate receptors targeting HY5 for polyubiquitination and proteasomal degradation (Chen *et al.*, 2010). Interestingly, UV-B triggers the dissociation of COP1-SPA core complexes from CUL4-DDB1, probably through UVR8, suggesting that this impairs ubiquitination and degradation of HY5 (Huang *et al.*, 2013; Huang *et al.*, 2014a). In effect, unique UVR8-COP1-SPA protein complexes are formed under UV-B that may have UV-B-specific signalling functions (Oravecz *et al.*, 2006; Favory *et al.*, 2009; Heijde *et al.*, 2013; Huang *et al.*, 2013; Huang *et al.*, 2014a).

It is notable that phytochromes and cryptochromes also inhibit the COP1-SPA complex through direct interaction. However, UVR8 interacts with COP1 UV-B-dependently and does not interact with SPA proteins, whereas phytochromes and cryptochromes interact light-dependently with SPAs but constitutively with COP1 (Fankhauser and Ulm, 2011; Heijde and Ulm, 2012; Heijde *et al.*, 2013; Menon *et al.*, 2016). Although SPA proteins remain in complex with COP1 upon UV-B perception (Heijde *et al.*, 2013; Huang *et al.*, 2013), early UV-B signalling does not require SPA proteins (Oravecz *et al.*, 2006). Thus, phytochrome, cryptochrome and UVR8 photoreceptors regulate COP1-SPA E3 ubiquitin ligase activity but in mechanistically different ways.

Alternatively or additionally, given that the UVR8 C-terminus mimics the COP1-interaction sequence of HY5 and other COP1 substrates, HY5 release through competition for the COP1 binding site could be the main mechanism leading to COP1 substrate stabilization (Cloix *et al.*, 2012; Yin *et al.*, 2015; Uljon *et al.*, 2016). Importantly, UVR8 itself is not destabilized upon interaction with COP1 and is thus probably not a substrate of polyubiquitination by the COP1-SPA complex (Favory *et al.*, 2009; Gruber *et al.*, 2010; Heijde and Ulm, 2013; Heilmann and Jenkins, 2013). Indeed, a similar mechanism of cryptochrome C-terminal activity and COP1 substrate competition has been proposed recently (Muller and Bouly, 2015).

RUP1 and RUP2 facilitate UVR8 ground state reversion by redimerization

UVR8 reverts to its ground state by redimerization (Heijde and Ulm, 2013; Heilmann and Jenkins, 2013). In contrast, UVR8 degradation and *de novo* protein synthesis do not contribute significantly to the regeneration of the UVR8 homodimer pool (Heijde and Ulm, 2013; Heilmann and Jenkins, 2013). Redimerized UVR8 may again be UV-B-activated (Heijde and Ulm, 2013; Heilmann and Jenkins, 2013). Whereas recovery of UVR8 dimer *in vitro* takes up to 48 hours, UVR8 fully reverts within a few hours *in vivo* (Christie *et al.*, 2012; Wu *et al.*, 2012; Heijde and Ulm, 2013; Heilmann and Jenkins, 2013). The WD40-repeat proteins RUP1 and RUP2 interact with UVR8 and facilitate its redimerization *in vivo*, which interferes negatively with the UVR8-COP1 interaction (Gruber *et al.*, 2010; Heijde and Ulm, 2013). The *RUP1* and *RUP2* genes are UV-B-induced in a UVR8- and COP1-dependent manner and thus constitute a negative feedback regulation (Gruber *et al.*, 2010). Similar to plants overexpressing the UVR8 photoreceptor (Favory *et al.*, 2009), *Arabidopsis* mutant plants lacking RUP1 and RUP2 show an exaggerated UV-B response that results in impaired growth and dwarfism (Gruber *et al.*, 2010). This correlates with elevated and sustained UVR8 monomer levels as well as UVR8-COP1 interaction (Heijde and Ulm, 2013). In contrast, *RUP2* overexpression was shown to render *Arabidopsis* 'UV-B blind', comparable to the *uvr8* null mutant, but without affecting UVR8 protein levels (Gruber *et al.*, 2010). Instead, UVR8 remains largely homodimeric in the presence of UV-B and the UVR8-COP1 interaction is prevented in *RUP2* overexpression lines (Heijde and Ulm, 2013). Indeed, RUP1 and RUP2 are important for the establishment of the UVR8 homodimer/monomer photo-equilibrium in wild-type plants under natural conditions (Findlay and Jenkins, 2016).

Interestingly, RUP1 and RUP2 interact, like COP1, with the C-terminus of UVR8 (Cloix *et al.*, 2012; Yin *et al.*, 2015) and the UVR8$^{VP\text{-}AA}$ mutation abolishes this interaction (Yin *et al.*, 2015). It is of note that RUP1 and RUP2, as WD40-repeat proteins, are predicted to form a β-propeller structure similar to UVR8 and COP1 (Xu and Min, 2011). However, in contrast to COP1, interaction with a UVR8 core lacking the C-terminal domain was not detected for RUP1 and RUP2 (Yin *et al.*, 2015). The fact that the same UVR8 domain is an interaction surface for COP1 and the negative regulators RUP1 and RUP2 indicates competition for binding to the C-terminal UVR8 domain (Cloix *et al.*, 2012; Ouyang *et al.*, 2014; Yin *et al.*, 2015). However, RUP1 and RUP2 facilitate UVR8 redimerization even in the absence of COP1, i.e. UVR8 redimerization through RUP1 and RUP2 is independent of COP1 and this process may release COP1 from UVR8, not the other way round (Heijde and Ulm, 2013). Competition between RUP1/RUP2 and COP1 for the apparently overlapping C-terminal UVR8 binding site probably also plays a role (Cloix *et al.*, 2012; Heijde and Ulm, 2013; Yin *et al.*, 2015). Thus, alternatively, UVR8 redimerization and disruption of the UVR8-COP1 interaction may also be independent activities of RUP1 and RUP2. In stark contrast to COP1, RUP1 and RUP2 interact with both the UVR8 monomer and the homodimer and UV-B may enhance the interaction (Gruber *et al.*, 2010; Cloix *et al.*, 2012; Huang *et al.*, 2014b; Yin *et al.*, 2015). However, the UVR8-RUP1/RUP2 interaction *in vivo* is affected indirectly by UV-B as *RUP1* and *RUP2* gene expression is UV-B-induced (Gruber *et al.*, 2010).

Subcellular localization and activity of UVR8

UVR8 is localized mainly in the cytosol in the absence of UV-B and to a lower level in the nucleus. The nuclear fraction increases upon UV-B activation (Kaiserli and Jenkins, 2007; Yin *et al.*, 2016). Nuclear localization of UVR8 is indeed required for UV-B signalling and the UV-B-dependent nuclear accumulation of UVR8 requires COP1 (Yin *et al.*, 2016). In contrast to UVR8, COP1 includes intrinsic nuclear localization signal (NLS) and nuclear export signal (NES) sequences and the nucleocytosolic partitioning of COP1 is known to be regulated by visible light, mainly towards its nuclear exclusion (von Arnim and Deng, 1994; Lau and Deng, 2012; Pacin *et al.*, 2014). In contrast, *COP1* gene expression induced in response to UV-B and COP1 protein is stabilized and accumulates in the nucleus (Oravecz *et al.*, 2006; Favory *et al.*, 2009; Huang *et al.*, 2012). It was proposed, therefore, that cytosolic UVR8-COP1 interaction leads to their combined COP1-NLS–mediated nuclear import (Yin *et al.*, 2016). Alternatively, UVR8 monomers may enter nuclei by a further unknown mechanism and nuclear COP1 may retain UVR8 in the nucleus by inhibition of its nuclear export (Yin *et al.*, 2016). Independent of the exact mechanism, COP1 has a dual action in the regulation of UV-B-induced UVR8 nuclear accumulation and in UVR8-mediated UV-B signalling (Yin *et al.*, 2016).

Although UVR8 activity in the nuclear compartment leads to UV-B-dependent changes in gene expression, just how UVR8 activation and interaction with COP1 result in transcriptional changes is not well understood. Different transcription factors are involved, including stabilization of the bZIP transcription factors HY5 and HYH. This is described in more detail below. In addition, the results of chromatin immunoprecipitation (ChIP) experiments have suggested that UVR8 binding to chromatin in the vicinity of its target genes via histone H2B interaction may be involved (Brown *et al.*, 2005; Kaiserli and Jenkins, 2007; Cloix and Jenkins, 2008; Favory *et al.*, 2009; Cloix *et al.*, 2012). The possibility of UVR8 chromatin association was inspired by sequence homology between UVR8 and animal chromatin-associated Regulator of Chromatin Condensation (RCC1) proteins (Kliebenstein *et al.*, 2002; Brown *et al.*, 2005). It was postulated that chromatin-associated UVR8 recruits and/or activates chromatin modifiers and/or transcription factors (Cloix and Jenkins, 2008). Recent data show that histone H3 lysine K9 and/or K14 acetylation increases at UVR8-regulated gene loci upon UV-B exposure in a UVR8-dependent manner, but a mechanistic link between UVR8 and chromatin modification has not been established (Velanis *et al.*, 2016). The proposed UVR8 chromatin association via histones (preferentially H2B) was found to be independent of UV-B (Brown *et al.*, 2005; Cloix and Jenkins, 2008), but the gel-based ChIP analyses may not be sufficiently quantitative (Velanis *et al.*, 2016). A more quantitative approach using ChIP analysed by real-time qPCR did not detect any chromatin association (Binkert *et al.*, 2016). Moreover, a structural comparison of *Drosophila* DmRCC1 and *Arabidopsis* UVR8 crystal structures revealed that critical histone- and DNA-interaction residues in DmRCC1 are not conserved in UVR8 and recombinant UVR8 did not bind nucleosomes *in vitro* (Binkert *et al.*, 2016). Thus, activity of UVR8 directly at the level of target gene chromatin remains controversial and, if true, its functional significance remains to be determined.

UVR8-mediated gene regulation

The bZIP transcription factor HY5 and its homologue HYH are central regulators of UV-B-mediated gene expression

The basic leucine zipper (bZIP) transcription factor HY5 plays a major role in UV-B signalling (Ulm *et al.*, 2004; Brown *et al.*, 2005; Stracke *et al.*, 2010; Huang *et al.*, 2012). The *Arabidopsis* genome encodes a protein, HY5 HOMOLOG (HYH), with 49% amino acid sequence identity to HY5 and the same structural domains (Holm *et al.*, 2002). HY5 and HYH form homo- and heterodimers and show partial functional redundancy in visible light-dependent and UV-B-dependent gene expression; however, HY5 is the major agent in both conditions (Holm *et al.*, 2002; Brown and Jenkins, 2008; Stracke *et al.*, 2010; Feher *et al.*, 2011). HY5 and HYH are both positive regulators of photomorphogenesis that are targeted for degradation in darkness by the COP1-SPA E3 ubiquitin ligase complex and stabilized in response to visible light and UV-B (Holm *et al.*, 2002; Favory *et al.*, 2009; Lau and Deng, 2012; Huang *et al.*, 2014a).

Together, HY5 and HYH are thought to govern the majority of UVR8-mediated UV-B transcriptional responses (Tilbrook *et al.*, 2013; Jenkins, 2014). *HY5* and *HYH* are rapidly and transiently induced in response to UV-B, in a UVR8- and COP1-dependent manner (Ulm *et al.*, 2004; Brown *et al.*, 2005; Oravecz *et al.*, 2006; Favory *et al.*, 2009). Interestingly, HY5 and HYH are both required for UV-B-activated *HY5* gene expression (Binkert *et al.*, 2014). Similarly, HY5 is required for UV-B-induced expression of the negative regulatory genes *RUP1* and *RUP2* and establishment of the negative feedback loop in UV-B signalling (Gruber *et al.*, 2010). HY5 also activates *COP1* gene expression in response to UV-B and thus contributes to the accumulation of COP1 protein, in addition to the UV-B-mediated post-transcriptional stabilization of COP1 (Favory *et al.*, 2009; Huang *et al.*, 2012). HY5 associates with the promoter regions of its target genes *HY5*, *RUP1*, *RUP2*, *MYB12*, *BBX24* and *COP1* (Stracke *et al.*, 2010; Binkert *et al.*, 2014). Moreover, UV-B promotes HY5 binding to UV-B-induced genes in a UVR8-dependent manner (Binkert *et al.*, 2014).

The results of ChIP experiments combined with microarray analyses suggest that HY5 binds to the promoter regions of over 9,000 potential target genes (Zhang *et al.*, 2011), of which transcription factors are overrepresented. HY5 may thus be a hierarchical regulator of transcriptional cascades. Binding of HY5 to chromatin occurs close to transcriptional start sites, in particular to ACGT-containing elements (ACEs) (Ang *et al.*, 1998; Song *et al.*, 2008; Zhang *et al.*, 2011; Binkert *et al.*, 2014). The enrichment of *cis*-regulatory elements in proximity to HY5 binding sites suggests that HY5 regulates many target genes as part of a regulatory network involving other transcription factors (Zhang *et al.*, 2011).

HY5 and HYH are required for the transcriptional activation of the *HY5* gene (Abbas *et al.*, 2014; Binkert *et al.*, 2014). A T/G-box in the *HY5* promoter (-90-CACGTT-85) is a HY5- and HYH-binding site required for *HY5* gene induction in response to UV-B (Binkert *et al.*, 2014). Given the absence of further major UV-B-responsive elements in the *HY5* promoter and the lack of an activation domain in HY5 and HYH (Ang *et al.*, 1998; Stracke *et al.*, 2010), other transcriptional regulators are expected to contribute in combination with HY5 to the activation of *HY5* gene expression. Several HY5-interacting proteins have been identified as candidates for such a function (Gangappa and Botto, 2016). Supporting this notion, a recent study revealed that B-BOX DOMAIN PROTEIN 21 (BBX21), a HY5-interacting transcription factor (Datta *et al.*, 2007), also binds to the T/G-box in the *HY5* promoter and positively regulates the visible-light inducibility of *HY5* (Xu *et al.*, 2016). Overexpression of *BBX21* results in a hypermorphogenic hypocotyl phenotype in light-grown seedlings that depends on functional HY5 protein (Xu *et al.*, 2016), indicating that BBX21 requires HY5 for its activity. BBX21 is one of four B-box family proteins that act as co-activators (BBX21 and BBX22) or co-repressors (BBX24 and BBX25) of HY5 in visible light signalling (Crocco and Botto, 2013). Interestingly, BBX24 was shown to negatively regulate photomorphogenic UV-B responses through interaction with both COP1 and HY5 (Jiang *et al.*, 2012). It is thus tempting to speculate that further B-box family members may also regulate UV-B responses.

Next to *HY5*, a UV-B-induced marker gene that is often used is *CHALCONE SYNTHASE* (*CHS*). CHS catalyses the first committed step in flavonoid biosynthesis and is regulated mainly at the transcriptional level (Wade *et al.*, 2001; Winkel-Shirley, 2002). *CHS* induction in response to UV-B and visible light depends on the binding of HY5 to an ACE element in the *CHS* promoter (Ang *et al.*, 1998; Hartmann *et al.*, 1998;

Shin *et al.*, 2007; Stracke *et al.*, 2010). Apart from this direct link, HY5 also regulates UV-B-induced expression of *MYB12* by binding to an ACE element in its promoter region (Stracke *et al.*, 2010). MYB12 is part of the PRODUCTION OF FLAVONOL GLYCOSIDES (PFG) family of MYB transcription factors that regulate *CHS* and other genes of the flavonoid biosynthesis pathway (Stracke *et al.*, 2007). Basal *CHS* expression depends on the binding of MYB12 to an MYB recognition element in the *CHS* promoter (MRE^{CHS}) (Mehrtens *et al.*, 2005). Even though UV-B-induced *CHS* expression is independent of MYB12, *CHS* expression in response to UV-B is strongly reduced in a triple mutant lacking MYB12 and two functionally related PFG family members (Stracke *et al.*, 2010). Even though MRE^{CHS} is not a UV-B-responsive *cis*-element per se, MYB12 accumulation and binding to MRE^{CHS} in response to UV-B contributes to establishing UV-B tolerance (Stracke *et al.*, 2010).

FHY3 contributes to the UV-B-induced expression of COP1

HY5 is implicated in a positive feedback loop promoting *COP1* gene expression in response to UV-B (Huang *et al.*, 2012). ACE elements, targeted by HY5 and a FAR-RED ELONGATED HYPOCOTYL 3 (FHY3) binding site (FBS) in the *COP1* promoter act as UV-B-responsive *cis*-elements (Huang *et al.*, 2012). FHY3 contributes to UV-B-induced expression of *HY5* (and other genes) and tolerance to UV-B damage (Huang *et al.*, 2012). Interestingly, HY5 and FHY3 interact physically and UV-B diminishes the HY5-FHY3 interaction (Huang *et al.*, 2012). HY5 and FHY3 were shown to regulate *COP1* gene expression in a combinatorial, non-competitive manner by binding independently to distinct ACE and FBS *cis*-elements, respectively (Huang *et al.*, 2012). Further mechanistic details of this regulation in the context of UV-B signalling and how COP1-UVR8 impinges on the HY5-FHY3 interaction remain to be determined.

Concluding Remarks

Despite recent major steps in understanding UV-B photoreceptor signalling, it is not yet clear exactly how this unique signalling pathway functions. In particular, it is not known how the UV-B-dependent UVR8-COP1 interaction results in UV-B-responsive gene expression involving HY5, HYH, BBX24 and FHY3 (Ulm *et al.*, 2004; Brown and Jenkins, 2008; Huang *et al.*, 2012; Jiang *et al.*, 2012) and probably other transcription factors. Similarly, it is not clear how this signalling pathway regulates chromatin changes that are broadly associated with UV-B-responsive gene expression (Casati *et al.*, 2006; Casati *et al.*, 2008; Schenke *et al.*, 2014; Velanis *et al.*, 2016). It also remains to be determined exactly how UVR8-COP1 signalling impinges on phytohormone signalling and provokes UV-B-responsive morphological changes (Vanhaelewyn *et al.*, 2016). First hints with regard to the antagonizing impact of UVR8 on auxin signalling have been recorded (Hayes *et al.*, 2014; Vandenbussche *et al.*, 2014). In summary, important steps have been made towards a better understanding of how plants respond to the UV-B intrinsic to sunlight, but many open questions remain.

Acknowledgements

Research in the Ulm laboratory is supported by the University of Geneva, the European Research Council (ERC) and the Swiss National Science Foundation.

References

Abbas, N., Maurya, J.P., Senapati, D., Gangappa, S.N. and Chattopadhyay, S. (2014) Arabidopsis CAM7 and HY5 physically interact and directly bind to the *HY5* promoter to regulate its expression and thereby promote photomorphogenesis. *The Plant Cell* 26, 1036–1052. DOI:10.1105/tpc.113.122515

Ang, L.H., Chattopadhyay, S., Wei, N., Oyama, T., Okada, K. *et al.* (1998) Molecular interaction between COP1 and HY5 defines a regulatory switch for light control of Arabidopsis development. *Molecular Cell* 1, 213–222. DOI:10.1016/S1097-2765(00)80022-2

Belhaj, K., Chaparro-Garcia, A., Kamoun, S., Patron, N.J. and Nekrasov, V. (2015) Editing plant genomes with CRISPR/Cas9. *Current Opinion in Biotechnology* 32, 76–84. DOI:10.1016/j.copbio.2014.11.007

Biever, J.J., Brinkman, D. and Gardner, G. (2014) UV-B inhibition of hypocotyl growth in etiolated *Arabidopsis thaliana* seedlings is a consequence of cell cycle arrest initiated by photodimer accumulation. *Journal of Experimental Botany* 65, 2949–2961. DOI:10.1093/jxb/eru035

Binkert, M., Kozma-Bognar, L., Terecskei, K., De Veylder, L., Nagy, F. *et al.* (2014) UV-B-responsive association of the Arabidopsis bZIP transcription factor ELONGATED HYPOCOTYL5 with target genes, including its own promoter. *The Plant Cell* 26, 4200–4213. DOI:10.1105/tpc.114.130716

Binkert, M., Crocco, C.D., Ekundayo, B., Lau, K., Raffelberg, S. *et al.* (2016) Revisiting chromatin binding of the Arabidopsis UV-B photoreceptor UVR8. *BMC Plant Biology* 16, 42. DOI:10.1186/s12870-016-0732-5

Brown, B.A. and Jenkins, G.I. (2008) UV-B signaling pathways with different fluence-rate response profiles are distinguished in mature Arabidopsis leaf tissue by requirement for UVR8, HY5, and HYH. *Plant Physiology* 146, 576–588. DOI:10.1104/pp.107.108456

Brown, B.A., Cloix, C., Jiang, G.H., Kaiserli, E., Herzyk, P. *et al.* (2005) A UV-B-specific signaling component orchestrates plant UV protection. *Proceedings of the National Academy of Sciences of the United States of America* 102, 18225–18230. DOI:10.1073/pnas.0507187102

Casati, P., Stapleton, A.E., Blum, J.E. and Walbot, V. (2006) Genome-wide analysis of high-altitude maize and gene knockdown stocks implicates chromatin remodeling proteins in response to UV-B. *The Plant Journal* 46, 613–627. DOI:10.1111/j.1365-313X.2006.02721.x

Casati, P., Campi, M., Chu, F., Suzuki, N., Maltby, D. *et al.* (2008) Histone acetylation and chromatin remodeling are required for UV-B-dependent transcriptional activation of regulated genes in maize. *The Plant Cell* 20, 827–842. DOI:10.1105/tpc.107.056457

Chen, D., Gibson, E.S. and Kennedy, M.J. (2013) A light-triggered protein secretion system. *The Journal of Cell Biology* 201, 631–640. DOI:10.1083/jcb.201210119

Chen, H., Huang, X., Gusmaroli, G., Terzaghi, W., Lau, O.S. *et al.* (2010) Arabidopsis CULLIN4-damaged DNA binding protein 1 interacts with CONSTITUTIVELY PHOTOMORPHOGENIC1-SUPPRESSOR OF PHYA complexes to regulate photomorphogenesis and flowering time. *The Plant Cell* 22, 108–123. DOI:10.1105/tpc.109.065490

Christie, J.M., Arvai, A.S., Baxter, K.J., Heilmann, M., Pratt, A.J. *et al.* (2012) Plant UVR8 photoreceptor senses UV-B by tryptophan-mediated disruption of cross-dimer salt bridges. *Science* 335, 1492–1496. DOI:10.1126/science.1218091

Cloix, C. and Jenkins, G.I. (2008) Interaction of the Arabidopsis UV-B-specific signaling component UVR8 with chromatin. *Molecular Plant* 1, 118–128. DOI:10.1093/mp/ssm012

Cloix, C., Kaiserli, E., Heilmann, M., Baxter, K.J., Brown, B.A. *et al.* (2012) C-terminal region of the UV-B photoreceptor UVR8 initiates signaling through interaction with the COP1 protein. *Proceedings of the National Academy of Sciences of the United States of America* 109, 16366–16370. DOI:10.1073/pnas.1210898109

Crefcoeur, R.P., Yin, R., Ulm, R. and Halazonetis, T.D. (2013) Ultraviolet-B-mediated induction of protein-protein interactions in mammalian cells. *Nature Communications* 4, 1779. DOI:10.1038/ncomms2800

Crocco, C.D. and Botto, J.F. (2013) BBX proteins in green plants: insights into their evolution, structure, feature and functional diversification. *Gene* 531, 44–52. DOI:10.1016/j.gene.2013.08.037

Datta, S., Hettiarachchi, G.H., Deng, X.W. and Holm, M. (2006) Arabidopsis CONSTANS-LIKE3 is a positive regulator of red light signaling and root growth. *The Plant Cell* 18, 70–84. DOI:10.1105/tpc.105.038182

Datta, S., Hettiarachchi, C., Johansson, H. and Holm, M. (2007) SALT TOLERANCE HOMOLOG2, a B-box protein in Arabidopsis that activates transcription and positively regulates light-mediated development. *The Plant Cell* 19, 3242–3255. DOI:10.1105/tpc.107.054791

Davey, M.P., Susanti, N.I., Wargent, J.J., Findlay, J.E., Quick, W.P., Paul, N.D. and Jenkins, G.I. (2012) The UV-B photoreceptor UVR8 promotes photosynthetic efficiency in *Arabidopsis thaliana* exposed to elevated levels of UV-B. *Photosynthesis Research* 114, 121–131. DOI:10.1007/s11120-012-9785-y

Demkura, P.V. and Ballare, C.L. (2012) UVR8 mediates UV-B-induced Arabidopsis defense responses against *Botrytis cinerea* by controlling sinapate accumulation. *Molecular Plant* 5, 116–126. DOI:10.1093/mp/sss025

Fankhauser, C. and Ulm, R. (2011) Light-regulated interactions with SPA proteins underlie cryptochrome-mediated gene expression. *Genes & Development* 25, 1004–1009. DOI:10.1101/gad.2053911

Fasano, R., Gonzalez, N., Tosco, A., Dal Piaz, F., Docimo, T. *et al.* (2014) Role of Arabidopsis UV RESISTANCE LOCUS 8 in plant growth reduction under osmotic stress and low levels of UV-B. *Molecular Plant* 7, 773–791. DOI:10.1093/mp/ssu002

Favory, J.J., Stec, A., Gruber, H., Rizzini, L., Oravecz, A. *et al.* (2009) Interaction of COP1 and UVR8 regulates UV-B-induced photomorphogenesis and stress acclimation in Arabidopsis. *The EMBO Journal* 28, 591–601. DOI:10.1038/emboj.2009.4

Feher, B., Kozma-Bognar, L., Kevei, E., Hajdu, A., Binkert, M. *et al.* (2011) Functional interaction of the circadian clock and UV RESISTANCE LOCUS 8-controlled UV-B signaling pathways in *Arabidopsis thaliana*. *The Plant Journal* 67, 37–48. DOI:10.1111/j.1365-313X.2011.04573.x

Fierro, A.C., Leroux, O., De Coninck, B., Cammue, B.P., Marchal, K. *et al.* (2015) Ultraviolet-B radiation stimulates downward leaf curling in *Arabidopsis thaliana*. *Plant Physiology and Biochemistry* 93, 9–17. DOI:10.1016/j.plaphy.2014.12.012

Findlay, K.M. and Jenkins, G.I. (2016) Regulation of UVR8 photoreceptor dimer/monomer photo-equilibrium in Arabidopsis plants grown under photoperiodic conditions. *Plant, Cell & Environment* 39, 1706–1714. DOI:10.1111/pce.12724

Galvao, V.C. and Fankhauser, C. (2015) Sensing the light environment in plants: photoreceptors and early signaling steps. *Current Opinion in Neurobiology* 34, 46–53. DOI:10.1016/j.conb.2015.01.013

Gangappa, S.N. and Botto, J.F. (2016) The multifaceted roles of HY5 in plant growth and development. *Molecular Plant* 9, 1353–1365. DOI:10.1016/j.molp.2016.07.002

González Besteiro, M.A. and Ulm, R. (2013) ATR and MKP1 play distinct roles in response to UV-B stress in Arabidopsis. *The Plant Journal* 73, 1034–1043. DOI:10.1111/tpj.12095

González Besteiro, M.A., Bartels, S., Albert, A. and Ulm, R. (2011) Arabidopsis MAP kinase phosphatase 1 and its target MAP kinases 3 and 6 antagonistically determine UV-B stress tolerance, independent of the UVR8 photoreceptor pathway. *The Plant Journal* 68, 727–737. DOI:10.1111/j.1365-313X.2011.04725.x

Gruber, H., Heijde, M., Heller, W., Albert, A., Seidlitz, H.K. *et al.* (2010) Negative feedback regulation of UV-B-induced photomorphogenesis and stress acclimation in Arabidopsis. *Proceedings of the National Academy of Sciences of the United States of America* 107, 20132–20137. DOI:10.1073/pnas.0914532107

Hartmann, U., Valentine, W.J., Christie, J.M., Hays, J., Jenkins, G.I. *et al.* (1998) Identification of UV/blue light-response elements in the *Arabidopsis thaliana chalcone synthase* promoter using a homologous protoplast transient expression system. *Plant Molecular Biology* 36, 741–754. DOI:10.1023/A:1005921914384

Hayes, S., Velanis, C.N., Jenkins, G.I. and Franklin, K.A. (2014) UV-B detected by the UVR8 photoreceptor antagonizes auxin signaling and plant shade avoidance. *Proceedings of the National Academy of Sciences of the United States of America* 111, 11894–11899. DOI:10.1073/pnas.1403052111

Heijde, M. and Ulm, R. (2012) UV-B photoreceptor-mediated signalling in plants. *Trends in Plant Science* 17, 230–237. DOI:10.1016/j.tplants.2012.01.007

Heijde, M. and Ulm, R. (2013) Reversion of the Arabidopsis UV-B photoreceptor UVR8 to the homodimeric ground state. *Proceedings of the National Academy of Sciences of the United States of America* 110, 1113–1118. DOI:10.1073/pnas.1214237110

Heijde, M., Binkert, M., Yin, R., Ares-Orpel, F., Rizzini, L. *et al.* (2013) Constitutively active UVR8 photoreceptor variant in Arabidopsis. *Proceedings of the National Academy of Sciences of the United States of America* 110, 20326–20331. DOI:10.1073/pnas.1314336110

Heilmann, M. and Jenkins, G.I. (2013) Rapid reversion from monomer to dimer regenerates the ultraviolet-B photoreceptor UV RESISTANCE LOCUS8 in intact Arabidopsis plants. *Plant Physiology* 161, 547–555. DOI:10.1104/pp.112.206805

Heilmann, M., Velanis, C.N., Cloix, C., Smith, B.O., Christie, J.M. *et al.* (2016) Dimer/monomer status and *in vivo* function of salt-bridge mutants of the plant UV-B photoreceptor UVR8. *The Plant Journal* 88, 71–81. DOI:10.1111/tpj.13260

Hideg, E., Jansen, M.A. and Strid, A. (2013) UV-B exposure, ROS, and stress: inseparable companions or loosely linked associates? *Trends in Plant Science* 18, 107–115. DOI:10.1016/j.tplants.2012.09.003

Holm, M., Hardtke, C.S., Gaudet, R. and Deng, X.W. (2001) Identification of a structural motif that confers specific interaction with the WD40 repeat domain of Arabidopsis COP1. *The EMBO Journal* 20, 118–127. DOI:10.1093/emboj/20.1.118

Holm, M., Ma, L.G., Qu, L.J. and Deng, X.W. (2002) Two interacting bZIP proteins are direct targets of COP1-mediated control of light-dependent gene expression in Arabidopsis. *Genes & Development* 16, 1247–1259. DOI:10.1101/gad.969702

Horak, E. and Farre, E.M. (2015) The regulation of UV-B responses by the circadian clock. *Plant Signaling & Behavior* 10, e1000164. DOI:10.1080/15592324.2014.1000164

Huang, X., Ouyang, X., Yang, P., Lau, O.S., Li, G. *et al.* (2012) Arabidopsis FHY3 and HY5 positively mediate induction of *COP1* transcription in response to photomorphogenic UV-B light. *The Plant Cell* 24, 4590–4606. DOI:10.1105/tpc.112.103994

Huang, X., Ouyang, X., Yang, P., Lau, O.S., Chen, L. *et al.* (2013) Conversion from CUL4-based COP1-SPA E3 apparatus to UVR8-COP1-SPA complexes underlies a distinct biochemical function of COP1 under UV-B. *Proceedings of the National Academy of Sciences of the United States of America* 110, 16669–16674. DOI:10.1073/pnas.1316622110

Huang, X., Ouyang, X. and Deng, X.W. (2014a) Beyond repression of photomorphogenesis: role switching of COP/DET/FUS in light signaling. *Current Opinion in Plant Biology* 21, 96–103. DOI:10.1016/j.pbi.2014.07.003

Huang, X., Yang, P., Ouyang, X., Chen, L. and Deng, X.W. (2014b) Photoactivated UVR8-COP1 module determines photomorphogenic UV-B signaling output in Arabidopsis. *PLoS Genetics* 10, e1004218. DOI:10.1371/journal.pgen.1004218

Jenkins, G.I. (2014) Structure and function of the UV-B photoreceptor UVR8. *Current Opinion in Structural Biology* 29, 52–57. DOI:10.1016/j.sbi.2014.09.004

Jiang, L., Wang, Y., Li, Q.F., Bjorn, L.O., He, J.X. *et al.* (2012) Arabidopsis STO/BBX24 negatively regulates UV-B signaling by interacting with COP1 and repressing HY5 transcriptional activity. *Cell Research* 22, 1046–1057. DOI:10.1038/cr.2012.34

Kaiserli, E. and Jenkins, G.I. (2007) UV-B promotes rapid nuclear translocation of the Arabidopsis UV-B specific signaling component UVR8 and activates its function in the nucleus. *The Plant Cell* 19, 2662–2673. DOI:10.1105/tpc.107.053330

Kliebenstein, D.J., Lim, J.E., Landry, L.G. and Last, R.L. (2002) Arabidopsis UVR8 regulates ultraviolet-B signal transduction and tolerance and contains sequence similarity to human regulator of chromatin condensation 1. *Plant Physiology* 130, 234–243. DOI:10.1104/pp.005041

Lau, O.S. and Deng, X.W. (2012) The photomorphogenic repressors COP1 and DET1: 20 years later. *Trends in Plant Science* 17, 584–593. DOI:10.1016/j.tplants.2012.05.004

Li, J., Yang, L., Jin, D., Nezames, C.D., Terzaghi, W. *et al.* (2013) UV-B-induced photomorphogenesis in Arabidopsis. *Protein & Cell* 4, 485–492. DOI:10.1007/s13238-013-3036-7

Li, N., Teranishi, M., Yamaguchi, H., Matsushita, T., Watahiki, M.K. *et al.* (2015) UV-B-induced CPD photolyase gene expression is regulated by UVR8-dependent and -independent pathways in Arabidopsis. *Plant & Cell Physiology* 56, 2014–2023. DOI:10.1093/pcp/pcv121

Mathes, T., Heilmann, M., Pandit, A., Zhu, J., Ravensbergen, J. *et al.* (2015) Proton-coupled electron transfer constitutes the photoactivation mechanism of the plant photoreceptor UVR8. *Journal of the American Chemical Society* 137, 8113–8120. DOI:10.1021/jacs.5b01177

Mazza, C.A. and Ballare, C.L. (2015) Photoreceptors UVR8 and phytochrome B cooperate to optimize plant growth and defense in patchy canopies. *The New Phytologist* 207, 4–9. DOI:10.1111/nph.13332

Mehrtens, F., Kranz, H., Bednarek, P. and Weisshaar, B. (2005) The Arabidopsis transcription factor MYB12 is a flavonol-specific regulator of phenylpropanoid biosynthesis. *Plant Physiology* 138, 1083–1096. DOI:10.1104/pp.104.058032

Menon, C., Sheerin, D.J. and Hiltbrunner, A. (2016) SPA proteins: SPAnning the gap between visible light and gene expression. *Planta* 244, 297–312. DOI:10.1007/s00425-016-2509-3

Morales, L.O., Brosche, M., Vainonen, J., Jenkins, G.I., Wargent, J.J. *et al.* (2013) Multiple roles for UV RESISTANCE LOCUS8 in regulating gene expression and metabolite accumulation in Arabidopsis under solar ultraviolet radiation. *Plant Physiology* 161, 744–759. DOI:10.1104/pp.112.211375

Muller, K., Engesser, R., Schulz, S., Steinberg, T., Tomakidi, P. *et al.* (2013) Multi-chromatic control of mammalian gene expression and signaling. *Nucleic Acids Research* 41, e124. DOI:10.1093/nar/gkt340

Muller, P., and Bouly, J.P. (2015) Searching for the mechanism of signalling by plant photoreceptor cryptochrome. *FEBS Letters* 589, 189–192. DOI:10.1016/j.febslet.2014.12.008

O'Hara, A., and Jenkins, G.I. (2012) *In vivo* function of tryptophans in the Arabidopsis UV-B photoreceptor UVR8. *The Plant Cell* 24, 3755–3766. DOI:10.1105/tpc.112.101451

Oravecz, A., Baumann, A., Mate, Z., Brzezinska, A., Molinier, J. *et al.* (2006) CONSTITUTIVELY PHOTOMORPHOGENIC1 is required for the UV-B response in Arabidopsis. *The Plant Cell* 18, 1975–1990. DOI:10.1105/tpc.105.040097

Ouyang, X., Huang, X., Jin, X., Chen, Z., Yang, P. *et al.* (2014) Coordinated photomorphogenic UV-B signaling network captured by mathematical modeling. *Proceedings of the National Academy of Sciences of the United States of America* 111, 11539–11544. DOI:10.1073/pnas.1412050111

Pacin, M., Legris, M. and Casal, J.J. (2014) Rapid decline in nuclear CONSTITUTIVE PHOTOMORPHOGENESIS1 abundance anticipates the stabilization of its target ELONGATED HYPOCOTYL5 in the light. *Plant Physiology* 164, 1134–1138. DOI:10.1104/pp.113.234245

Rizzini, L., Favory, J.J., Cloix, C., Faggionato, D., O'Hara, A. *et al.* (2011) Perception of UV-B by the Arabidopsis UVR8 protein. *Science* 332, 103–106. DOI:10.1126/science.1200660

Robson, T.M., Klem, K., Urban, O. and Jansen, M.A. (2015) Re-interpreting plant morphological responses to UV-B radiation. *Plant, Cell & Environment* 38, 856–866. DOI:10.1111/pce.12374

Schenke, D., Cai, D. and Scheel, D. (2014) Suppression of UV-B stress responses by flg22 is regulated at the chromatin level via histone modification. *Plant, Cell & Environment* 37, 1716–1721. DOI:10.1111/pce.12283

Shin, J., Park, E. and Choi, G. (2007) PIF3 regulates anthocyanin biosynthesis in an HY5-dependent manner with both factors directly binding anthocyanin biosynthetic gene promoters in Arabidopsis. *The Plant Journal* 49, 981–994. DOI:10.1111/j.1365-313X.2006.03021.x

Song, Y.H., Yoo, C.M., Hong, A.P., Kim, S.H., Jeong, H.J. *et al.* (2008) DNA-binding study identifies C-box and hybrid C/G-box or C/A-box motifs as high-affinity binding sites for STF1 and LONG HYPOCOTYL5 proteins. *Plant Physiology* 146, 1862–1877. DOI:10.1104/pp.107.113217

Stracke, R., Ishihara, H., Huep, G., Barsch, A., Mehrtens, F. *et al.* (2007) Differential regulation of closely related R2R3-MYB transcription factors controls flavonol accumulation in different parts of the *Arabidopsis thaliana* seedling. *The Plant Journal* 50, 660–677. DOI:10.1111/j.1365-313X.2007.03078.x

Stracke, R., Favory, J.J., Gruber, H., Bartelniewoehner, L., Bartels, S. *et al.* (2010) The Arabidopsis bZIP transcription factor HY5 regulates expression of the *PFG1/MYB12* gene in response to light and ultraviolet-B radiation. *Plant, Cell & Environment* 33, 88–103. DOI:10.1111/j.1365-3040.2009.02061.x

Tilbrook, K., Arongaus, A.B., Binkert, M., Heijde, M., Yin, R. *et al.* (2013) The UVR8 UV-B photoreceptor: perception, signaling and response. *The Arabidopsis Book* 11, e0164. DOI:10.1199/tab.0164

Tilbrook, K., Dubois, M., Crocco, C.D., Yin, R., Chappuis, R. *et al.* (2016) UV-B perception and acclimation in *Chlamydomonas reinhardtii*. *The Plant Cell* 28, 966–983. DOI:10.1105/tpc.15.00287

Tossi, V., Lamattina, L., Jenkins, G.I. and Cassia, R.O. (2014) Ultraviolet-B-induced stomatal closure in Arabidopsis is regulated by the UV RESISTANCE LOCUS8 photoreceptor in a nitric oxide-dependent mechanism. *Plant Physiology* 164, 2220–2230. DOI:10.1104/pp.113.231753

Uljon, S., Xu, X., Durzynska, I., Stein, S., Adelmant, G. *et al.* (2016) Structural basis for substrate selectivity of the E3 ligase COP1. *Structure* 24, 687–696. DOI:10.1016/j.str.2016.03.002

Ulm, R. (2006) UV-B perception and signalling in higher plants. In: Schäfer, E. and Nagy, F. (eds) *Photomorphogenesis in Plants*, 3rd Edition, Kluwer Academic Publishers, Dordrecht, The Netherlands, pp. 279–304. DOI:10.1007/1-4020-3811-9_14

Ulm, R. and Jenkins, G.I. (2015) Q&A: How do plants sense and respond to UV-B radiation? *BMC Biology* 13, 45. DOI:10.1186/s12915-015-0156-y

Ulm, R., Baumann, A., Oravecz, A., Mate, Z., Adam, E. *et al.* (2004) Genome-wide analysis of gene expression reveals function of the bZIP transcription factor HY5 in the UV-B response of Arabidopsis. *Proceedings of the National Academy of Sciences of the United States of America* 101, 1397–1402. DOI:10.1073/pnas.0308044100

Vandenbussche, F., Tilbrook, K., Fierro, A.C., Marchal, K., Poelman, D. *et al.* (2014) Photoreceptor-mediated bending towards UV-B in Arabidopsis. *Molecular Plant* 7, 1041–1052. DOI:10.1093/mp/ssu039

Vanhaelewyn, L., Prinsen, E., Van Der Straeten, D. and Vandenbussche, F. (2016) Hormone-controlled UV-B responses in plants. *Journal of Experimental Botany* 67, 4469–4482. DOI:10.1093/jxb/erw261

Velanis, C.N., Herzyk, P. and Jenkins, G.I. (2016) Regulation of transcription by the Arabidopsis UVR8 photoreceptor involves a specific histone modification. *Plant Molecular Biology* 92, 425–443. DOI:10.1007/s11103-016-0522-3

von Arnim, A.G. and Deng, X.W. (1994) Light inactivation of Arabidopsis photomorphogenic repressor COP1 involves a cell-specific regulation of its nucleocytoplasmic partitioning. *Cell* 79, 1035–1045. DOI:10.1016/0092-8674(94)90034-5

Wade, H.K., Bibikova, T.N., Valentine, W.J. and Jenkins, G.I. (2001) Interactions within a network of phytochrome, cryptochrome and UV-B phototransduction pathways regulate *chalcone synthase* gene expression in Arabidopsis leaf tissue. *The Plant Journal* 25, 675–685. DOI:10.1046/j.1365-313x.2001.01001.x

Wargent, J.J., Gegas, V.C., Jenkins, G.I., Doonan, J.H. and Paul, N.D. (2009) UVR8 in *Arabidopsis thaliana* regulates multiple aspects of cellular differentiation during leaf development in response to ultraviolet B radiation. *The New Phytologist* 183, 315–326. DOI:10.1111/j.1469-8137.2009.02855.x

Wellmann, E. (1971) Phytochrome-mediated flavone glycoside synthesis in cell suspension cultures of *Petroselinum hortense* after preirradiation with ultraviolet light. *Planta* 101, 283–286. DOI:10.1007/BF00386835

Wellmann, E. (1983) UV radiation in photomorphogenesis. In: Shropshire, W. Jr and Mohr, H. (eds) *Encyclopedia of Plant Physiology Vol. 16B*. Springer, Heidelberg, Germany, pp. 745–756. DOI:10.1007/978-3-642-68918-5_29

Winkel-Shirley, B. (2002) Biosynthesis of flavonoids and effects of stress. *Current Opinion in Plant Biology* 5, 218–223. DOI:10.1016/S1369-5266(02)00256-X

Wu, D., Hu, Q., Yan, Z., Chen, W., Yan, C. *et al.* (2012) Structural basis of ultraviolet-B perception by UVR8. *Nature* 484, 214–219. DOI:10.1038/nature10931

Wu, M., Eriksson, L.A. and Strid, A. (2013) Theoretical prediction of the protein–protein interaction between *Arabidopsis thaliana* COP1 and UVR8. *Theoretical Chemistry Accounts* 132, 1371. DOI:10.1007/s00214-013-1371-7

Xie, L., Lang-Mladek, C., Richter, J., Nigam, N. and Hauser, M.T. (2015) UV-B induction of the E3 ligase ARIADNE12 depends on CONSTITUTIVELY PHOTOMORPHOGENIC 1. *Plant Physiology and Biochemistry* 93, 18–28. DOI:10.1016/j.plaphy.2015.03.006

Xu, C. and Min, J. (2011) Structure and function of WD40 domain proteins. *Protein & Cell* 2, 202–214. DOI:10.1007/s13238-011-1018-1

Xu, D., Jiang, Y., Li, J., Lin, F., Holm, M. *et al.* (2016) BBX21, an Arabidopsis B-box protein, directly activates HY5 and is targeted by COP1 for 26S proteasome-mediated degradation. *Proceedings of the National Academy of Sciences of the United States of America* 113, 7655–7660. DOI:10.1073/pnas.1607687113

Yang, X., Montano, S. and Ren, Z. (2015) How does photoreceptor UVR8 perceive a UV-B signal? *Photochemistry and Photobiology* 91, 993–1003. DOI:10.1111/php.12470

Yin, R., Arongaus, A.B., Binkert, M. and Ulm, R. (2015) Two distinct domains of the UVR8 photoreceptor interact with COP1 to initiate UV-B signaling in Arabidopsis. *The Plant Cell* 27, 202–213. DOI:10.1105/tpc.114.133868

Yin, R., Skvortsova, M.Y., Loubery, S. and Ulm, R. (2016) COP1 is required for UV-B-induced nuclear accumulation of the UVR8 photoreceptor. *Proceedings of the National Academy of Sciences of the United States of America* 113, E4415–4422. DOI:10.1073/pnas.1607074113

Zeng, X., Ren, Z., Wu, Q., Fan, J., Peng, P.-P. *et al.* (2015) Dynamic crystallography reveals early signalling events in ultraviolet photoreceptor UVR8. *Nature Plants* 1, 14006. DOI:10.1038/nplants.2014.6

Zhang, H., He, H., Wang, X., Yang, X., Li, L. *et al.* (2011) Genome-wide mapping of the HY5-mediated gene networks in Arabidopsis that involve both transcriptional and post-transcriptional regulation. *The Plant Journal* 65, 346–358. DOI:10.1111/j.1365-313X.2010.04426.x

10 The Effects of Ultraviolet-B on *Vitis vinifera* – How Important Is UV-B for Grape Biochemical Composition?

Brian R. Jordan*

Faculty of Agriculture and Life Sciences, Department of Wine, Food and Molecular Biosciences, Centre for Viticulture & Oenology, Lincoln University, Lincoln, Canterbury, New Zealand

Dedicated to Gillian Barbara Jordan who passed away on 25 December 2012. Gill was everything to me for 35 years and supported my career throughout.

Introduction

The light environment, including visible (400–700 nm) and ultraviolet radiation (UV: 280–380 nm) is a major determinant of plant growth and development (Whitelam and Halliday, 2007). Ultraviolet radiation-B (UV-B: 280–315 nm) is part of this natural radiation that plants are exposed to. It is a highly energetic form of radiation and is generally associated with detrimental effects upon the biosphere. This association with harmful outcomes is largely attributed to UV-B being absorbed by and causing damage to a wide variety of important molecules, such as DNA, proteins and lipids (Jordan, 1996). This absorption of UV-B can, for instance, lead to changes in gene function, loss of enzyme activity and production of highly reactive oxygen species (ROS) (Jordan, 2002; Jordan, 2011; Hideg *et al.*, 2013). Furthermore, the potential for damaging consequences of UV-B was previously heightened by the reduction in the ozone layer as a consequence of manmade chlorofluorocarbon compounds entering the stratosphere (Jordan, 1996). More recently, a more balanced perspective of the role of UV-B has developed (Ballare, 2012; Wargent and Jordan, 2013; Robson *et al.*, 2015). Thus, UV-B can induce more subtle responses, such as those associated with photomorphogenesis and can also have effects at a number of different trophic levels (Rozema *et al.*, 1997; Day, 2001). It has recently been suggested that a specific UV-B photoreceptor may be responsible for these photomorphogenetic responses and UV RESISTANCE LOCUS 8 (UVR8) has been identified as the photoreceptor (Rizzini *et al.*, 2011; Wu *et al.*, 2012). From a number of recent studies two types of response to UV-B can be identified: an acclimation response that is caused by low fluence UV-B and a 'defence response' to UV-B stress that is due to higher UV-B fluence levels and has commonality to mechanisms involved in other plant defence responses. Both types of response require changes in gene expression, but use different signal transduction and biochemical

*Corresponding author: Brian.Jordan@lincoln.ac.nz

pathways. These recent insights provide the potential for a greater understanding of how UV-B can influence plant growth and development. Indeed, Wargent and Jordan (2013) have proposed that understanding and manipulating UV-B responses may well have potential benefits in horticultural and agricultural production (see also Chapter 11, this volume).

There are, unfortunately, relatively limited examples of UV-B responses in the natural environment, particularly relating to commercial crops. One commercial crop, grapevine, *Vitis vinifera*, is, however, frequently subjected to UV-B exposure through vine canopy management practice (see below). The response to UV-B is particularly significant because many potential UV-B-induced changes alter chemical composition, which is important in the wine-making process. For instance, UV-B can change the levels and types of flavonoids that can influence the colour, astringency, bitterness and mouth feel of the wine. This is particularly relevant in many parts of the southern hemisphere where grapes are grown for wine production (e.g. Australia, Chile, New Zealand and South Africa) and the vines are exposed to relatively high levels of UV-B. For instance, Christchurch in New Zealand is at an equivalent latitude to the major wine-producing region of Bordeaux in France, but UV-B levels can be 30% higher in Christchurch, exacerbated by the clear unpolluted New Zealand atmosphere (Richard McKenzie, personal communication; Seckmeyer *et al.*, 2008; and also McKenzie *et al.*, 2006 for a similar comparison to the USA). The level of UV-B will also be influenced by altitude and the aspect of the crop to solar exposure. Understanding the UV-B-induced response is particularly complex because of the predicted global climate change as UV-B effects are strongly influenced by other environmental parameters (Jordan, 1996; and see Chapter 1, this volume). Thus the response of grapevine to UV-B must be considered in a world scenario of potentially increasing temperature and CO_2 with declining water resources. The interaction, however, between global climate change and UV-B levels still remains uncertain (Andrady *et al.*, 2012; and see Chapter 1, this volume).

The importance of a UV-B-induced response for *Vitis vinifera* is particularly significant when the management practice of canopy leaf removal is considered. Leaf removal around the fruiting zone is frequently used to reduce humidity and hence reduce disease pressure on the grape clusters. Thus, not only are the remaining canopy leaves exposed to ambient UV-B, but the berries themselves will also be directly exposed. This can have at least three potential consequences: (i) the physiology and metabolism of the leaves is altered and consequently the assimilates going to the fruit can be qualitatively and quantitatively changed; (ii) the berries respond to the direct exposure to UV-B by modification of their chemical composition; (iii) the microflora are affected either directly or indirectly by the UV-B environment.

A major development in advancing the understanding of grapevine research and the response to UV-B has come about by the application of molecular biological techniques. Major contributions have been made by the sequencing of the Pinor noir genome (Jaillon *et al.*, 2007), transcriptomic approaches (Pilati *et al.*, 2007; Pontin *et al.*, 2010) and characterization of functional gene activity and their regulation (Downey *et al.*, 2003; Dunlevy *et al.*, 2010, 2013; Azuma *et al.*, 2012; Guillaumie *et al.*, 2013; Liu *et al.*, 2015).

In this chapter the effects of UV-B on the biosynthesis of a number of important biochemical compounds (flavonoids, amino acids, aroma and lipid/carotenoids) that have implications for winemaking will be discussed. It is also the intention to discuss the molecular mechanism regulating their biosynthesis and how they compare to those described for commonly used model plant systems. In addition, the interaction of UV-B with pathogens and other environmental parameters will be considered. For a complementary review of grape berry biochemical changes during development see Conde *et al.*, 2007.

The Effect of UV-B on Vine Leaves

The leaves of plants are the primary location of photosynthesis and therefore the source of carbohydrate biosynthesis. Ultimately this carbohydrate is then distributed to growing points that act as sinks, such as new shoots and fruits. Leaves are also a major site of nitrogen assimilation, fatty acid biosynthesis and synthesis of other important compounds, such as carotenoids and terpenoids. It is therefore very important to understand how UV-B radiation impinges upon leaf function. This is particularly significant for grapevines as assimilates from the leaf are critical to fruit development and ripeness. Furthermore, predicting the outcome of light and UV responses is particularly complex as they can change with vine development (Dokoozlian *et al.*, 1996); row orientation (Grifoni *et al.*, 2008) and cultivar (Nunez-Oliveral *et al.*, 2006).

Many studies have focused on UV-B-induced responses in leaves and primarily on photosynthesis itself (Jordan 1996; Jordan *et al.*, 2016). These studies have shown a large range of detrimental effects on the biochemistry and molecular mechanisms associated with photosynthesis (Strid *et al.*, 1990, 1994; Jordan *et al.*, 1992, 1994, 2016 and references therein). In grapevine leaves exposed to natural UV, photosynthesis was inhibited and UV-induced damage to photosystem 2 was considered to be a main factor in this inhibition (Pfündel, 2003). The electron transport function represented by photosystem 2 was more capable of recovery than CO_2 fixation (Kolb *et al.*, 2001). In these studies, the inhibition of photosystem 2 was dependent on the amount of epidermal screening and an efficient repair mechanism, whereas inhibition of CO_2 was not diminished by UV-B screening and the recovery was slower than for photosystem 2. There are a number of factors that provide protection to the photosynthetic apparatus from UV-B and these include the levels of UV-B absorbing compounds, the potential to reflect the UV-B, antioxidants and DNA repair activity. UV radiation is generally considered to induce two forms of compounds in the leaf epidermis capable of protecting the plant: the hydroxycinnamates and flavonoids. The effectiveness of these compounds must, however, be considered in the context of the physical distribution and morphology of the leaf. For instance the vacuoles of the epidermal cells contain UV-absorbing compounds and may protect 90% of the leaf area (Jordan, 1996). However, passage between cells through the anticlinal cell walls will allow UV-B to damage the exposed mesophyll and palisade cells (Day *et al.*, 1992, 1993; Jordan, 1996). These cells may also contain UV-absorbing compounds, but not necessarily reduce UV-B damage. Indeed, the relationship between UV protection and levels of UV-absorbing compounds may be overestimated due to the action of non-epidermal compounds (Kolb and Pfündel, 2005). Some of these UV-B absorbing compounds also add to the antioxidant capacity of the tissue. In addition, UV-B causes an increase in a range of other antioxidants, such as glutathione and ascorbate (Hideg *et al.*, 2013). Martínez-Lüscher *et al.* (2013) exposed grapevines to two doses of biologically effective UV-B radiation (5.98 and 9.66 kJ m^{-2} day^{-1}) over two separate periods, from fruit set until harvest ripeness (75 days) and from veraison to harvest ripeness (20 days). The major effects were seen after the short-term exposure at higher dose with inhibition of a wide range of photosynthetic functions and a significant increase in antioxidant capacity. However, exposure to a long period of UV-B at either dose showed that most photosynthetic and biochemical parameters were unaffected, with no sign of oxidative damage. These results show that grapevine has a long-term acclimation response to UV-B, probably due to the accumulation of UV-B absorbing compounds and antioxidant capacity. In addition, UV-B may also cause the formation, or induce complex cross linking, of compounds such as cuticular wax. This would lead to some minor reflection of the UV-B (about 5% for most leaf tissue, Jordan, 1996), but would also change the cell surface characteristic in terms of thickness and permeability. The responses to UV-B at the cell surface will undoubtedly vary between grape varieties and could be very significant in terms of changing the skin properties of the fruit in relation to the winemaking process. Furthermore, the nature of the cuticular wax is thought to play a role in plant defence and disease susceptibility (Rozema *et al.*, 1997; Keller *et al.*, 2003). Consequently, UV-B exposure will influence disease progression and herbivory (see below and Chapter 3, this volume).

Another complex aspect of UV-B protection is the role of high photosynthetically active radiation (PAR). High PAR is well known to provide protection to the photosynthetic apparatus from UV damage (Jordan, 1996). This protection is frequently ascribed to the higher light providing more DNA protection through photo-repair mechanisms. This, however, is not the case as experiments in which the effects of photo-repair were excluded by using low-pressure sodium lamps (which had a wavelength different from that which activates photo-repair through photolyase, but which still maintains photosynthesis) still showed high light protection against UV-B (Mackerness, *et al.*, 1996). The authors concluded that some component of photosynthesis itself was involved, probably photophosphorylation.

In addition to the chemical changes that are induced in the leaves by UV-B, there are well-characterized changes in leaf morphology (Wargent *et al.*, 2009a; Wargent and Jordan, 2013). These changes frequently include a reduction in leaf area and increase in leaf thickness, leaf mass per unit area and a reduction in total leaf number. The mechanism involved in these changes is uncertain, but UVR8 regulates multiple aspects of cell differentiation in the leaf in response to UV-B (Wargent *et al.*, 2009b). Interestingly many of these changes in leaf morphology appear similar to the changes that take place in leaves in response to high irradiance (Davies *et al.*, 1986). The changes in morphology in high light are reflected in physiological function, such as increased photosynthesis which may drive greater assimilation of amino acids, etc. As a consequence of this knowledge, the impact on grape berry biochemistry may be elucidated. As leaves that are exposed to high light probably receive the highest UV-B it is clear that increased leaf exposure through viticultural management may have consequences for grape development. It would certainly be very interesting to know if the mechanisms involved in UV-B and high light responses are similar or even the same. This is particularly interesting when the concept of 'gap' sensing through UV-B signalling is considered (see Chapter 3 for discussion). This may have consequences for the biochemical composition of the grape berries and the pathology and herbivory of the fruit.

Changes in the Chemical Composition of Grape Berries

Flavonoids

Most attention has focused on the influence of UV radiation on the phenylpropanoid pathway in grapes, as this pathway is involved in colour, tannin formation and the production of compounds with antioxidant capacity. The three major classes of flavonoids in grapes are flavonols, flavan-3-ols (condensed tannins/proanthocyanidins) and anthocyanins.

In white grape varieties, UV-B exposure leads to increased pigmentation as spots on the fruit skin. One of the most interesting aspects of this UV-B-induced pigmentation is that the response takes place after veraison when the berries soften, irrespective of UV-B exposure levels being normally higher pre-veraison. Thus there is a clear developmental regulation that overrides the response to UV-B (Gregan *et al.*, 2012; and see below in relation to anthocyanin formation and PR proteins). This regulation has been linked to the levels of carbohydrate within the fruit (Lenk *et al.*, 2007). Thus, the sugar accumulates within the berry until a point is reached after which biosynthesis of flavonoid compounds takes place. This may be the case for coloured flavonoids, such as anthocyanins, but certainly colourless flavonols can also be synthesized pre-veraison (Liu *et al.*, 2015; and see below). This is particularly important when the formation of another related chemical is considered, resveratrol (trans-3,5,4'-trihydroxystilbene). Resveratrol is thought to play an important role in plant defence and to have a beneficial role in human health (Pan *et al.*, 2009). The pivotal enzyme in the biosynthesis of resveratrol is stilbene synthase, STS (Pan *et al.*, 2009). This enzyme and the subsequent synthesis of resveratrol takes place mostly pre-veraison in contrast to the biosynthesis of the pathway to anthocyanins (post-veraison) through the first committed enzyme step, CHS (chalcone synthase). Both CHS and STS use coumaroyl CoA as a substrate. Although more understanding is needed, it is likely that this split in the pathway to either anthocyanin or resveratrol is a key to the outcome of significant grape chemistry. Most importantly, UV radiation has a strong influence on both of these pathways although at different stages of development (pre- and post-veraison). This is also consistent with studies that show that the overall response to UV-B is determined by the stage of development (Jordan, 1996 and references therein; Jordan *et al.*, 1998; Mackerness *et al.*, 1998).

UV absorbance increases rapidly in grape berries in response to UV-B exposure, primarily in the skins (Fig. 10.1a, b), and this is reflected in increases in individual flavonoid compounds as determined by HPLC analysis (primarily flavonols, Fig. 10.1c). The qualitative composition, however, of individual

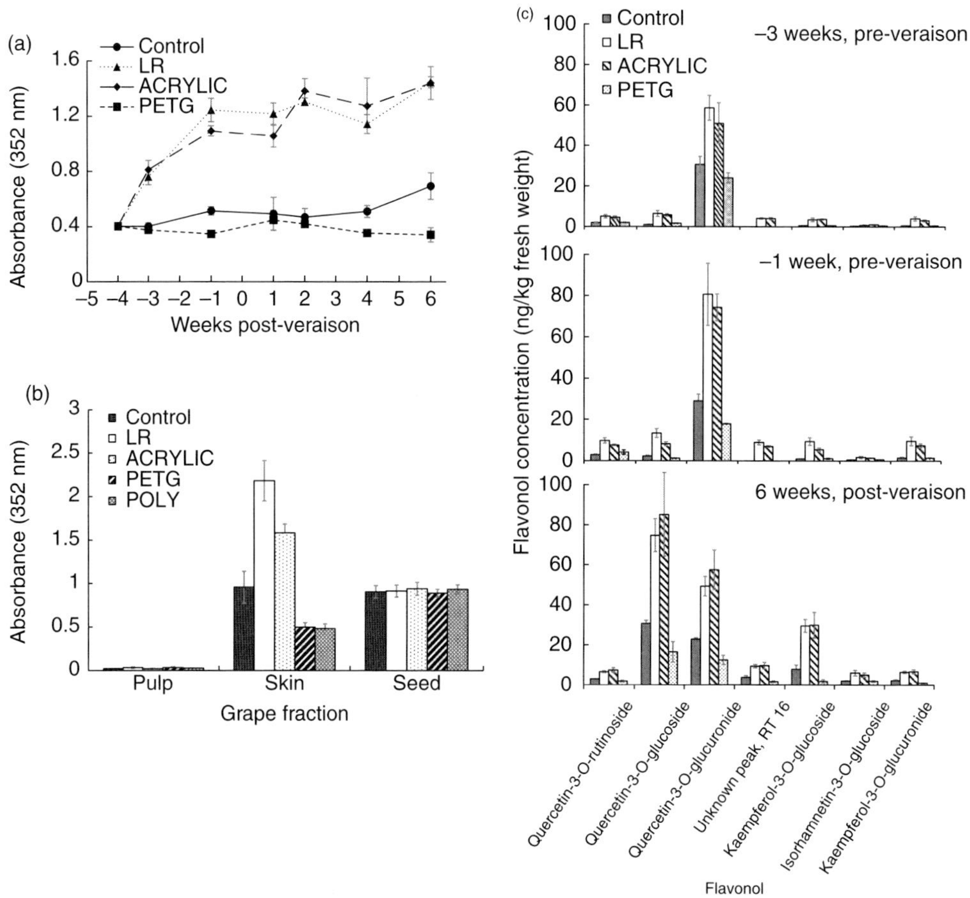

Fig. 10.1. The influence of UV-B on flavonol biosynthesis in grapevine. Fig.10.1a shows the increase in absorbance in grape berries when exposed to UV-B compared to UV-B exclusion. Fig 10.1b shows absorbance in different grape fractions. Fig.10.1c shows the changes in individual flavonols during development and in response to UV-B exposure or exclusion. Control and PETG (UV-B exclusion). LR, leaf removal and ACRYLIC (UV-B exposure). Data adapted from Liu *et al.* (2015), Figs 1, 2 and Table 1.

flavonols changes between veraison and harvest. For instance, in Sauvignon blanc preveraison, quercetin-3-0-glucoside was increased to a greater level than other flavonols, but after veraison a number of other flavonols, notably quercetin-3-0-glucoside and kaempherol-3-0-glucoside are relatively enhanced (Fig.10.1c). This is consistent with many other studies that show the chemical composition of flavonoids can be changed by UV-B (Ryan *et al.*, 1997; Markham *et al.*, 1998; Hofmann *et al.*, 2000; Liu *et al.*, 2015). This qualitative change has also recently been seen in a latitudinal analysis of *Vitis vinifera* cv. Pinot noir across Europe (Del-Castillo-Alonso, M.-A. *et al.*, 2016). In this study, the influence of latitude (between Spain and Germany) and associated environmental parameters was determined on the metabolite composition of Pinot noir berry skins. The ratio between trihydroxylated and monohydroxylated flavonols, which was positively correlated with antioxidant capacity, was the berry-skin variable best correlated with latitudinal solar variables.

The biosynthesis of flavonoids is complex and has been extensively studied in many species (Schwinn and Davies, 2004). In respect to UV-B it is generally acknowledged that the biosynthesis of flavonoids is to provide protection, both by absorption of the UV-B and as antioxidants (Jordan, 2004). In grapes, as in other species, flavonols are produced by flavonol synthase (Downey *et al.*, 2003); its gene expression is specifically increased by UV-B exposure (Fujita *et al.*, 2006 and see below). The biosynthesis of flavonols involves the regulation of flavonol synthase by a complex of proteins including a R2R3-MYB transcription factor (*Vv*MYB12/MYBF1), a basic helix-loop-helix domain protein (*Vv*MYCA1/bHLH) and a WD40 repeat protein (*Vv*WDR1 and WDR2). These regulatory complexes have been characterized in grapes in relation to their role in flavonoid biosynthesis (Czemmel *et al.*, 2009; Matus *et al.*, 2009; Hichri *et al.*, 2010; Matus *et al.*, 2010).

As mentioned in the introduction, the understanding of UV-B regulation of plant cell function has taken a 'step change' with the characterisation of UVR8 as the UV-B photoreceptor (Jenkins and Brown, 2007; Rizzini *et al.*, 2011; Wu *et al.*, 2012; Jenkins, 2013; see also Chapters 8 and 9, this volume). To date these studies have largely been carried out on the model plant *Arabidopsis thaliana*. This research has pointed towards a specific low UV-B fluence response perceived by UVR8 and mediated through transcription factors such as HY5 and COP1 (Brown and Jenkins, 2008; Stracke *et al.*, 2010; see also Chapter 9, this volume). The regulation is then mediated through the MYB/bHLH/WD40 complex to determine the activity of key genes such as flavonol synthase (FLS; Fig.10.2). In addition to the UVR8 pathway a non-specific mechanism exists that overlaps with other known signal transduction pathways, commonly including ROS, jasmonic acid and salicylic acid (Surplus *et al.*, 1998; Mackerness *et al.*, 1999; Jordan, 2002). This pathway has been shown to induce gene expression for a variety of enzymes including the pathogen-related proteins and has usually been related to high UV fluence and potentially damaging stress (see Hideg *et al.*, 2013 for discussion). This pathway has been shown to be mediated by mitogen activated protein kinase signalling cascades (González Besteiro *et al.*, 2011).

Jordan and colleagues have recently studied the pathways of gene expression from UVR8 in *Vitis vinifera var.* Sauvignon blanc in vineyard trials (Liu *et al.*, 2015). To examine the UV-B signalling in detail they studied the gene expression of UVR8 and reaction partners that could reflect the high and low fluence signal transduction pathways responding to UV-B. Thus, they selected reaction partners for UVR8 thought to be involved in the low fluence response, such as HY5, COP1, the MYB/bHLH/WD40 transcription factor complex and functional genes known to be UV-B-induced in grapes, such as flavonol synthase (Table 10.1 shows gene expression for the major genes tested, indicating up or down regulation). Five flavonol synthase genes (FLS) were tested and only two (FLS 4 and 5) showed expression. These genes were expressed predominantly at different times during development and both were UV-B up-regulated. Of the signal transduction and regulatory genes, HY5 and

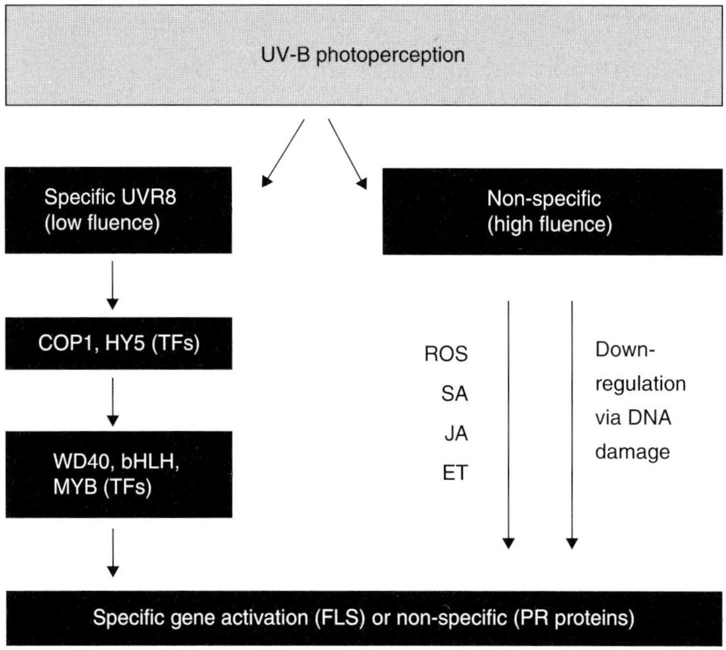

Fig. 10.2. Schematic diagram to represent the specific and non-specific UV-B signal transduction pathways. Adapted from Jordan (2011), Fig. 22.2.

Table 10.1. UV-B-induced gene expression in *Vitis vinifera*.

Gene signalling pathway	Gene activity
UVR8-COP1-HY5	
VvUVR8	X
VvCOP1	X
VvHY5	↑
MYB12-bHLH-WD40	
VvMYB12	↑
VvbHLH	X
VvWDR1/2	X
PR proteins	
VvTL1-3	X
VvChi4A/4B	↑
Flavonoid biosynthetic	
VvFLS4/5	↑
VvCHS1/2	↑

MYB were up-regulated in response to UV-B in these vineyard trials. Genes that are thought to be involved in the non-specific high fluence response, PR genes and MAKP kinase were also tested. Interestingly, the UV-B-induced response of PR genes was limited and they showed a much clearer developmental response. MAKP kinase was not UV-B-induced. To complement the vineyard field trials, Liu *et al.* (2017) studied UV-B induction of the same genes in controlled environments with Sauvignon blanc 'Mullins' vines at two developmental time points (equivalent to pre-veraison and just pre-harvest). The results were consistent with the findings in the vineyard, except that PR genes were more consistently up-regulated by UV-B. Overall, these results suggest that in a natural environment (which has routinely high fluence rates), a UV-B-induced pathway that has been considered low-fluence specific (from model plant studies), is active in a commercial crop (Liu *et al.*, 2017).

Anthocyanins are also very important flavonoids for the wine industry as they are visual and textural cues in wine. Anthocyanin levels are strongly regulated by development, and increase from veraison through to harvest. They are also influenced by environmental factors, including light/UV-B and temperature (higher temperatures reducing anthocyanin accumulation). The response to light/UV-B is not consistent and apparently varies between varieties of grapevine. For instance, in Shiraz that was shaded, anthocyanin increased in response to development, but not the reduction in light (Downey *et al.*, 2004). The gene expression of UDP-glucose flavonoid glucosyl transferase (UFGT), which is an important gene in anthocyanin biosynthesis was not regulated by UV-B, but only regulated by development. Similarly, UV-B did not have an effect on Pinot noir anthocyanin concentrations (Price *et al.*, 1995). In contrast, anthocyanin levels were related to UV-B in Tempranillo grapes and UFGT gene expression was up-regulated (Martínez-Lüscher *et al.*, 2014). Martínez-Lüscher *et al.* (2016) also studied the interaction of potential climate change on anthocyanin synthesis. Higher temperatures and CO_2 enhance grape development resulting in increased sugar accumulation while decreasing anthocyanin biosynthesis. This uncoupling of sugar accumulation from anthocyanin was found to be alleviated by UV-B exposure as it negated the CO_2/temperature response while increasing anthocyanin biosynthesis. Another important flavonoid group that is important for the wine industry is the proanthocyanidins (PAs; sometimes referred to as condensed tannins) which are polymers of flavanol-3-ol units and provide bitter attributes, astringency and complexity to the wine. The biosynthesis of proanthocyanidins in the skins of grapevine berries has been found to be primarily dependent on visible light and contrasted with flavonols whose biosynthesis is strongly UV-B-dependent (Koyama *et al.*, 2012). Shading in lightproof boxes of Cabernet sauvignon from flowering to 49 days after treatment partially reduced the levels of PAs and transcript levels. This treatment had a much more profound effect on flavonol levels and related transcript abundance. Light exclusion also decreased the mean degree of polymerisation of the PAs, which is very significant for wine quality. UV-B per se did not affect the concentration and composition of PAs in contrast to the major effect on flavonol concentrations. Natural shading by leaf foliage has produced wines with lower flavanol-3-ol and PAs compared to leaf removal from the fruiting zone (Kemp *et al.*, 2011). There is therefore a very different light regulation of these flavonoids (visible light regulating PAs and UV-B regulating flavonols) and this may reflect the different biochemical pathways (Kemp *et al.*, 2011, and references therein).

Finally, flavonoids are known to have a wide range of different effects on grapevine (see Czemmel *et al.*, 2012, and references therein). For example, one important effect of a flavonoid increase is that it can lead to poor fruit set. This is because high levels of flavonols can inhibit auxin transport, which in turn will inhibit fruit set. This effect can also be enhanced by nitrogen deficiency. As UV-B clearly is a major regulator of flavonoid biosynthesis and ultimately will be a determinant of flavonoid levels, understanding the wider consequences of UV-B in grapevine biology is therefore critical.

Amino acids and nitrogen assimilation

Another important area of vine biochemistry is the metabolism of amino acids (Bell and Henschke, 2005; Ugliano *et al.*, 2008). The assimilation of amino acids takes place primarily in leaves through the GS-GOGAT pathway in the chloroplasts (Lam *et al.*, 1996; Forde and Lea, 2007). This pathway is light-dependent and the assimilate is redistributed from the leaves to other parts of the plant, including the berries. Amino acids are very important for viticulture and oenology as they are precursors for a variety of important compounds associated with winemaking (Bell and Henschke, 2005; Ugliano *et al.*, 2008; Gregan *et al.*, 2012, 2017). For example, the phenolic compounds are synthesized from phenylalanine through the activity of phenylalanine ammonia lyase. Aromatic compounds such as methoxypyrazines are synthesized from valine, leucine and isoleucine through a pathway that has not been fully characterized in grape (Dunlevy *et al.*, 2010). The formation of sulphur containing aroma compounds called thiols, including 3-mercaptohexan-1-ol (3MH) and 3-mercaptohexyl acetate (3MHA), require cysteine and glutathione to attach to hexanyl carbon chains in the final stages of their synthesis during winemaking. The amino acid composition in the must also has an effect on the final wine characteristics and can also be manipulated by the use of different microbes in the fermentation. Importantly, most amino acids are utilised during the fermentation process (YAN: yeast assimilable nitrogen). It is also significant that a major amino acid at harvest, proline, is non-YAN. Therefore a large percentage of the final amino acid concentration is not readily available for fermentation (Stines *et al.*, 1999; Stines *et al.*, 2000; Gregan *et al.*, 2012, 2017). Low levels of YAN amino acids can cause the ferment to become 'stuck' and lead to artificial nitrogen being added to complete the ferment, which is not a particularly satisfactory alternative for natural winemaking. Despite the importance of amino acids and the role that light has been shown to play in assimilation, there is not that much known about the control and regulation of the various amino acids in grapevine. In particularly there is very little known about their regulation by UV radiation. Schultz *et al.*, (1998) showed that over two seasons in Riesling grapes the total amino acid levels were higher when UV-B was reduced by screens to approximately 10% of ambient. Furthermore, there was a qualitative change in amounts of amino acids that contribute substantially to YAN, such as arginine and glutamine which increased in lower UV-B conditions. The non-YAN amino acid, proline, also increased under lower UV-B. Proline is an amino acid that frequently shows a response to stress, so it is no surprise that it responded to UV-B (but see below for further discussion). In contrast to these results, Keller and Torres-Martinez (2004) found no change in amino acid composition under UV-B treatments, although they did find differences in polyphenolics (as discussed above). The differences in response could be due to significantly different experimental treatments (vineyard trials versus potted vines) and varieties (Riesling versus Chardonnay). In another series of trials, UV screens were placed over rows of Sauvignon blanc grapevines to study the effects of UV-B and UV-A radiation in 'leaf plucked' vines and in comparison to control vines (no screens and no leaf removal). Both qualitative and quantitative changes to amino acids were detected in the grape berries (Gregan *et al.*, 2012, 2017). Significantly, the level of these amino acids in the berries related to the presence of the proximal leaves remaining over the fruiting zone; leaf removal reducing the total amino acids. Despite the fruit being exposed to high natural UV levels, no changes to amino acids could be related to a UV response. These results contrast with those of Schultz *et al.* (1998) even though a similar screening approach was used. Recently, Martínez-Lüscher *et al.* (2014) used fluorescent UV-B lamps in a glasshouse to study amino acid and flavonoid accumulation in Tempranillo grapes. Grapes exposed to the highest fluence of UV-B from fruit set to ripeness showed no difference in total free amino acids. Major amino acids such as arginine and glutamine also showed no difference (in agreement with Gregan *et al.*, 2012). However, some reduction was seen between veraison and full ripeness, both at the higher fluence and a medium exposure of UV-B. One amino acid that did increase was gamma-amino-butyrate. It was suggested this may be due to a possible role in preventing accumulation of reactive oxygen intermediates and cell death under UV-B exposure. From a variety of different methodologies and using different grape varieties, the evidence points to little direct effect of UV-B radiation on amino acid metabolism, though there are some indications of a UV-B response. This positive response, however, may also be indirect (impact on leaf photosynthesis) which leads to reduced production and or export of the amino acids to the fruit. With the advent of transcriptomic approaches etc., the regulation of amino acid biosynthesis can be addressed in more detail. For instance, in our present research we are using Nanostring nCounter transcriptomics technology which will enable us to simultaneously investigate multiple known and putative genes involved in amino acid metabolism, for instance glutamate/glutamine and proline/arginine metabolism. This will provide expression profiles for biosynthesis and degradation of amino acids throughout development and in response to the environment. One observation is that using this approach, flavonol synthase 4 clearly responds to UV-B in agreement with the results of Liu *et al.*, 2015 using qPCR. In contrast, no genes related to amino acid biosynthesis or turnover show an overt UV-B response/induction in the grape berry (Gregan *et al.*, unpublished observations). It is possible that the regulation of amino acid metabolism is largely at the level of protein activation and enzyme activity.

Aroma compounds, lipids and isoprenoids

Methoxypyrazines (mostly 3-isobutyl-2-methoxypyrazine (IBMP) and 3-isopropyl-2-methoxypyrazine (IPMP)) are important aroma compounds giving an herbaceous/grassy aroma to wine (Allen *et al.*, 1991; Lacey *et al.*, 1991; Lakso and Sacks, 2010). Their presence is considered beneficial in some wines, such as New Zealand Sauvignon blanc (Parr *et al.*, 2007a, b), but detrimental for the aroma of red wines, such as Cabernet sauvignon (Dunlevy *et al.*, 2010; Martinson and Scheiner, 2010; Dunlevy *et al.*, 2013; Guillaumie *et al.*, 2013). A variety of studies have suggested that light has an effect either on the biosynthesis or degradation of these compounds (Gregan and Jordan, 2012, 2016 and references therein). The research has frequently inferred that UV-B has a role in this light response. Recently, methoxypyrazines were determined in grape berries throughout development from vines with canopy leaves retained, with leaves removed over the fruiting zone (no screen), and with leaves removed and PETG screens used to block UV-B radiation (Gregan *et al.*, 2012, 2016 and unpublished data). Over three seasons the highest level of methoxypyrazine was achieved pre-veraison. Using screening material to exclude UV-B, some indications were obtained that early in development UV-B did reduce methoxypyrazine levels. However, this was not sustained and no effect could be seen at harvest in juice made from these berries. Furthermore, specific analysis of berry skins showed no changes to methoxypyrazine levels at harvest when UV-B was removed, compared to fully exposed berries.

Grapevines comprise a mixture of hydrophobic compounds such as waxes, fatty acids, membrane lipids and sterols, as would normally be expected within the various cellular compartments of plants. Their leaves contain polyunsaturated fatty acids, such as linolenic acid and linoleic acid (C18:3 and C18:2) as the major fatty acid constituents of galactolipids which are the predominant lipids of leaf chloroplasts. These polyunsaturated fatty acids within the lipid fractions are potentially susceptible to UV-B-induced oxidation and the subsequent action of enzymes such as lipases and lipoxygenase (LOX) to produce a cascade of damaging reactive oxygen species (Podolyan *et al.*, 2010). In addition, the hexanyl carbon chains formed from this enzymatic breakdown are the precursors for combination with sulphur-containing amino acids to form thiols. The limited research, however, shows little evidence that UV-B causes a breakdown of these fatty acids (Giordano *et al.*, 2004) and in the normal growing environment lipids are stable due to a wide range of protective mechanisms, which include cellular antioxidants and physical barriers such as changes to surface wax composition and formation of trichomes. However, it is likely that the leaves will become more susceptible to UV-B-induced lipid breakdown with senescence. Furthermore, the leaf fall at senescence will expose the fruit to more UV-B and potential lipid degradation (see below for potential consequences).

The berries themselves also contain plastids, but their morphology and function changes throughout development. Specifically the grape pericarp plastids change from a small amyloplast stage at anthesis to a larger pleomorphic form containing lipid-like granules (Hardie *et al.*, 1996). The lipid-like globules are the cellular compartments associated with the biosynthesis of isoprenoid secondary metabolites, such as monoterpenes and damascenone (see below). Within the lipid fractions of the grape berry, the major fatty acids are saturated fatty acids, such as palmitic acid (C16) and stearic acid

(C18), although there are still substantial levels of polyunsaturated fatty acids present at harvest. There is also a large LOX gene family that is capable of breaking down these fatty acids (Podolyan *et al.*, 2010). From the wine industry perspective, the breakdown of these fatty acids pre- and post-harvest is very significant as it leads to the formation of important aroma compounds called thiols (3MH and 3MHA: Podolyan *et al.*, 2010). The role of UV-B on this process at this latter stage of grape berry development is unknown, but clearly there is a significant potential to both cause the lipid breakdown and induce the expression of genes for enzymes such as LOX which could completely alter the wine aroma profile. This is particularly relevant as, at this late stage of development, leaf fall will expose the berry fruit to additional UV-B exposure and potentially cause further change in lipid composition *in vivo*. Subsequent tissue disruption during harvesting will then bring the lipid/fatty acid substrate in contact with enzymes of degradation such as LOX.

Another major protective mechanism involves carotenoids which are hydrophobic molecules found in plastids and act as antioxidants and photo-protectants. These compounds are very important in grapes as they are precursors of volatile norisoprenoids, notably beta-damascenone and beta-ionone that are contributors to wine flavour and aroma. The changes in carotenoid levels in response to UV-B has been variable depending largely on the species or variety (Jordan, 1996). In grapes the response has usually been a lowering of the carotenoid levels (Schultz *et al.*, 1998; Steel and Keller, 2000). Steel and Keller (2000) showed a decrease in leaf and berry carotenoids in response to UV-B in Cabernet sauvignon vines. This reduction in carotenoids has also been linked to a reduction in the photoprotective efficiency of the xanthophyll cycle. In addition, the norisoprenoids change in response to the light/UV-B environment (see Jug and Rusjan (2012) for further discussion and references therein). Another important group of hydrophobic compounds are the terpenoids, which include diterpenoids such as tocopherol and triterpenes such as sitosterol and stigmasterol. The triterpenes are synthesized by a cytosolic MAV pathway while the diterpenes are produced in plastids by the methylerytritol phosphate pathway. Recent studies suggest that these pathways are induced by different fluences of UV-B radiation and may have different roles (Gil *et al.*, 2012). For instance, low fluence UV-B causes the induction of sterols that are involved in membrane stabilisation, whereas high fluence UV-B increases the diterpenes which have an antioxidant role to protect against damaging ROS. Diterpenes, such as tocopherol, are also lipophilic and therefore protect against lipid oxidation damage within the hydrophobic regions of membranes. High UV-B also increases the levels of the stress-related sesquiterpenoid hormone ABA. Taken together the results suggest the involvement of terpenoids at two levels. At low fluence UV-B they are involved in an adaptive mechanism involving the production of sterols to stabilize cell membranes and at higher UV-B fluence, terpenoids with antioxidant properties are produced to be part of a defence-related response (Gil *et al.*, 2012).

The Effects of UV-B on Grape Pathogens

UV-B is highly energetic radiation and therefore has potentially damaging consequences for a wide variety of plant pathogens (Jordan, 1996). There are a number of ways that UV-B radiation can influence disease progression in grapes:

1. Direct impact of the UV-B radiation on the pathogens.
2. Enhancement of the plant's endogenous biochemical defence.
3. Formation of physical barriers such as waxes.

In addition, different cultivars of grapes show different susceptibility to UV-B, and cultivars will respond to the UV-B exposure differently at different developmental stages.

UV-B radiation is potentially harmful to any microbe (bacteria, fungi, bacteriophage or virus) present on the surface of the grapevine. However, the response of the microbe will depend on its individual tolerance or susceptibility to UV-B, which will be dependent on a number of factors, such as the life cycle stage, degree of protective pigmentation and effectiveness of DNA repair mechanisms. In addition, it is important to consider the impact of multiple environmental parameters on the pathogens since a combination of stress factors can have a different outcome compared to the effect of a single stress (Jordan, 1996). The microbes, themselves can frequently be adapted to particular UV environments, such as the soil or phyllosphere. The variety of individual responses will in turn influence the ecological community effect. The general response is, however, for microbes to be affected negatively by UV-B radiation. *Uncinula necator* (powdery mildew) is a major pathogen of grapevine and tends to be most prevalent within the shade inside the vine canopy (Austin and Wilcox, 2010). Exposure to UV-B both inhibited conidial spore germination and mycelial growth of this pathogen (Willocquet *et al.*, 1996). UV-B radiation is also known to share a number of common signal transduction pathways with pathogens and elicit pathogen defence pathways (Surplus *et al.*, 1998). These signal transduction pathways induce changes in gene expression to produce pathogen protective compounds. Evidence that this may be an important factor in protection of grapes comes from studies on infection of abaxial and adaxial sides of leaves (Keller *et al.*, 2003). These studies showed that a highly significant effect of UV was found on abaxial sides of leaves, despite the fact that this surface was protected from direct UV exposure. These results strongly suggest that a physiological response of the plants cellular defence system is a major determinant of infection. In addition, the same workers looked at the interaction of UV and nitrogen status in different varieties of grapes (Chardonnay and Cabernet sauvignon) that show different susceptibility to powdery mildew. Incidence of powdery mildew infection was substantially increased by high nitrogen status and low UV, particularly in Chardonnay. Under these conditions, the response was related to leaf succulence, low phenolics and cuticular wax deposition, high nitrogen and photosynthesis.

A well-reported response of plants to UV-B is the production of pathogen-related proteins (PR proteins). These PR proteins are a mixture of acidic and basic enzymatic proteins that can inhibit pathogen infection, through degrading cell wall components for instance (e.g. chitinase). In Sauvignon blanc grapevines, PR proteins increased in the skin and pulp with the severity of powdery mildew infection (Tian *et al.*, 2015). UV exclusion (UV-B and UV-A) reduced the PR proteins in the skin, but not the pulp. In similar experiments on Sauvignon blanc, but specifically separating UV-B and UV-A responses, PR gene expression was considered (Liu *et al.*, 2015). In these studies, PR gene expression was very dependent on the developmental stage of the grapes, being predominantly expressed towards harvest. Some of the PR genes were induced by UV-B, but not all of those tested. In another series of experiments in controlled environment cabinets (Liu *et al.*, 2017), PR genes were routinely UV-B-induced which is consistent with previous findings (Mackerness *et al.*, 1999). For further related discussion on UV-B effects on ecosystems, including herbivory, see Chapter 3, this volume.

Conclusions and Going Forward

Grapevine is a major international crop and there is a long heritage of successful viticulture and winemaking. However, there are significant challenges for the future. A major challenge has to be the impact of global climate change on a crop which is frequently produced as a monoculture. The general consensus is that global climate change is taking place and there are potential challenges for viticulture in the predicted climate scenarios (Schultz, 2000; see also Chapter 1, this volume). The predicted changes include increased temperature, increased CO_2 and reduced water availability. The increase in CO_2 scenarios that have been predicted have been experimentally tested with a number of species. For most species the results are relatively consistent with an increase in net photosynthesis, biomass production and water use efficiency. These findings are also consistent for grape, however, long-term acclimation may reduce these positive aspects (Schultz, 2000).

In addition to these environmental parameters, UV-B levels are unlikely to decline substantially and in some latitudes may even increase in response to other climatic factors changing (see also Chapter 1, this volume). The role of UV-B becomes very important as it is well known that the response of plants to UV-B radiation is particularly sensitive to interaction with other environmental parameters (Jordan, 1996). Thus there is a general expectation that with latitudinal climate changes taking place, some varieties of grapes will not cope in their existing environments (e.g. Merlot in Bordeaux). Furthermore it is predicted that the environment of whole regions may be essentially moved to a different location (e.g. the climate of the major wine region of New Zealand, Marlborough, will be found further south in Canterbury). It is therefore essential that the consequences of a changing environment be taken into account, with multiple environmental parameters being considered. For instance, the predicted increase in temperature and CO_2 is expected to increase photosynthesis and hence productivity. However, UV-B may counteract this response (see Martínez-Lüscher *et al.*, 2016). Similarly, the UV-B response of plants is strongly influenced by the level of water availability – a reduced water level generally decreases the UV-B response. This could be because lower levels of water slow growth, meaning that there is less potential impact of UV-B on cell division, etc. Even the colour of red grape varieties may be changed by the alteration in temperature and UV-B brought about by climate change.

From this short review we can make a number of particular observations on the biochemical and molecular response to UV-B in grapevine:

1. Flavonols respond specifically to UV-B radiation and the regulation is at the level of gene activity. The role of different signal transduction pathways need further investigation, but it is highly likely that the UVR8/HY5/COP1/MYB pathway is involved (Liu *et al.*, 2015).
2. Flavanol-3-ols (proanthocyanidins/condensed tannins) respond mostly to visible light.
3. Anthocyanins show some response to UV-B in different varieties, but this requires further investigation to determine the molecular mechanism.
4. The interactions between the flavonoids (1, 2 and 3 above) in response to UV-B/visible light are very significant in relation to wine quality (e.g. oxidation status and co-pigmentation).
5. Amino acid levels may be affected in certain varieties of grapevine by UV-B, but this is likely to be indirect.
6. Aroma compounds do show changes in response to UV-B, be they direct or indirect. These changes are wide ranging and reflect the diverse nature of the biochemical pathways that are involved.
7. The effect of UV-B on pathogens has only been investigated to a limited extent in grapevine. This remains an opportunity for further understanding and exploitation through viticulture practice.

Acknowledgements

My UV-B research on grapevine has been supported by a joint investment by Plant and Food Research and New Zealand winegrowers. I would like specifically to acknowledge New Zealand wine industry colleagues for their open discussions and interest in my research. I would also like to acknowledge the major contributions of Scott Gregan and Linlin Liu to my research in recent years. Special thanks to all my great colleagues in the UV-B field in particular, Ake Strid, Marcel Jansen, Jason Wargent and Rainer Hofmann. Two individual plant scientists, Daphne Vine Prue (photomorphogenesis) and the late Professor Jan Anderson (photosynthesis) had a profound influence on my career and I am forever grateful.

References

Allen, M.S., Lacey, M.J., Harris, R.L.M. and Brown, W.V. (1991) Contribution of methoxypyrazines to Sauvignon blanc wine aroma. *American Journal of Enology and Viticulture* 42, 109–112.

Austin, C. and Wilcox, W. (2010) Heat and UV radiation from sunlight exposure inhibit powdery mildew. Cornell Researchers tackle green flavours in red wines. *Research Focus 2010-2: Cornell Viticulture and Enology*, 1–6.

Andrady, A.L., Aucamp, P.J., Austin, A.T., Bais, A.F., Ballaré, C.L. *et al.* (2012) Environmental effects of ozone depletion and its interactions with climate change: Progress report, 2011. United Nations Environment Programme, Environmental Effects Assessment Panel. *Photochemical and Photobiological Sciences* 11, 13–27.

Azuma, A., Yakushiji, H., Koshita, Y. and Kobayashi, S. (2012) Flavonoid biosynthesis-related genes in grape skin are differentially regulated by temperature and light conditions. *Planta* 236, 1067–1080.

Ballaré, C.L., Mazza, C.A., Austin, A.T. and Pierik, R. (2012) Canopy light and plant health. *Plant Physiology* 160, 145–155.

Bell, S.-J. and Henschke, P.A. (2005) Implications of nitrogen nutrition for grapes, fermentation and wine. *Australian Journal of Grape and Wine Research* 11, 242–295.

Brown, B.A. and Jenkins, G.I. (2008) UV-B signaling pathways with different fluence-rate response profiles are distinquished in mature Arabidopsis leaf tissue by requirements for UVR8, HY5, and HYH. *Plant Physiology* 146, 576–588.

Conde, C., Silva, P., Fontes, N., Dias, A.C.P., Tavares, R.M. *et al.* (2007) Biochemical changes throughout grape berry development and fruit and wine quality. *Food* 1, 1–22.

Czemmel, S., Stracke, R., Weisshaar, B., Cordon, N., Harris, N.N. *et al.* (2009) The grapevine R2R3-MYB transcription factor VvMYBF1 regulates flavonol synthesis in developing grape berries. *Plant Physiology* 151, 1513–1530.

Czemmel, S., Heppel, S.C. and Bogs, J. (2012) R_2R_3 MYB transcription factors: key regulators of the flavonoid biosynthetic pathway in grapevine. *Protoplasma* 249, 109–118.

Davies, E.C., Chow, J.M., Le Fay, J.M. and Jordan, B.R. (1986) Acclimation of tomato leaves to changes in light intensity: effects on the function of the thylakoid membrane. *Journal of Experimental Botany* 37, 211–220.

Day, T.A. (2001) Multiple trophic levels in UV-B assessments-completing the ecosystem. *New Phytologist* 152, 181–186.

Day, T.A., Vogelmann, T.C. and DeLucia, E.H. (1992) Are some plant life forms more effective than others in screening out ultraviolet-B radiation? *Oecologia* 92, 513–519.

Day, T.A., Martin, G. and Vogelmann, T.C. (1993) Penetration of UV-B radiation in foliage – evidence that the epidermis behaves as a nonuniform filter. *Plant, Cell & Environment* 16, 735–741.

Del-Castillo-Alonso, M.-A., Castagna, A. and Csepregi, K. *et al.* (2016) Environmental factors correlated with the metabolite profile of *Vitis vinifera* cv. Pinot noir berry skins along a European latitudinal gradient. *Journal of Agricultural and Food Chemistry* 64, 8722–8734.

Dokoozlian, N.K. and Kliewer, W.M. (1996) Influence of light on grape berry growth and composition varies during fruit development. *Journal of American Society of Horticultural Science* 121, 869–874.

Downey, M.O., Harvey, J.S. and Robinson, S.P. (2003) Synthesis of flavonols and expression of flavonol synthase genes in developing grape berries of Shiraz and chardonnay (*Vitis vinifera* L.). *Australian Journal of Grape and Wine Research* 9, 110–121.

Downey, M.O., Harvey, J.S. and Robinson, S.P. (2004) The effect of bunch shading on berry development and flavonoid accumulation in Shiraz grapes. *Australian Journal of Grape and Wine Research* 10, 55–73.

Dunlevy, J.D., Soole, K.L., Perkins, M.V., Dennis, E.G., Keyzers, R.A., Kalua, C.M. and Boss, P.A. (2010) Two O-methyltransferases involved in the biosynthesis of methoxypyrazines: grape-derived aroma compounds important to wine flavour. *Plant Molecular Biology* 74, 77–89.

Dunlevy, J.D., Dennis, E.G., Soole, K.L., Perkins, M.V., Davies, C. and Boss, P.K.A. (2013) Methyltransferase essential for the methoxypyrazine-derived flavour of wine. *Plant Journal* 75, 606–617.

Forde, B.G. and Lea, P.J. (2007) Glutamate in plants: metabolism, regulation and signalling. *Journal of Experimental Botany* 58, 2339–2358.

Fujita, A., Goto-Yamamoto, N., Aramaki, I. and Hashizume, K. (2006) Organ-specific transcription of putative flavonol synthase genes of grapevine and effects of plant hormones and shading on flavonol biosynthesis in grape berry skins. *Bioscience, Biotechnology, and Biochemistry* 70, 632–638.

Gil, M., Pontin, M., Berli, F., Bottini, R. and Piccoli, P. (2012) Metabolism of terpenes in the response of grape (*Vitis vinifera* L.) leaf tissues to UV-B radiation. *Phytochemistry* 77, 89–98.

Giordano, C.V., Galactro, A., Puntarulo, S. and Ballares, C.L. (2004) The inhibitory effects of UV-B radiation (280–315nm) on *Gunnera magellanica* growth correlate with increased DNA damage not with oxidative damage to lipids. *Plant, Cell & Environment* 27, 1415–1423.

González Besteiro, M., Bartels, S., Albert, A. and Ulm, R. (2011) Arabidopsis MAP kinase phosphatase 1 and its target MAP kinases 3 and 6 antagonistically determine UV-B stress tolerance, independent of the UVR8 photoreceptor pathway. *The Plant Journal* 68, 727–737.

Gregan, S.M. and Jordan, B. (2016) Methoxypyrazine accumulation and O-methyltransferase gene expression in sauvignon blanc grapes: the role of leaf removal, light exposure, and berry development. *Journal of Agricultural and Food Chemistry* 64, 2200–2208.

Gregan, S.M., Wargent, J.J., Liu, L., Shinkle, J., Hofmann, R. *et al.* (2012) Effects of solar ultraviolet radiation and canopy manipulation on the biochemical composition of Sauvignon Blanc grapes. *Australian Journal of Grape and Wine Research* 18, 227–238.

Gregan, S.M., Winefield, C. and Jordan, B. (2017) Amino acid metabolism and accumulation in Sauvignon blanc: investigating berry composition in response to canopy manipulation. *Acta Horticulturae.* In press.

Grifoni, D., Carreras, G., Zipoli, G., Sabatini, F., Dalla Marta, A. and Orlandini, S. (2008) Row orientation effect on UV-B, UV-A and PAR solar irradiation components in vineyards at Tuscany, Italy. *International Journal of Biometeorology* 52, 755–763.

Guillaumie, S., Ilg, A., Réty, S., Brette, M., Trossat-Magnin, C. *et al.* (2013) Genetic analysis of the biosynthesis of 2-methoxy-3-isobutylpyrazine, a major grape-derived aroma compound impacting wine quality. *Plant Physiology* 162, 604–615.

Hardie, W.J., Aggenbach, S.J. and Jaudzems, V.G. (1996) The plastids of the grape pericarp and their significance in isoprenoid synthesis. *Australian Journal of Grape and Wine Research* 2, 144–154.

Hichri, I., Heppel, S.C., Pillet, J., Leon, C., Czemmel, S. *et al.* (2010) The basic Helix-Loop-Helix transcription factor MYC1 is involved in the regulation of the flavonoid biosynthetic pathway in grapevine. *Molecular Plant* 3, 509–523.

Hideg, É., Jansen, M. and Strid, Å. (2013) UV-B radiation, ROS and stress: inseparable companions or loosely linked associates? *Trends in Plant Science* 18, 107–115.

Hofmann, R.W., Swinny, E.E., Bloor, S.J., Markham, K.R., Ryan, K.G. *et al.* (2000) Response of nine *Trifolium repens* L. populations to ultraviolet-B radiation: differential flavonol glycoside accumulation and biomass production. *Annals of Botany* 86, 527–537.

Jaillon, O., Aury, J.M., Noel, B., Policriti, A., Clepet, C. *et al.* (2007) The grapevine genome sequence suggests ancestral hexaploidization in major angiosperm phyla. *Nature* 449, 463–467.

Jenkins, G.I. (2013) Photomorphogenic responses of plants to UV-B radiation. Available at: http://photobiology.info/Jenkins.html (accessed 21 April 2017).

Jenkins, G.I. and Brown, B.A. (2007) UV-B perception and signal transduction. In: Whitelam, G.C. and Halliday, K.J. (eds) *Annual Plant Reviews Volume 30: Light and Plant Development.* Blackwell Publishing Ltd, Oxford, UK, pp. 155–182.

Jordan, B.R. (1996) The effects of ultraviolet-B radiation on plants: a molecular perspective. In: Callow, J.A. (ed.) *Advances in Botanical Research Volume 22.* Academic Press. Cambridge, Massachusetts, pp. 97–162.

Jordan, B.R. (2002) Molecular response of plant cells to UV-B stress. *Functional Plant Biology* 29, 909–916.

Jordan, B.R. (2004) Plant pigments and protection against UV-B radiation. In: Davies, K. (ed.) *Annual Plant Reviews Vol. 14: Plant Pigments and Their Manipulation.* CRC Press/Blackwell Publishing, Oxford, UK, pp. 275–292.

Jordan, B. (2009) The effects of light on grape berry and leaf biochemistry: a new insight into canopy management and quality wine production. *New Zealand Winegrower*, October/November 13 (2).

Jordan, B. (2011) The effects of UV-B radiation on plants: molecular mechanisms involved in UV-B response. In: Pessarakli, M. (ed.) *The Handbook of Plant and Crop Stress*, 3rd Edition. CRC Press, Boca Raton, Florida, pp. 565–576.

Jordan, B.R., He, J., Chow, W.S. and Anderson, J.M. (1992) Changes in mRNA levels and polypeptide subunits of ribulose 1,5-bisphosphate carboxylase in response to supplementary ultraviolet-B radiation. *Plant, Cell & Environment* 15, 91–98.

Jordan, B.R., James, P., Strid, Å. and Anthony, R. (1994) The effect of supplementary UV-B radiation on gene expression and pigment composition in etiolated and green pea leaf tissue: UV-B-induced changes are gene-specific and dependent upon the developmental stage. *Plant, Cell & Environment* 17, 45–54.

Jordan, B.R., James, P. and Mackerness, S.A.H. (1998) Factors affecting UV-B induced changes in *Arabidopsis thaliana* gene expression: role of development, protective pigments and the chloroplast signal. *Plant and Cell Physiology* 39, 769–778.

Jordan, B.R., Strid, Å. and Wargent, J. (2016) What role does UV-B play in determining photosynthesis? In: Pesslarakli, M. (ed.) *Handbook of Photosynthesis*, 3rd Edition. CRC Press, Boca Raton, Florida, pp. 275–286.

Jug, T. and Rusjan, D. (2012) Advantages and disadvantages of UV-B radiations on grapevine (*Vitis* sp.). *Emir Journal of Agriculture* 24, 576–585.

Keller, M. and Torres-Martinez, N. (2004) Does UV radiation affect winegrape composition? *Acta Horticulturae* 640, 313–319.

Keller, M., Rogiers, S.Y. and Schultz, H.R. (2003) Nitrogen and ultraviolet radiation modify grapevines' susceptibility to powdery mildew. *Vitis* 42, 87–94.

Kemp, B.S., Harrison, R. and Creasy, G.L. (2011) Effect of mechanical leaf removal and its timing on flavan-3-ol composition and concentration in *Vitis vinifera* L. cv. Pinot noir wine. *Australian Journal of Grape and Wine Research* 17, 270–279.

Kennedy, J. (2002) Understanding grape berry development. *Winegrowing* July/August, 1–5.

Koch, A., Doyle, C.L., Matthews, M.A., Williams, L.E. and Ebeler, S.E. (2010) 2-Methoxy-3-isobutlypyrazine in grape berries and its dependence on genotype. *Phytochemistry* 1–9.

Kolb, C.A. and Pfündel, E.E. (2005) Origins of non-linear and dissimilar relationships between epidermal UV absorbance and UV absorbance of extracted phenolics in leaves of grapevine and barley. *Plant, Cell & Environment* 25, 580–590.

Kolb, C.A., Kaser, M.A., Kopecký, J., Zotz, G., Riederer, M. and Pfündel, E.E. (2001) Effects of natural intensities of visible and ultraviolet radiation on epidermal ultraviolet screening and photosynthesis in grape leaves. *Plant Physiology* 127, 863–875.

Kolb, C.A., Kopecký, J., Riederer, M. and Pfündel, E.E. (2003) UV screening by phenolics in berries of grapevine (*Vitis vinifera*). *Functional Plant Biology* 30, 1177–1186.

Koyama, K., Ikeda, H., Poudel, P.R. and Goto-Yamamoto, N. (2012) Light quality affects flavonoid biosynthesis in young berries of Cabernet Sauvignon grapes. *Phytochemistry* 78, 54–64.

Lacey, M.J., Allen, M.S., Harris, R.L.N. and Brown, W.V. (1991) Methoxypyrazines in Sauvignon blanc grapes and wines. *American Journal of Enology and Viticulture* 42, 103–108.

Lakso, A.N. and Sacks, G.L. (2010) Growing season dynamics – vine balance impact on methoxypyrazines. *Winegrowing* May/June, 35–49.

Lam, H.-M., Coschigano, K.T., Oliveira, I.C., Melo-Oliveira, R. and Coruzzi, G.M. (1996) The molecular genetics of nitrogen assimilation into amino acids in higher plants. *Annual Review of Plant Physiology and Plant Molecular Biology* 47, 569–593.

Lenk, S., Buschmann, C. and Pfündel, E.E. (2007) *In vivo* assessing flavonols in white grape berries (*Vitis vinifera* L. cv. Pinot Blanc) of different degrees of ripeness using chlorophyll fluorescence imaging. *Functional Plant Biology* 34, 1092–1104.

Liu, L., Gregan, S., Winefield, C. and Jordan, B. (2015) From UVR8 to flavonol synthase: UV-B-induced gene expression in Sauvignon blanc grape berry. *Plant, Cell & Environment* 38, 905–919.

Liu, L., Gregan, S., Winefield, C. and Jordan, B. (2017) A comparison between controlled environment and vineyard trials to study the effects of UV-B on flavonoid biosynthesis in *Vitis vinefera* L. Sauvignon blanc berries. *Plant Biochemistry and Physiology*. Submitted.

Mackerness, S.A.H., Butt, J.P. and Jordan, B.R. (1996) Amelioration of ultraviolet-B-induced downregulation of mRNA transcripts for chloroplast proteins, by high irradiance, is mediated by photosynthesis. *Journal of Plant Physiology* 148, 100–106.

Mackerness, S.A.H., Surplus, S.L., Jordan, B.R. and Thomas, B. (1998) Effects of supplementary UV-B radiation on photosynthetic transcripts at different stages of leaf development and light levels in pea: role of active oxygen species and antioxidant enzymes. *Photochemistry and Photobiology* 68, 88–96.

Mackerness, S.A.H., Surplus, S.L., Blake, P., John, C.F., Buchanan-Wollaston, V., Jordan, B.R. and Thomas, B. (1999) Ultraviolet-B-induced stress and changes in gene expression in *Arabidopsis thaliana*: role of signalling pathways controlled by jasmonic acid, ethylene and reactive oxygen species. *Plant, Cell & Environment* 22, 1413–1423.

Markham, K.R., Ryan, K.G., Bloor, S.J. and Mitchell, K.A. (1998) An increase in luteolin: Apigenin ratio in *Marchantia polymorpha* on UV-B enhancement. *Phytochemistry* 8, 191–194.

Martínez-Lüscher, J., Morales, F., Delrot, S., Sánchez-Díaz, M., Gomés, E., Aguirreolea, J. and Pascual, I. (2013) Short- and long-term physiological responses of grapevine leaves to UV-B radiation. *Plant Science* 213, 114–122.

Martínez-Lüscher, J., Sánchez-Díaz, M., Delrot, S., Aguirreolea, J., Pascual, I. and Gomés, E. (2014) Ultraviolet-B radiation and water deficit interact to alter flavonol and anthocyanin profiles in grapevine berries through transcriptomic regulation. *Plant and Cell Physiology* 55, 1925–1936.

Martínez-Lüscher, J., Sánchez-Díaz, M., Delrot, S., Aguirreolea, J., Pascual, I. and Gomés, E. (2016) Ultraviolet-B alleviates the uncoupling effect of elevated CO_2 and increased temperature on grape berry (*Vitis vinifera* cv. Tempranillo) anthocyanin and sugar accumulation. *Australian Journal of Grape and Wine Research* 22, 87–95.

Martinson, T. and Scheiner, J. (2010) Cornell researchers tackle green flavours in red wines. *Research Focus 2010-11: Cornell Viticulture and Enology* 1–5.

Matus, J.T., Loyola, R., Vega, A., Peña-Neira, A., Bordeu, E., Arce-Johnson, P. and Alcalde, J.A. (2009) Post-veraison sunlight exposure induces MYB-mediated transcriptional regulation of anthocyanin and flavonol synthesis in berry skins. *Journal of Experimental Botany* 60, 853–867.

Matus, J.T., Poupin, M.J., Cañón, P., Bordeu, E., Alcalde, J.A. and Arce-Johnson, P. (2010) Isolation of WDR and bHLH genes related to flavonoid synthesis in grapevine (*Vitis vinifera* L.). *Plant Molecular Biology* 72, 607–620.

McKenzie, R.L., Bodeker, G.E., Scott, G. and Slusser, J. (2006) Geographical differences in erythemally-weighted UV measured at mid-latitude USDA sites. *Photochemistry and Photobiology Science* 5, 343–352.

Nunez-Oliveral, E., Martínez-Abaigar, J., Tomas, R., Otero, S. and Arroniz-Crespo, M. (2006) Physiological effects of solar ultraviolet-B exclusion on two cultivars of *Vitis vinfera* L. from La Rioja, Spain. *American Journal Enology and Viticulture* 57, 441–448.

Pan, Q.-H., Wang, L. and Li, J.-M. (2009) Amounts and subcellular localization of stilbene synthase in response of grape berries to UV irradiation. *Plant Science* 176, 360–366.

Parr, W.V., Sherlock, R.R. and Green, J.A. (2007a) The importance of green: sensory and chemical analysis of New Zealand Sauvignon blanc. *Australian and New Zealand Grapegrower and Winemaker* Annual Technical Issue 62–69.

Parr, W.V., Green, J.A., White, G.K. and Sherlock, R.R. (2007b) The distinctive flavour of New Zealand Sauvignon blanc: sensory characterisation by wine professionals. *Food Quality and Preference* 18, 849–861.

Pfündel, E.E. (2003) Action of UV and visible radiation on chlorophyll fluorescence from dark-adapted grape leaves (*Vitis vinifera* L.). *Photosynthesis Research* 75, 29–39.

Pilati, S., Perazzolli, M., Malossini, A., Cestaro, A., Dematte, L., Fontana, P. *et al.* (2007) Genome-wide transcriptional analysis of grapevine berry ripening reveals a set of genes similarly modulated during three seasons and the occurrence of an oxidative burst at veraison. *BMC Genomics* 8, 422.

Podolyan, A., White, J., Jordan, B. and Winefield, C. (2010) Identification of the lipoxygenase gene family from *Vitis vinifera* and biochemical characterisation of two 13-lipoxygenases expressed in grape berries of Sauvignon blanc. *Functional Plant Biology* 37, 1–18.

Pontin, M.A., Piccoli, P.N., Francisco, R., Bottini, R., Martinez-Zapater, J.M. and Lijavetzky, D. (2010) Transcriptome changes (*Vitis vinifera* L.) cv. Malbec leaves induced by ultraviolet-B radiation. *BMC Plant Biology* 10, 224–236.

Price, S.F., Breen, P.J., Valladao, M. and Watson, B.T. (1995) Cluster sun exposure and quercetin in Pinot noir grapes and wine. *American Journal of Enology and Viticulture* 46, 187–194.

Robson, M.T., Karel, K., Otmar, U. and Jansen, M.A.K. (2015) Re-interpreting plant morphological responses to UV-B radiation. *Plant, Cell & Environment* 38, 856–866.

Rizzini, L., Favory, J.-J., Cloix, C., Faggionato, D., O'Hara, A. *et al.* (2011) Perception of UV-B by the Arabidopsis UVR8 protein. *Science* 332, 103–106.

Rozema, J., Staaij, J., van de Bjorn, L.O. and Caldwell, M. (1997) UV-B as an environmental factor in plant life: stress and regulation. *Tree* 12, 22–28.

Ryan, K.G., Markham, K.R., Bloor, S.J., Bradley, J.M., Mitchell, K.A. and Jordan, B.R. (1997) UV-B radiation induced increase in Quercetin: Kaempferol ratio in normal and transgenic lines of Petunia. *Photochemistry and Photobiology* 68, 323–330.

Schultz, H.R. (2000) Climate change and viticulture: a European perspective on climatology, carbon dioxide and UV-B effects. *Australian Journal of Grape and Wine Research* 6, 2–12.

Schultz, H.R., Lohnertz, O., Bettner, W., Balo, B., Linsenmeier, A. *et al.* (1998) Is grape composition affected by current levels of UV-B radiation? *Vitis* 37, 191–192.

Schwinn, K.E. and Davies, K.M. (2004) Flavonoids. In: Davies, K. (ed.) *Annual Plant Reviews Vol. 14: Plant Pigments and Their Manipulation.* CRC Press/Blackwell Publishing, Oxford, UK, pp. 92–149.

Seckmeyer, G., Glandorf, M., Wichers, C., McKenzie, R., Henriques, D. *et al.* (2008) Europe's darker atmosphere in the UV-B. *Photochemical and Photobiological Sciences* 7, 925–930.

Steel, C.C. and Keller, M. (2000) Influence of UV-B irradiation on the carotenoid content of *Vitis vinifera* tissues. *Biochemical Society Transactions* 28, 883–885.

Stines, A.P., Naylor, D.J., Hoj, P.B. and Heeswijck, R.V. (1999) Proline accumulation in developing grapevine fruit occurs independently of changes in the levels of delta 1-pyrroline-5-carboxylate synthetase mRNA or protein. *Plant Physiology* 120, 923–931.

Stines, A.P., Grubb, J., Gockowiak, H., Henschke, P.A., Hoj, P.B. and Heeswijck, R.V. (2000) Proline and arginine accumulation in developing berries of *Vitis vinifera* L. in Australian vineyards: influence of vine cultivar, berry maturity and tissue type. *Australian Journal of Grape and Wine Research* 6, 150–158.

Stracke, R., Favory, J.-J., Gruber, H., Bartelniewoehner, L., Bartels, S. *et al.* (2010) The Arabidopsis bZIP transcription factor HY5 regulates expression of the PFG1/MYB12 gene in response to light and ultraviolet-B radiation. *Plant, Cell & Environment* 33, 88–103.

Strid, Å., Chow, W.S. and Anderson, J.M. (1990) Effects of supplementary UV-B radiation on photosynthesis in *Pisum sativum. Biochimica Biophysica Acta* 1020, 260–268.

Strid, Å., Chow, W.S. and Anderson, J.M. (1994) UV-B damage and protection at the molecular level in plants. *Photosynthesis Research* 39, 475–489.

Surplus, S.L., Jordan, B.R., Murphy, A.M., Carr, J.P., Thomas, B. and Mackerness, S.A.H. (1998) Ultraviolet-B-induced responses in *Arabidopsis thaliana*: role of salicylic acid and reactive oxygen species in the regulation of transcripts encoding photosynthetic and acidic pathogenesis-related proteins. *Plant, Cell & Environment* 21, 685–694.

Tian, B., Harrison, R., Jaspers, M. and Morton, J. (2015) Influence of ultraviolet exclusion and of powdery mildew infection on sauvignon blanc grape composition and on extraction of proteins into juice. *Australian Journal of Grape and Wine Research* 21, 417–424.

Ugliano, M., Henschke, P.A., Henderich, M.J. and Pretorius, I.S. (2008) Nitrogen management-critical for wine flavour and style. *Winemaking* May/June, 6–25.

Wargent, J.J. and Jordan, B.R. (2013) From ozone depletion to agriculture: understanding the role of solar UV radiation in sustainable crop production. *New Phytologist Tansley Review* 197, 1058–1076.

Wargent, J.J., Gegas, V.C., Jenkins, G.I., Doonan, J.H. and Paul, N.D. (2009a) UVR8 in *Arabidopsis thaliana* regulates multiple aspects of cellular differentiation during leaf development in response to ultraviolet-B radiation. *New Phytologist* 183, 315–326.

Wargent, J.J., Moore, J.P., Ennos, A.R. and Paul, N.D. (2009b) Ultraviolet radiation as a limiting factor in leaf expansion and development. *Photochemistry and Photobiology* 85, 279–286.

Whitelam, G.C. and Halliday, K.J. (2007) Light and plant development. In: Whitelam, G.C. and Halliday, K.J. (eds) *Annual Plant Reviews Vol. 30.* Blackwell Publishing, Oxford, UK.

Willocquet, L., Colombet, D., Rougier, M., Fargues, J. and Clerjeau, M. (1996) Effects of radiation, especially ultraviolet-B, on conidial germination and mycelial growth of grape powdery mildew. *European Journal of Plant Pathology* 102, 441–449.

Wu, D., Hu, Q., Yan, Z. *et al.* (2012) Structural basis of ultraviolet-B perception by UVR8. *Nature* 484, 214–219.

11 Turning UV Photobiology into an Agricultural Reality

Jason Wargent*

Institute of Agriculture and Environment, Massey University, Palmerston North, New Zealand

Introduction

Global agriculture faces significant challenges in order to provide sustainable, nutritious, high-quality food for our growing population. During the last 50 years, agricultural productivity has expanded at a pace that has typically exceeded increases in human population, but this phase of expansion has now reached an end. The rising tide of food 'insecurity' demands that the pipeline of agricultural innovation is hyper-accelerated, delivering new solutions for food production at a previously unprecedented rate. Such a task calls for the integration of new plant biological knowledge into agricultural practice, and reduced reliance on purely conventional approaches to increase crop yields.

The Green Revolution of the 20th century was underpinned by waves of technology transfer that led to large changes in agronomic methods, including the introduction of high-yielding grain varieties, increased use of agrochemicals and increased use of mechanization. The result of the adoption of such approaches was an almost threefold increase in grain yields from 1961 to 2010 (Godfray *et al.*, 2010). However, the world now faces a range of threats to agricultural improvement, including a predicted rise in human population to 9 billion by 2050, the consequences of climate change, and increased land consumption demand for meat farming, dairy production and other non-food land use such as biofuel feedstock cultivation. Taken together, it is estimated that global crop yields need to increase by 100–110% by 2050 in order to meet required demands (Tilman *et al.*, 2011). It is currently clear that achieving such a quantum in crop yield increase will be highly challenging, and it has been estimated that the current rates of improvement for key grains indicate that a doubling of yields will be very unlikely by the middle of this century (Ray *et al.*, 2013). The challenges associated with achieving large increases in agricultural productivity are significant, and more so when combined with the threat of climate change. For example, it has been suggested that a potential outcome of increased atmospheric CO_2 levels is a fertilization effect in terms of an increase in net photosynthetic rates of many crop plants (Ainsworth and Rogers, 2007). However, such potential benefits of increased CO_2 supply to plants could be limited by other consequences of climate change, such as increases in tropospheric

*Corresponding author: J.Wargent@massey.ac.nz

ozone, which can be injurious to plant photosynthetic function (Lobell and Burke, 2009). Equally, crop plants such as maize and sorghum, which utilize C4 photosynthesis, could be less sensitive to increased CO_2 in terms of photosynthetic improvement (Ainsworth and Long, 2005), yet contradictory evidence also suggests C4 crops may indeed be sensitive to CO_2 enrichment (Leakey *et al.*, 2004). The interactions between various outcomes of climate change and net crop productivity are still not well understood, but even predicted climatic variability alone is a factor that underpins a significant proportion of crop yield variability (Ray *et al.*, 2015). In addition to the marked concerns regarding crop productivity from a food insecurity standpoint, other related challenges exist for agriculture; namely food quality and provision for a burgeoning middle-class global population. By 2030, it is estimated that the number of Global Middle Class (i.e. those households with a daily expenditure ranging from US$10 to $100 per person) could increase to 4.9 billion (from 1.8 billion in 2009) (Kharas, 2010). Such an increase in the number of middle class consumers will act as drivers for agricultural change, not in terms of nourishment itself, but with regards to increased demand for certain food types (e.g. dairy, meat) and higher-quality ('premium') food products, such as fresh produce with optimized taste, increased health-giving properties, and shelf-life characteristics. As well as direct and tangible improvements in food quality, a growing demand for premium food will also lead to increased pressure to further reduce agrochemical inputs, and progression of the use of sustainable agronomic practices in general.

The pathway to increasing food production to such a scale requires significant innovation, particularly with regard to the use of 'sustainable intensification' (Pretty, 2008), e.g. achieving large increases in crop yield, but in the absence of large expansion of production land area. Therefore, closing the yield gaps in agriculture requires the use of numerous agronomic and plant biological tools (Foley *et al.*, 2011), utilising techniques based on the principles of precision agriculture (Wang *et al.*, 2006), integrated pest management (Vet and Dicke, 1992), and sustainable nutrient management (Spiertz, 2010), in addition to the use of genetic modification. But no single tool or approach will single-handedly achieve the yield increases or innovations in premium food production that are needed, and an integrated model of agricultural evolution will be the only route to achieve such goals. One of the goals to achieving sustainable intensification is the increased substitution of chemical crop inputs with biologically based inputs (Alexandratos and Bruinsma, 2012), and increased use of biologically based crop improvement knowledge in general. The exploitation of plant-crop responses to light have been critical throughout the history of crop cultivation, not least due to the requirement of light for photoautotrophic nutrition via photosynthesis, but also due to our understanding of how light is perceived as an 'informational' cue by plants. The direct agronomic value of increasing our plant photobiological knowledge is significant; for example, our understanding of plant responses to shade that is imposed by neighbouring plants forms a vital aspect of crop plant density optimization practices within modern agriculture (as discussed by Ballare *et al.*, 1997). Indeed, it could be argued that the exploitation of photobiology in agriculture has constituted a 'silent innovation' in the past, despite our enormous reliance on plant-light interactions. In this chapter, we will explore the exciting potential to use our understanding of UV-B plant response for maximum future gain in food cultivation.

UV-B Photomorphogenesis and Crop Plants

Plant responses to UV-B radiation [280–315 nm] are diverse and numerous, yet there is much we still do not understand regarding the 'impact' of UV-B plant responses upon natural ecosystems and cultivated crop production systems. If the advent of concerns in the early 1980s regarding the potential consequences of stratospheric ozone depletion is considered in many ways the birth of modern UV plant photobiological research, it is not a coincidence that many early studies of plant UV response were focused on crop plants, largely due to the concerns regarding the deleterious effects of increased UV-B exposure to crops. Many of those early studies indeed attributed observed reductions in plant growth and/or productivity to enhanced UV exposure, and UV-B wavelengths in particular. As discussed earlier (see Chapter 2, this volume), UV-B plant photobiological experimental design has evolved substantially over the last three decades, and changes in experimental technique have inarguably influenced many of the observations made in the past. For example, a number of studies from the 1980s and early 1990s that focused on crop plant species concluded that crop plants were vulnerable to UV-B exposure: Teramura *et al.* (1991) observed reductions in biomass across a number of rice cultivars; He *et al.* (1993) noted intraspecific differences in UV response in rice. In the latter work, the researchers attributed sensitivities in rice and pea plants to UV on the basis of decreases in the ratio of variable to maximum chlorophyll fluorescence yield, and in the quantum yield of photosynthetic oxygen evolution. Yet, as our knowledge of UV–plant interactions grew, and experimental technique developed further, an increasing number of studies questioned the likelihood that ambient solar UV-B levels could induce significant reductions in crop productivity. Moreover, our understanding of UV photomorphogenesis was expanding, whereby UV-induced changes (e.g. in leaf expansion rate) did not necessarily indicate a net productivity loss to a particular crop plant. For example, a review of experiments where realistic UV-B exposures of plants under modified or ambient field conditions were carried out (i.e. no use of hyper-ambient UV fluxes; no use of non-earth surface incident UV wavelengths), concluded that ambient levels of UV-B at that time (or in the predicted near future of ozone depletion) were very unlikely to lead to reductions in photosynthetic competency (Allen *et al.*, 1998). Simultaneously, there was increasing interest in the breadth of interactions that could be mediated by UV between crop plants and other biotic/abiotic factors, including plant pathogens (Paul, 2000), and insect pests of crops (Antignus *et al.*, 2001).

Two key developments have since occurred in UV-B plant photobiological research, which have significant consequences for the agronomic impact of UV photobiological knowledge. Firstly, our understanding of the molecular basis of UV-B perception and early-stage signalling events has grown considerably since the early 2000s, most notably in the discovery of UV RESISTANCE LOCUS 8 [UVR8] as a UV-B specific photoreceptor protein (Rizzini *et al.*, 2011), and the subsequent elucidation of UVR8's structure and likely mode of activity (Christie *et al.*, 2012). Such advances in our understanding provide tools to dissect the mechanistic drivers of previously well-known physiological and biochemical responses to UV. Moreover, a key observation arising from the discovery of UVR8, is that UV-B specific photoperception and downstream responses can be activated by relatively lower, regulatory fluence rates of UV-B, such as the inhibition of hypocotyl elongation for example (Favory *et al.*, 2009). Such studies are in stark contrast to those experimental approaches used to simulate hyper-ambient UV environments in relation to past concerns around ozone depletion. Second, as our interest in the regulatory effects of UV plant response has increased, particularly with regard to decreased focus on the solely deleterious consequences of UV exposure, an increasing number of studies have focused on the intriguing overlap between key plant responses to UV as agronomically desirable traits. For example, plant growth regulation (UV morphogenesis), crop product taste or flavour (UV sunscreening responses), and plant hardiness (increased resistance to abiotic/biotic stress), are all arguably desirable traits in agricultural production. Indeed, as has been noted by other authors, there are marked opportunities to consider the agricultural benefits of UV-B plant response (Ballare *et al.*, 2012; Wargent and Jordan, 2013; Andrady *et al.*, 2016). This now-significant interest in the agronomic exploitation of UV biology is underpinned by studies that have grown this new revolution in UV-B plant photobiology, based on the integration of new research tools and a refreshed perspective regarding the range of possibilities to alter the UV environment for many cultivated crops.

Opportunities to Exploit UV Response in Agriculture

Agricultural environments and UV light exposure

Studies of UV plant response have historically relied upon a range of experimental approaches, from completely controlled (indoor) lighting environments, to outdoor/field systems that utilize supplementation with artificial UV lighting sources, to selective filtration or transmission of key spectral regions from ambient sunlight at a given location. The use of spectral filtration as a consideration for plant cultivation is not hugely new per se, since initial studies of responses of crop species under modified red:far-red (R:FR) environments took place more than two decades ago (McMahon *et al.*, 1991; Rajapakse *et al.*, 1993; van Haeringen *et al.*, 1998). Crop covers, such as polyethylene films used to cover polytunnels or 'hoop-houses', and/or various cladding materials used for glasshouse covering, have evolved a great deal over the last 20 years, whereby the spectral transmission of crop covers can now be specified to quite a high extent (Krizek *et al.*, 2005). For example, a 'standard' horticultural crop cover often exhibits good transmission within the visible spectrum, some transmission in the long-wave UV-A wavelengths, and no transmission in the UV-B waveband. It is also clear that crop cover specifications can vary quite widely. The innovation of crop covers that allow for greater or lesser transmission of UV (dating to the early 2000s) has allowed researchers to study how modifying the ambient UV component of sunlight can potentially introduce a new level of crop quality and yield control. A large and growing proportion of the world's crops are grown under some form of protected environment (e.g. polytunnel/hoop-house, glasshouse), and in countries such as Spain and The Netherlands, the land-area use of protected cultivation practices is very significant. While a number of crops may be grown under protection for an entire production cycle (i.e. from seed/planting to harvest), protected cultivation is often used as a 'nursery' phase for crops that are routinely raised to a young age, and then transplanted into an outdoor field environment for onward growth. The practice of the transplantation of plants from a nursery environment is arguably in an expansion phase within global agriculture, largely due to the likely positive effects of protected propagation on plant health, relative growth and harvested produce quality. Currently, typical crops that are transplanted include vegetables in particular (e.g. lettuce, broccoli), in contrast to the traditional 'row crops' such as corn, soy and maize. The rise in protected cultivation, combined with those new technologies related to crop cover films with differing UV radiation transmission profiles, has led to a number of studies which have attempted to assess the consequences for crops grown under modified UV environments. For example, Paul *et al.* (2005) provided an initial demonstration that exposing propagation lettuce (*Lactuca sativa*) plants to a modified UV tunnel environment can have significant consequences for plant growth, including increases in plant fresh weight and leaf area, when comparing plants grown in a UV-excluding tunnel environment to those in a UV-inclusive environment (i.e. the addition of short-wave UV-A and long-mid wave UV-B). However, this response when excluding UV was traded off against an impact on the preferable 'bitter, strong' taste of lettuce, which was dramatically enhanced in plants grown (for 14 days) under a UV-inclusive environment. In this study, plants remained under the UV-modifying films for the duration of the experiments. A follow-on study quantified the responses of *L. sativa* plants, not only during the time that plants were exposed to varied UV regimes under polytunnel filters, but also tracked plants following a field transplantation event. The mean final harvestable fresh weight of lettuces which were briefly propagated under a UV+ filter environment was increased by 69%, as compared to plants originally cultivated under a UV-blocking filter, with UV+ plants increased in fresh weight by 31% compared to plants maintained at that young age under a 'standard' horticultural polytunnel cladding material (Wargent *et al.*, 2011). The mechanisms by which plants exposed to UV at a young age could exhibit increased biomass accumulation by the end of the field-growing phase remain somewhat elusive. However, following studies within artificially lit controlled environments, it was shown that enhanced exposure to UV can lead to increases in photosynthetic efficiency and capability, following a period of high light (Photosynthetically Active Radiation (PAR): 400–700 nm) and high temperature stress. Such resilience could possibly provide benefits to crop performance in the long run. These examples emphasize the paradigm shift that UV-B plant photobiology is arguably undergoing at this time, where those well-described morphogenic responses to UV-B may translate into desirable agricultural crop characteristics, e.g. plants with an increased tolerance to the various unpredictable stresses of the commercial food growing environment. UV morphogenic responses are clearly responsive to changes in the transmission of ambient UV, even at mid-latitudinal locations, but given the spectral and dose-based specificity of many observed UV responses to date, 'broadband' modifications to ambient UV-B may present agronomic challenges in terms of reliable manipulation in a year-round crop production environment. There are still quite limited examples of commercial uptake of UV-modifying films for horticultural production – this may be as a result of the cost of implementation, and/or the reliance on local climate/sunlight levels to achieve any previous reported responses. Nonetheless, some larger-scale uses of commercially available glasshouse cladding materials which transmit high levels of UV-B have been described, with morphological benefits (e.g. stem shortening) attributed to increased UV-B exposure (Torre *et al.*, 2012), and in other cases, limited benefits to enhanced UV-B transmission, i.e. flower colour (Sakai *et al.*, 2010). The dynamic nature of the vast majority of crop production environments (e.g. exposure to different climates, soils and critical events, such as drought or pathogen attack) also highlights the likely challenges in exploiting morphological responses to stimuli such as UV for agronomic gain. In other words, a much greater level of understanding of UV response would be needed to predict interactions with different components of an agronomic system (Fig. 11.1).

Plant secondary metabolism could be UV-modified for consumer benefit

In terms of secondary metabolic responses (arguably a good proxy for crop colour and/or taste), lettuce has been used as a model crop a number of times, with increases in anthocyanins and other phenolic compounds regularly observed under a UV-enhanced environment (Krizek *et al.*, 1998; Tsormpatsidis *et al.*, 2008) and, interestingly, in response to both enhanced UV-B and/or UV-A light. The fact that both overlapping and diverse roles in secondary metabolism induction have been identified for UV-B and UV-A radiation is of relevance to the consideration of agronomic interactions. While the typical focus of UV morphogenesis research is on UV-B responses, UV-A fluxes are significant in terms of solar radiation, and may also be subject to perturbation as a result of climate change (McKenzie *et al.*, 2011). At the same time, the majority of protected cropping environments are likely to allow more transmission of (long-wave) UV-A than UV-B. However, responses to UV-A radiation remain somewhat poorly defined, and there is limited evidence for exposure to UV-A alone leading to potential agronomically desirable traits, with the exception of secondary metabolite induction.

For example, in a recent study *L. sativa* seedlings cultivated under glasshouse (GH) conditions (which included mid-long wave exposure to UV-A but not UV-B) were compared with seedlings grown in a controlled environment (CE) where a modest flux of

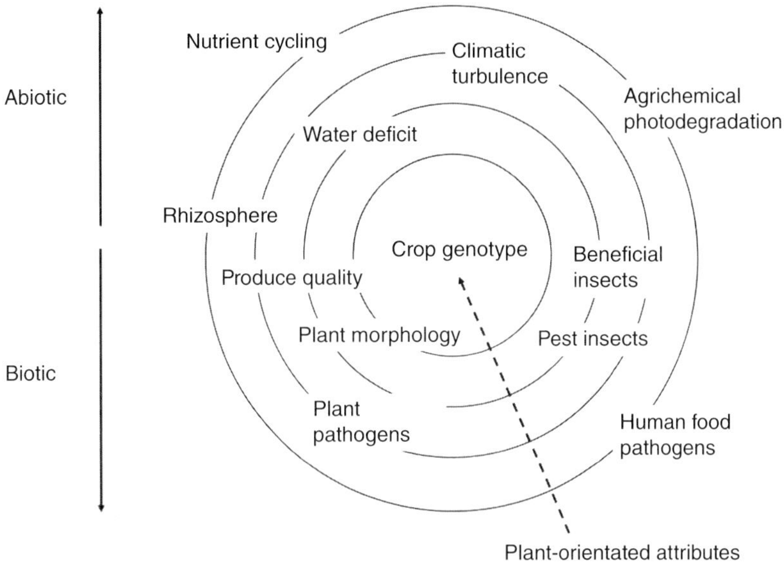

Fig. 11.1. Agricultural targets for UV response. Crop production exists as an interactive system, composed of both biotic and abiotic aspects. The diverse range of identified agri-outcomes related to UV exposure can be described as strongly plant-orientated (e.g. relative level of UV response in a given crop genotype; produce quality such as taste, colour), or somewhat less plant-orientated (e.g. persistence of human food pathogens on crop surfaces; interactions with soil rhizosphere). It can be argued that building a more holistic understanding of how numerous components of the agri-ecosystem interact in response to varying UV environments, will lead to future opportunities to exploit such endpoints for agronomic gain.

UV-A was present, but UV-B wavelengths were also included. The CE (UV-B+) plants exhibited the largest increases in net photosynthetic rate, and epidermal UV shielding levels (Wargent *et al.*, 2015).

Moreover, CE (UV-B+) plants were more elevated in levels of the most abundant secondary metabolites identified in the seedlings via LC-MS, such as quercetin 3-(6″-malonyl-glucoside). It has been shown that simultaneous UV-A and UV-B exposure can be required for maximal acclimation of plants under ambient sunlight (Barnes *et al.*, 2013), which further highlights our limited current knowledge regarding signalling cross-talk between the UVR8-COP1-HY5 pathway and UV-A signalling. Management of foliar colour of leafy salad vegetables, such as red lettuce, is a known crop quality attribute and the use of UV modification to regulate colour would arguably be desirable if a stable and cost-effective technology could be offered to achieve this. At the same time, the predominant focus of many secondary metabolite studies has centred on short-term experiments, where plants are exposed using UV-B (or UV-A) sources that were developed for laboratory use. Such approaches cannot necessarily be used in an identical form for commercial treatment of crops, and technological advances may further extend the possibilities of exploiting plant biochemical responses to UV. The expanding use of fully controlled environments to cultivate crops (using so-called ‘vertical farming’ approaches for example) also indicates an opportunity to completely control the light environment for crops (Yeh and Chung, 2009), yet technological and photobiological challenges certainly remain.

The manipulation of the UV response for control of produce colour is not the only potential endpoint of modifying plant secondary metabolism. The development of tools to increase the health-giving properties of fresh produce has been a high-profile aim of many in the food industry for some time now. This may be in the form of increased antioxidant availability (Gulcin, 2012), dietary fibre (Slavin and Lloyd, 2012) and consideration of the broader role of dietary phenolics (Parr and Bolwell, 2000). There is certainly an abundance of studies which have examined the effects of UV exposure on numerous aspects of dietary health related plant components. However, it is not clear at the present time if such traits as consumer health benefits can be routinely up-regulated by UV enhancement (Jansen *et al.*, 2008), and most importantly, if consumer/market acceptability would embrace such outcomes. While a 'willingness-to-pay' model has certainly been tested with regard to, for example, new varieties of kiwifruit with non-green flesh colour (Jaeger and Harker, 2005), there is still little understanding of the value proposition associated with a 'health-boosted' version of a particular type of produce. If no strong value proposition exists for the use of UV to induce desirable health properties in produce, then the commercial impetus will not exist. However, there is growing evidence of the potential for UV enhancement to increase the content of various health-associated compounds in indigenous horticultural crops (Pandey and Pandey-Rai, 2014; Takshak and Agrawal, 2016). Consideration of the ambient UV environment in which many such crops are cultivated could be used to further develop opportunities for solar UV to be maximally exploited for health-promoting gain in the future.

Crop density and UV interactions

Further agronomic integration of our understanding of plant UV response can also be considered in terms of plant–plant interactions such as shade. All crops are sown and/or transplanted at cropping densities determined on the basis of numerous physiological factors (e.g. minimizing plant-to-plant competition, increasing PAR exposure for production of assimilates), but also with the overall goal of maximised yield at the forefront of decision making, for which factors such as access for mechanization and land costs can also be highly influential. The impact of UV morphogenesis has a diverse range of implications for cropping density, across the entire range of agricultural environments. For example, 'stretching' of plant seedlings is a condition which is associated with poor onward crop growth (Carriedo *et al.*, 2016), and increased susceptibility to pest and pathogen attack (Izaguirre *et al.*, 2006). Where a particular propagation or nursery environment may result in stem elongation, due to a reduced red to far-red light ratio (R:FR) as perceived by the plant population, exposure to UV-B could arguably ameliorate such an undesirable response, both in a direct physiological sense, and with regard to induction of plant defences. Our pre-UVR8 understanding of photomorphogenesis has often been compartmentalized with regard to interactions between UV-B and R/FR light; i.e. a perceived mechanistic separation of the UV-B inhibition of stem elongation, and induction of stem elongation in response to a reduction in R:FR. Yet, as discussed in detail by other authors (Ballare *et al.*, 2012; Mazza and Ballare, 2015), the light-related interactions in a plant canopy are far from simple and singular in nature. While our understanding of the molecular basis for shade-avoidance related signalling is now quite well defined (i.e. a lower R:FR ratio as perceived by phytochrome (phy) resulting in inactivation of the active phy Pfr form, thus leading to increased levels of PHYTOCHROME INTERACTING FACTORS (PIFs), degradation of DELLA proteins, increased auxin biosynthesis, and stem elongation as a result), our understanding of the cooperative influence of UV-B on this signalling complex is now expanding. Hayes *et al.* (2014) demonstrated that under shade mimicking conditions (R:FR ratio = 0.05), stem elongation is significantly reduced when plants were co-exposed to UV-B light, and that this attenuation of shade avoidance response was part-mediated by UVR8-dependent induction of gibberellic acid oxidases (e.g. GA2-oxidases), resulting in stabilization of DELLA proteins. The relevance of such cross-talk is further underlined when we consider that UV-B:PAR ratios can be elevated in shaded areas of vegetation (Flint and Caldwell, 1998). Hence, despite UV-B fluxes being predictably lower in shade, the significance of UV-B signalling as a co-regulator of shade avoidance could be profound in agricultural canopies. An increased understanding of the actual light environment in a range of crop production scenarios, and further elucidation of how such signalling elements interact in different genotypes, will be valuable in bridging such photobiology with agricultural gain.

Pests, pathogens, and UV radiation

The agricultural ecosystem is not solely composed of plants, with crop success tightly linked to a complex web of biotic interactions. Equally, the attention of UV-B biologists has also focused on understanding the influence of UV on such interactions, predominantly the consequences of UV response for pest (insects) and pathogen (disease) attack. Regarding UV–pest interactions, our understanding to date can be viewed from two viewpoints: insect vision and perception of UV, and indirect responses of insects, based on UV plant photomorphogenic response. There is a good understanding of the spectral basis for vision in many insect taxa, where both perception and behavioural responses are sensitive to UV-A wavelengths (Tovee, 1995). There are also examples of UV-B sensitivity in insects, such as in the herbivore *Caliothrips phaseoli* (Mazza *et al.*, 2010). Colleagues in Israel have pioneered much of our existing knowledge related to this aspect of pest protection, driven by the exploration and exemplification of UV-opaque crop cover netting materials. The role of UV-opaque crop covers in obscuring normal vision (and thus optimal flight navigation) has been demonstrated in phloem-feeding insects, e.g. *Bemisia argentifolii* (whitefly) (Antignus *et al.*, 2001). In terms of indirect consequences of UV morphogenesis for pest insects (i.e. UV-mediated plant–insect interactions), there is good, albeit still somewhat limited understanding regarding how prior or concurrent UV exposure can influence pest dynamics. Foraging patterns of herbivorous pests upon plants previously exposed to UV-B have been examined a number of times, with a consistent 'deterrence' effect observed, i.e. reduced consumption of UV-exposed foliar material (Ballare *et al.*, 1996; Hatcher and Paul, 1994). However, the consequences of plant-mediated UV response for pest behaviour are not confined solely to herbivory. Using the herbivorous larvae–adult moth experimental model, adult *Plutella xylostella* larvae oviposited preferentially on *Arabidopsis* or *Brassica* plants which had been grown in the absence of UV-B, as compared to plants offered which had been exposed previously to UV-B (Caputo *et al.*, 2006; Foggo *et al.*, 2007). From a mechanistic perspective, the potential for overlapping signalling networks between UV-B response and resistance to herbivory is significant. For example, exposure to UV-B under field and supplemental light conditions in *Nicotiana* plants triggered the induction of certain phenylpropanoid compounds which were also induced by the application of oral secretions from the herbivore *Manduca sexta* (Izaguirre *et al.*, 2007), thus indicating some potential overlap in UV-B and herbivory-related signalling. The continuing evolution of our understanding of the molecular basis for both herbivory resistance and UV-B perception has led to some new knowledge of underlying signalling components. Morales *et al.* (2012) carried out microarray analysis of wild-type *Arabidopsis* and the *uvr8* mutant following exposure to modified ambient UV-B in a northern latitude, and UV-B-mediated transcript accumulation of several jasmonic acid (JA) biosynthesis and signalling genes was shown to be acting in a UVR8-dependent manner, after 12 h of solar UV exposure. However, to date there is still relatively little evidence that UV-B exposure leads to accumulation of endogenous JA; there is however evidence that solar UV-B can increase sensitivity to jasmonates, leading to enhanced expression of insect wounding response genes, such as trypsin proteinase inhibitor (TPI), as demonstrated in jasmonate-deficient *Nicotiana attenuata* plants (Demkura *et al.*, 2010). The single-system nature of research experimentation naturally has a tendency to oversimplify complex dynamics such as plant–insect interactions, and when our current knowledge regarding UV-mediated consequences for pests is considered in the context of agriculture, further questions are raised. For example, what are the indirect effects of UV upon pests beyond insect vision? How does UV affect pest–predator dynamics, and could certain UV-induced outcomes lead to consequences for biocontrol strategies, for example? There are few studies to date which have examined insect–insect interactions in this context, but the information that is available is intriguing. For example, in a series of choice assays, the egg parasitoid *Trichogramma* parasitized more eggs of the model pest organism *Manduca sexta* under UV-B illumination, as opposed to under the absence of UV-B (Van Atta *et al.*, 2015). In a tritrophic study, Foggo *et al.* (2007) observed that when *Plutella xylostella* larvae were allowed to forage on cabbage plants which had either been previously treated with UV-B, or not exposed to UV-B, adult *Cotesia plutellae* parasitoid wasps chose more frequently to visit those plants previously exposed to UV-B. It is tempting to speculate that such a response could be driven by UV-B-mediated increases in plant volatile cues, particularly given the role of the host plant in the attraction of parasitoids and predators of plant pests (Vet and Dicke, 1992). In terms of pest dietary consequences of plant UV exposure, a study of two genotypes of *Trifolium repens*, which were exposed to supplementary UV-B and fed to two herbivore species in short- and long-term feeding trials, indicated that both plant response in differing genotypes and dietary outcomes in two insect species were affected differently by identical UV-B treatments (Lindroth *et al.*, 2000). In terms of considering agricultural consequences or opportunities, such understanding clearly demonstrates the multi-modal nature of considering UV modification in agriculture. For example, for insect 'deterrence', where navigational interruption is targeted as above, the role of UV-A radiation is arguably paramount (Ben-Yakir *et al.*, 2012). Yet, in terms of plant-mediated outcomes for pest protection strategies, the influence of UV-B morphogenesis is likely to be significant. However, sizeable knowledge gaps still exist regarding the influence of UV-B on plant–pest interactions within end-to-end crop cycles or seasons, and regarding how manipulable such dynamics may be, in order to maximise sustainable pest management.

Plant pathogens represent a very significant threat to the closing of yield gaps in agriculture. Again as a result of concerns regarding ozone depletion, there has been long-standing interest in the effects of UV on plant pathogens, certainly from a terrestrial ecosystem perspective, with positive and negative consequences observed or hypothesized for fungal species, e.g. growth, reproduction (Paul *et al.*, 1997). When considering evidence to date for the role of UV modification for plant pathogen suppression, the importance of different UV wavebands is certainly complex. As with pest control, several studies have indicated that UV modification may offer a future means to reduce pathogen persistence. For example, in a study where two populations of lettuce plants were deliberately infected with the biotrophic fungal pathogen *Bremia lactucae*, and the necrotroph *Botrytis cinerea* respectively, the separate use of both UV-inclusive and UV-blocking polytunnel crop covers led to some level of disease suppression (Paul *et al.*, 2012). Such findings are very likely to be driven by the quite opposing effects of different UV and blue light wavebands in terms of fungal biology; e.g. that UV-B exposure can kill fungal spores, that UV-A wavelengths can be necessary for fungal sporulation, and that blue light exposure may suppress sporulation (Elad, 1997, Wu *et al.*, 2000). Furthermore, studies from Israel have shown that this outlook is further complicated by the indirect protection offered against insect-borne plant viruses by using a UV-opaque crop cover (Raviv and Antignus, 2004). From a molecular perspective, our current understanding regarding the UV-mediated basis for plant–pathogen interactions is even more limited than that of plant–insect interactions. While a number of studies have shown intriguing overlaps between UV-B-induced plant signalling, and that of known components of pathogen response (e.g. pathogenesis-related proteins; Surplus *et al.*, 1998; Mackerness *et al.*, 2001), our post-UVR8 understanding of UV–pathogen interactions is still quite constrained. Demkura *et al.* (2012) showed that a reduction in Arabidopsis infection by the necrotroph *Botrytis cinerea* was UVR8-dependent, and that hydroxycinnamic acids such as the sinapates were implicated in the defence response. However, a fully resolved mechanistic model of the role of UV in pathogen resistance remains elusive. There are many questions yet to be solved regarding the potential integration of UV-mediated pathogen control into agriculture, and as with many aspects of our current understanding of UV morphogenesis, there is limited knowledge regarding the long-term consequences or persistence of such UV-induced traits.

Post-harvest UV treatments in horticulture

Notwithstanding the scientific advances that have been initially inspired by concerns around ozone depletion, the use of UV treatment technology in food and crop production has also been explored at the post-harvest level for some time. Basic approaches have focused on disinfestation or related attempts to extend shelf life of harvested produce via increased mortality of spoilage organisms, such as moulds (Lu *et al.*, 1991), predominantly using short-wave UV treatments, e.g. UV-C radiation. More recently, attempts have been made to use post-harvest UV treatments as a means to extend shelf life (Bu *et al.*, 2013), and to enhance anti-oxidative potential of food and beverage products (Cantos *et al.*, 2003; Avena-Bustillos *et al.*, 2012). Yet at the same time, challenges do remain over the cost-effectiveness and process implementation aspects of UV treatments for food and produce (Shama, 2007).

UV modification technology: UV LEDs

LEDs have steadily become more accepted and utilized in horticulture in recent years, as supported by a growing research community (Mitchell, 2015). One of the challenges for the increased use of LEDs for crop production, and for researchers, includes the need to develop biological and technical frameworks which can provide end users with clear decision making steps in terms of the ideal light environment they should compose (or purchase off-the-shelf) when using LEDs. Because visible LEDs are now such a large part of our everyday lives, the availability, wavelength ranges, shelf life, and cost of, e.g. red and blue LEDs, has become much more competitive in recent years (Shur and Zukauskas, 2005). However, due to the much lower level of everyday applications for UV light compared to visible, the development of UV LEDs has been slower and slighter, with cost implications also significant, albeit compared to visible LEDs. UV LEDs require different base materials from visible LEDs (Hu *et al.*, 2006), and thus there are certain differences in manufacturing processes. To date, such considerations have slowed the rate of scientific development regarding UV LED implementation into horticultural research and production. However, some opportunities exist regarding UV LED use, and embryonic research advances have been achieved. The potential applications for UV LEDs in agriculture and horticulture have certainly been subject to recent debate and discussion (Wargent and Jordan, 2013; Huche-Thelier *et al.*, 2016), yet few published studies currently exist, particularly in the pre-harvest application space. However, it has been demonstrated that induced increases in crop colour, again in lettuce, can be achieved using UV-A LED sources (Li and Kubota, 2009), as opposed to using spectral filters or fluorescent UV tubes. In addition, increased suppression of tomato mosaic virus was observed in response to exposure of plant material with 7–14 kJ m^{-2} day^{-1} at wavelengths of 280–290 nm supplied by UV LEDs (Matsuura and Ishikura, 2014). Equally, the potential applications of UV LEDs for extension of shelf-life of fresh produce has also been explored by Britz *et al.* (2013), who recently showed that cold-storage shelf life of strawberries could be extended twofold when the fruit was exposed to low doses of UV LED radiation. As more studies are completed in the future, our understanding of the potential of UV LEDs in horticulture will no doubt expand.

Conclusions

While many of the now well-described outcomes of UV crop exposure can be seen as valuable to agricultural production, it is also clear that our increasing knowledge of UV morphogenesis has to expand considerably further in order to achieve significant technology transfer in the future. One complex challenge we face is related to the deep variation of different UV wavelengths in initiating (or inhibiting) differing biological responses. This multi-wavelength orchestration of the biological effectiveness of most aspects of UV response will require enhanced understanding of the underlying biology of desirable (and undesirable) crop traits, and over a temporal scale. The same is also true regarding the evolution of UV lighting technology. For example, conventional research tools such as high-pressure lamps, or fluorescent tubes, are unlikely to offer a sturdy or efficient means of providing industrial UV treatments to crops, now or in the future. One main reason for this is the lack of wavelength control or specificity that can be achieved using broadband light sources. Spectrally modifying films can offer some potential to manipulate the UV environment, but are limited to predictably changeable ambient conditions. Moreover, spectral filters do not currently offer more discrete wavelength control either. UV LEDs are still a very new technology, with much more limited LED shelf life and higher costs at the present time, as well as more limited studies of photobiological efficacy when applied to horticultural crops. However, the potential to control the UV spectrum more tightly, and to have flexibility within end-use should make UV LEDs a desirable option for horticultural crop technology integration in the future.

In summary, different UV environments provide different outcomes for different crops; it is not enough to develop scientific advances *ex situ*, but future photobiological innovations need to be implementable into production agriculture, meeting agronomic, economic and consumer-based needs. Such approaches have the potential to significantly strengthen our efforts in the feeding of a hungry planet.

References

Ainsworth, E.A. and Long, S.P. (2005) What have we learned from 15 years of free-air CO_2 enrichment (FACE)? A meta-analytic review of the responses of photosynthesis, canopy properties and plant production to rising CO_2. *New Phytologist* 165, 351–371.

Ainsworth, E.A. and Rogers, A. (2007) The response of photosynthesis and stomatal conductance to rising [CO_2]: mechanisms and environmental interactions. *Plant, Cell and Environment* 30, 258–270.

Alexandratos, N. and Bruinsma, J. (2012) *World Agriculture Towards 2030/2050: The 2012 revision.* ESA Working Paper No. 12-03. Agricultural Development Economics Division, Food and Agriculture Organization of the United Nations, Rome.

Allen, D.J., Nogues, S. and Baker, N.R. (1998) Ozone depletion and increased UV-B radiation: Is there a real threat to photosynthesis? *Journal of Experimental Botany* 49, 1775–1788.

Andrady, A., Aucamp, P.J., Austin, A.T., Bais, A.F., Ballare, C.L. *et al.* (2016) Environmental effects of ozone depletion and its interactions with climate change: Progress report, 2015. United Nations Environmental Programme and Environmental Effects Assessment Programme. *Photochemical and Photobiological Sciences* 15, 141–174.

Antignus, Y., Nestel, D., Cohen, S. and Lapidot, M. (2001) Ultraviolet-deficient greenhouse environment affects whitefly attraction and flight-behavior. *Environmental Entomology* 30, 394–399.

Avena-Bustillos, R.J., Du, W.X., Woods, R., Olson, D., Breksa, A.P. and McHugh, T.H. (2012) Ultraviolet-B light treatment increases antioxidant capacity of carrot products. *Journal of the Science of Food and Agriculture* 92, 2341–2348.

Ballare, C.L., Scopel, A.L., Stapleton, A.E. and Yanovsky, M.J. (1996) Solar ultraviolet-B radiation affects seedling emergence, DNA integrity, plant morphology, growth rate, and attractiveness to herbivore insects in *Datura ferox*. *Plant Physiology* 112, 161–170.

Ballare, C.L., Scopel, A.L. and Sanchez, R.A. (1997) Foraging for light: photosensory ecology and agricultural implications. *Plant Cell and Environment* 20, 820–825.

Ballare, C.L., Mazza, C.A., Austin, A.T. and Pierik, R. (2012) Canopy light and plant health. *Plant Physiology* 160, 145–155.

Barnes, P.W., Kersting, A.R., Flint, S.D., Beyschlag, W. and Ryel, R.J. (2013) Adjustments in epidermal UV-transmittance of leaves in sun-shade transitions. *Physiologia Plantarum* 149, 200–213.

Ben-Yakir, D., Antignus, Y., Offir, Y. and Shahak, Y. (2012) Optical manipulations of insect pests for protecting agricultural crops. *Acta Horticulturae* 956, 609–615. DOI:10.17660/ActaHortic.2012.956.72.

Britz, S., Gaska, I., Shturm, I., Bilenko, Y., Shatalov, M. and Gaska, R. (2013) Deep ultraviolet (DUV) light-emitting diodes (LEDs) to maintain freshness and phytochemical composition during postharvest storage. 2013 Conference on Lasers and Electro-Optics (CLEO), 9–14 June 2013, San Jose, California. IEEE, New York, pp. 1–2. DOI:10.1364/CLEO_AT.2013.ATh3N.3

Bu, J.W., Yu, Y.C., Aisikaer, G. and Ying, T.J. (2013) Postharvest UV-C irradiation inhibits the production of ethylene and the activity of cell wall-degrading enzymes during softening of tomato (*Lycopersicon esculentum* L.) fruit. *Postharvest Biology and Technology* 86, 337–345.

Cantos, E., Espin, J.C., Fernandez, M.J., Oliva, J. and Tomas-Barberan, F.A. (2003) Postharvest UV-C-irradiated grapes as a potential source for producing stilbene-enriched red wines. *Journal of Agricultural and Food Chemistry* 51, 1208–1214.

Caputo, C., Rutitzky, M. and Ballare, C.L. (2006) Solar ultraviolet-B radiation alters the attractiveness of *Arabidopsis* plants to diamondback moths (*Plutella xylostella* L.): impacts on oviposition and involvement of the jasmonic acid pathway. *Oecologia* 149, 81–90.

Carriedo, L.G., Maloof, J.N. and Brady, S.M. (2016) Molecular control of crop shade avoidance. *Current Opinion in Plant Biology* 30, 151–158.

Christie, J.M., Arvai, A.S., Baxter, K.J., Heilmann, M., Pratt, A.J. *et al.* (2012) Plant UVR8 photoreceptor senses UV-B by tryptophan-mediated disruption of cross-dimer salt bridges. *Science* 335, 1492–1496.

Demkura, P.V. and Ballare, C.L. (2012) UVR8 mediates UV-B-induced *Arabidopsis* defense responses against *Botrytis cinerea* by controlling sinapate accumulation. *Molecular Plant* 5, 642–652.

Demkura, P.V., Abdala, G., Baldwin, I.T. and Ballare, C.L. (2010) Jasmonate-dependent and -independent pathways mediate specific effects of solar ultraviolet B radiation on leaf phenolics and antiherbivore defense. *Plant Physiology* 152, 1084–1095.

Elad, Y. (1997) Effect of filtration of solar light on the production of conidia by field isolates of *Botrytis cinerea* and on several diseases of greenhouse-grown vegetables. *Crop Protection* 16, 635–642.

Favory, J.J., Stec, A., Gruber, H., Rizzini, L., Oravecz, A. *et al.* (2009) Interaction of COP1 and UVR8 regulates UV-B-induced photomorphogenesis and stress acclimation in *Arabidopsis*. *EMBO Journal* 28, 591–601.

Flint, S.D. and Caldwell, M.M. (1998) Solar UV-B and visible radiation in tropical forest gaps: measurements partitioning direct and diffuse radiation. *Global Change Biology* 4, 863–870.

Foggo, A., Higgins, S., Wargent, J.J. and Coleman, R.A. (2007) Tri-trophic consequences of UV-B exposure: plants, herbivores and parasitoids. *Oecologia* 154, 505–512.

Foley, J.A., Ramankutty, N., Brauman, K.A., Cassidy, E.S., Gerber, J.S. *et al.* (2011) Solutions for a cultivated planet. *Nature* 478, 337–342.

Godfray, H.C.J., Beddington, J.R., Crute, I.R., Haddad, L., Lawrence, D. *et al.* (2010) Food security: the challenge of feeding 9 billion people. *Science* 327, 812–818.

Gulcin, I. (2012) Antioxidant activity of food constituents: an overview. *Archives of Toxicology* 86, 345–391.

Hatcher, P.E. and Paul, N.D. (1994) The effect of elevated UV-B radiation on herbivory of pea by *Autographa gamma*. *Entomologia Experimentalis et Applicata* 71, 227–233.

Hayes, S., Velanis, C.N., Jenkins, G.I., and Franklin, K.A. (2014) UV-B detected by the UVR8 photoreceptor antagonizes auxin signaling and plant shade avoidance. *Proceedings of the National Academy of Sciences of the United States of America* 111, 11894–11899.

He, J., Huang, L.K., Chow, W.S., Whitecross, M.I. and Anderson, J.M. (1993) Effects of supplementary ultraviolet-B radiation on rice and pea plants. *Australian Journal of Plant Physiology* 20, 129–142.

Hu, X., Deng, J., Zhang, J.P., Lunev, A., Bilenko, Y. *et al.* (2006) Deep ultraviolet light-emitting diodes. *Physica Status Solidi A: Applications and Materials Science* 203, 1815–1818.

Huche-Thelier, L., Crespel, L., Le Gourrierec, J., Morel, P., Sakr, S. and Leduc, N. (2016) Light signaling and plant responses to blue and UV radiations: perspectives for applications in horticulture. *Environmental and Experimental Botany* 121, 22–38.

Izaguirre, M.M., Mazza, C.A., Biondini, M., Baldwin, I.T. and Ballare, C.L. (2006) Remote sensing of future competitors: impacts on plant defenses. *Proceedings of the National Academy of Sciences of the United States of America* 103, 7170–7174.

Izaguirre, M.M., Mazza, C.A., Svatos, A., Baldwin, I.T. and Ballare, C.L. (2007) Solar ultraviolet-B radiation and insect herbivory trigger partially overlapping phenolic responses in *Nicotiana attenuata* and *Nicotiana longiflora*. *Annals of Botany* 99, 103–109.

Jaeger, S.R. and Harker, F.R. (2005) Consumer evaluation of novel kiwifruit: willingness-to-pay. *Journal of the Science of Food and Agriculture* 85, 2519–2526.

Jansen, M.A.K., Hectors, K., O'Brien, N.M., Guisez, Y. and Potters, G. (2008) Plant stress and human health: Do human consumers benefit from UV-B acclimated crops? *Plant Science* 175, 449–458.

Kharas, H. (2010) *The Emerging Middle Class in Developing Countries*. The Brookings Institution. Available at: www.oecd.org/development/pgd/44798225.pdf (accessed 23 April 2017).

Krizek, D.T., Britz, S.J. and Mirecki, R.M. (1998) Inhibitory effects of ambient levels of solar UV-A and UV-B radiation on growth of cv. New Red Fire lettuce. *Physiologia Plantarum* 103, 1–7.

Krizek, D.T., Clark, H.D. and Mirecki, R.M. (2005) Spectral properties of selected UV-blocking and UV-transmitting covering materials with application for production of high-value crops in high tunnels. *Photochemistry and Photobiology* 81, 1047–1051.

Leakey, A.D.B., Bernacchi, C.J., Dohleman, F.G., Ort, D.R. and Long, S.P. (2004) Will photosynthesis of maize (*Zea mays*) in the US Corn Belt increase in future CO_2 rich atmospheres? An analysis of diurnal courses of CO_2 uptake under free-air concentration enrichment (FACE). *Global Change Biology* 10, 951–962.

Li, Q. and Kubota, C. (2009) Effects of supplemental light quality on growth and phytochemicals of baby leaf lettuce. *Environmental and Experimental Botany*, 67, 59–64.

Lindroth, R.L., Hofmann, R.W., Campbell, B.D., McNabb, W.C. and Hunt, D.Y. (2000) Population differences in *Trifolium repens* L – response to ultraviolet-B radiation: foliar chemistry and consequences for two lepidopteran herbivores. *Oecologia* 122, 20–28.

Lobell, D.B. and Burke, M. (2009) *Climate Change and Food Security: Adapting Agriculture to a Warmer World*. Springer, Dordecht, The Netherlands.

Lu, J.Y., Stevens, C., Khan, V.A., Kabwe, M. and Wilson, C.L. (1991) The effect of ultraviolet irradiation on shelf-life and ripening of peaches and apples. *Journal of Food Quality* 14, 299–305.

Mackerness, S.A.H., John, C.F., Jordan, B. and Thomas, B. (2001) Early signaling components in ultraviolet-B responses: distinct roles for different reactive oxygen species and nitric oxide. *FEBS Letters* 489, 237–242.

Matsuura, S. and Ishikura, S. (2014) Suppression of *Tomato mosaic virus* disease in tomato plants by deep ultraviolet irradiation using light-emitting diodes. *Letters in Applied Microbiology* 59, 457–463.

Mazza, C.A. and Ballare, C.L. (2015) Photoreceptors UVR8 and phytochrome B cooperate to optimize plant growth and defense in patchy canopies. *New Phytologist* 207, 4–9.

Mazza, C.A., Izaguirre, M.M., Curiale, J. and Ballare, C.L. (2010) A look into the invisible: ultraviolet-B sensitivity in an insect (*Caliothrips phaseoli*) revealed through a behavioural action spectrum. *Proceedings of the Royal Society B: Biological Sciences* 277, 367–373.

McKenzie, R.L., Aucamp, P.J., Bais, A.F., Bjorn, L.O., Ilyas, M. and Madronich, S. (2011) Ozone depletion and climate change: impacts on UV radiation. *Photochemical and Photobiological Sciences* 10, 182–198.

McMahon, M.J., Kelly, J.W., Decoteau, D.R., Young, R.E. and Pollock, R.K. (1991) Growth of *Dendranthema* x *grandiflorum* (Ramat) Kitamura under various spectral filters. *Journal of the American Society for Horticultural Science* 116, 950–954.

Mitchell, C.A. (2015) Academic research perspective of LEDs for the horticulture industry. *Hortscience* 50, 1293–1296.

Morales, L.O., Brosché, M., Vainonen, J., Jenkins, G.I., Wargent, J.J. *et al.* (2013) Multiple roles for UV RESISTANCE LOCUS8 in regulating gene expression and metabolite accumulation in Arabidopsis under solar ultraviolet radiation. *Plant Physiology* 161, 744–759.

Pandey, N. and Pandey-Rai, S. (2014) Modulations of physiological responses and possible involvement of defense-related secondary metabolites in acclimation of *Artemisia annua* L. against short-term UV-B radiation. *Planta* 240, 611–627.

Parr, A.J. and Bolwell, G.P. (2000) Phenols in the plant and in man: the potential for possible nutritional enhancement of the diet by modifying the phenols content or profile. *Journal of the Science of Food and Agriculture* 80, 985–1012.

Paul, N.D. (2000) Stratospheric ozone depletion, UV-B radiation and crop disease. *Environmental Pollution* 108, 343–355.

Paul, N.D., Rasanayagam, S., Moody, S.A., Hatcher, P.E. and Ayres, P.G. (1997) The role of interactions between trophic levels in determining the effects of UV-B on terrestrial ecosystems. *Plant Ecology* 128, 296–308.

Paul, N.D., Jacobson, R.J., Taylor, A., Wargent, J.J. and Moore, J.P. (2005) The use of wavelength-selective plastic cladding materials in horticulture: understanding of crop and fungal responses through the assessment of biological spectral weighting functions. *Photochemistry and Photobiology* 81, 1052–1060.

Paul, N.D., Moore, J.P., McPherson, M., Lambourne, C., Croft, P., Heaton, J.C. and Wargent, J.J. (2012) Ecological responses to UV radiation: interactions between the biological effects of UV on plants and on associated organisms. *Physiologia Plantarum* 145, 565–581.

Pretty, J. (2008) Agricultural sustainability: concepts, principles and evidence. *Philosophical Transactions of the Royal Society B: Biological Sciences* 363, 447–465.

Rajapakse, N.C., McMahon, M.J. and Kelly, J.W. (1993) End of day far-red light reverses height reduction of Chrysanthemum induced by $CuSO_4$ spectral filters. *Scientia Horticulturae* 53, 249–259.

Raviv, M. and Antignus, Y. (2004) UV radiation effects on pathogens and insect pests of greenhouse-grown crops. *Photochemistry and Photobiology* 79, 219–226.

Ray, D.K., Mueller, N.D., West, P.C. and Foley, J.A. (2013) Yield trends are insufficient to double global crop production by 2050. *PLoS ONE* 8, e66428.

Ray, D.K., Gerber, J.S., Macdonald, G.K. and West, P.C. (2015) Climate variation explains a third of global crop yield variability. *Nature Communications* 6, 5989.

Rizzini, L., Favory, J.-J., Cloix, C., Faggionato, D., O'Hara, A. *et al.* (2011) Perception of UV-B by the Arabidopsis UVR8 protein. *Science* 332, 103–106.

Sakai, W.S., Band, E., Edwards, J., Shioshita, R. and Fang, J. (2010) A preliminary study of the effect of ultraviolet radiation on visible flower coloration in two orchids. *Acta Horticulturae* 878, 367–371.

Shama, G. (2007) Process challenges in applying low doses of ultraviolet light to fresh produce for eliciting beneficial hormetic responses. *Postharvest Biology and Technology* 44, 1–8.

Shur, M.S. and Zukauskas, A. (2005) Solid-state lighting: toward superior illumination. *Proceedings of the IEEE* 93, 1691–1703. DOI:10.1109/JPROC.2005.853537

Slavin, J.L. and Lloyd, B. (2012) Health benefits of fruits and vegetables. *Advances in Nutrition* 3, 506–516.

Spiertz, J.H.J. (2010) Nitrogen, sustainable agriculture and food security: a review. *Agronomy for Sustainable Development* 30, 43–55.

Surplus, S.L., Jordan, B.R., Murphy, A.M., Carr, J.P., Thomas, B. and Mackerness, S.A.H. (1998) Ultraviolet-B-induced responses in *Arabidopsis thaliana*: role of salicylic acid and reactive oxygen species in the regulation of transcripts encoding photosynthetic and acidic pathogenesis-related proteins. *Plant, Cell and Environment* 21, 685–694.

Takshak, S. and Agrawal, S.B. (2016) The role of supplemental ultraviolet-B radiation in altering the metabolite profile, essential oil content and composition, and free radical scavenging activities of *Coleus forskohlii*, an indigenous medicinal plant. *Environmental Science and Pollution Research* 23, 7324–7337.

Teramura, A.H., Ziska, L.H. and Sztein, A.E. (1991) Changes in growth and photosynthetic capacity of rice with increased UV-B radiation. *Physiologia Plantarum* 83, 373–380.

Tilman, D., Balzer, C., Hill, J. and Befort, B.L. (2011) Global food demand and the sustainable intensification of agriculture. *Proceedings of the National Academy of Sciences of the United States of America* 108, 20260–20264.

Torre, S., Roro, A.G., Bengtsson, S., Mortensen, L.M., Solhaug, K.A., Gislerod, H.R. and Olsen, J.E. (2012) Control of plant morphology by UV-B and UV-B-temperature interactions. *Acta Horticulturae* 956, 207–214.

Tovee, M.J. (1995) Ultraviolet photoreceptors in the animal kingdom – their distribution and function. *Trends in Ecology and Evolution* 10, 455–460.

Tsormpatsidis, E., Henbest, R.G.C., Davis, F.J., Battey, N.H., Hadley, P. and Wagstaffe, A. (2008) UV irradiance as a major influence on growth, development and secondary products of commercial importance in Lollo Rosso lettuce 'Revolution' grown under polyethylene films. *Environmental and Experimental Botany* 63, 232–239.

Van Atta, K.J., Potter, K.A. and Woods, H.A. (2015) Effects of UV-B on environmental preference and egg parasitization by *Trichogramma* wasps (Hymenoptera: Trichogrammatidae). *Journal of Entomological Science* 50, 318–325.

Van Haeringen, C.J., West, J.S., Davis, F.J., Gilbert, A., Hadley, P. *et al.* (1998) The development of solid spectral filters for the regulation of plant growth. *Photochemistry and Photobiology* 67, 407–413.

Vet, L.E.M. and Dicke, M. (1992) Ecology of infochemical use by natural enemies in a tritrophic context. *Annual Review of Entomology* 37, 141–172.

Wang, N., Zhang, N.Q. and Wang, M.H. (2006) Wireless sensors in agriculture and food industry – recent development and future perspective. *Computers and Electronics in Agriculture* 50, 1–14.

Wargent, J.J. and Jordan, B.R. (2013) From ozone depletion to agriculture: understanding the role of UV radiation in sustainable crop production. *New Phytologist* 197, 1058–1076.

Wargent, J.J., Elfadly, E.M., Moore, J.P. and Paul, N.D. (2011) Increased exposure to UV-B radiation during early development leads to enhanced photoprotection and improved long-term performance in *Lactuca sativa*. *Plant, Cell and Environment* 34, 1401–1413.

Wargent, J.J., Nelson, B.C.W., McGhie, T.K. and Barnes, P.W. (2015) Acclimation to UV-B radiation and visible light in *Lactuca sativa* involves up-regulation of photosynthetic performance and orchestration of metabolome-wide responses. *Plant Cell and Environment* 38, 929–940.

Wu, B.M., Subbarao, K.V. and Van Bruggen, A.H.C. (2000) Factors affecting the survival of *Bremia lactucae* sporangia deposited on lettuce leaves. *Phytopathology* 90, 827–833.

Yeh, N.C. and Chung, J.P. (2009) High-brightness LEDs – energy efficient lighting sources and their potential in indoor plant cultivation. *Renewable and Sustainable Energy Reviews* 13, 2175–2180.

Index